DINOSAUR IN A HAYSTACK

Reflections in Natural History

Stephen Jay Gould

PENGUIN BOOKS

PENGUIN BOOKS

Published by the Penguin Group
Penguin Books Ltd, 27 Wrights Lane, London W8 5TZ, England
Penguin Books USA Inc., 375 Hudson Street, New York, New York 10014, USA
Penguin Books Australia Ltd, Ringwood, Victoria, Australia
Penguin Books Canada Ltd, 10 Alcorn Avenue, Toronto, Ontario, Canada M4V 3B2
Penguin Books (NZ) Ltd, 182–190 Wairau Road, Auckland 10, New Zealand

Penguin Books Ltd, Registered Offices: Harmondsworth, Middlesex, England

First published by Jonathan Cape 1996
Published in Penguin Books 1997
1 3 5 7 9 10 8 6 4 2

The essays in this book were previously published in *Natural History* magazine

Printed in England by Clays Ltd, St Ives plc

For my only brother, Peter (1944–1994)

My dearest and constant companion
May we someday, somewhere and somehow
live together in that two-family house
of our lifelong dreams.

CONTENTS

PART FIVE

The Glory of Museums

PART SIX

Disparate Faces of Eugenics

PART SEVEN

Evolutionary Theory, Evolutionary Stories

THEORY

STORIES

PART EIGHT

Linnaeus and Darwin's Grandfather

Come Seven

MICHEL DE MONTAIGNE, traditional founder of the essay as a literary genre, wrote a short letter as a preface for his *Essays* (1580). He stated "to the Readers":

> I desire therein to be viewed as I appear in mine own genuine, simple and ordinary manner . . . If I had lived among those nations, which (they say) yet dwell under the sweet liberty of nature's primitive laws, I assure thee I would most willingly have painted myself quite fully and quite naked.

I have been writing the monthly essays that construct these books since January 1974. This volume, the seventh in a continuing series, includes the piece that I wrote to mark the completion of twenty years, with never a month missed. I should therefore use this preface to celebrate what our founder Montaigne defined as crucial to the genre in his statement quoted above—ordinary things (with deeper messages).

I have always seen myself as a meat-and-potatoes man. You can take your ravioli stuffed with quail and . . . well, stuff it somewhere (I am also quite capable of releasing my own ground pepper from a shaker). I wrote, in the preface to the first volume of this series (*Ever Since Dar-*

win, 1977), that I viewed myself as a tradesman, not a polymath—and that my business was evolutionary biology.

As I have written with active passion, I have also watched with odd detachment—as my own essays have grown, shifted outward, and expanded focus throughout the seven volumes, across my own transition from rebellious youth to iconoclastic middle age. I trust that I have properly mixed my joy in new and challenging ideas with fierce fealty to the great and honorable traditions (not always followed, hence the need for iconoclasm) of "the ancient and universal company of scholars" (to cite the delightfully archaic line pronounced by Harvard's president each year as he confers degrees upon new Ph.D.s). But still, the essays always come home to their centering theme of evolution, the most exciting and the most portentous natural truth that science has ever discovered—and surely, as Freud noted in equating the most significant intellectual revolutions with the most troubling and penetrating assaults upon human arrogance, the most discombobulating intellectual change that science has ever compelled us to accept. What transition could be more profound than "created in God's image to rule a young world of stable entities made for our delectation" to "a fortuitous twig, budding but yesterday on an ancient and copious bush of ever changing, interrelated forms."

The first volume, *Ever Since Darwin*, covered the basic principles of Darwinian theory. The second, *The Panda's Thumb*, featured extensions and criticisms of these central ideas. The third, *Hen's Teeth and Horse's Toes*, expanded to the social implications of evolutionary thought, particularly to our profession's struggle with oxymoronic "creation science," then so threatening (we fought and won all the way to the Supreme Court), and now somewhat muted (though still dangerous), thanks to our vigilance and the power of our ideas and information. The fourth, *The Flamingo's Smile*, stressed the random and unpredictable character of life's history. The fifth, *Bully for Brontosaurus*, expanded this theme to a multifarious disquisition on the nature of history. *Eight Little Piggies*, the sixth volume, then added environmental subjects (curiously underplayed, I must say in self-criticism, in earlier contributions to the series).

This seventh volume, *Dinosaur in a Haystack*, features further extensions of evolutionary thought to subjects both primarily academic (literature, a personal love that has grown upon and through me over time, in parts 2 and 4; astronomy and calendrics, in continuity with a childhood passion that once ranked second only to paleontology, or maybe fourth after baseball and philately as well, in part 1) and primarily social and political (the vital role of museums in part 5, and the disparate faces of eugenics in part 6).

But I also remain true to my centering upon evolution in the two longest sections, treating the two great themes of my focal science—evolutionary theory, with emphasis on issues in Darwinism (part 7), and patterns (often quite surprising and upsetting to traditional ideas) in the recorded history of life (part 3). In other words, process and pattern, or mechanism and pathway; the how and the what of evolution's four-billion-year course on our planet. Finally, and further true to my love of history (not only of life itself, but of science trying to understand life), I have illustrated many of my themes by their expression in the life and work of fascinating individuals, emphasizing either the unfairly unknown, or unappreciated byways in careers of celebrated people: Pierre-Simon Laplace, Mary Shelley, Alfred, Lord Tennyson, Jonathan Swift, Edgar Allan Poe's work on shells, the unknown Mary Roberts, the equally unknown Gotthelf Fischer von Waldheim, Luther Burbank, R. A. Fisher, and J. B. S. Haldane. In the book's last section of three essays, I explore some wonderful connections between Linnaeus and Erasmus Darwin, with inevitable comments on Erasmus's far more famous grandson.

My dinosaur in the haystack may represent the gem of a detail always sought to ground a generality, dinosaurs as the champion details for public fascination, the haystack as an encompassing generality, and their conjunction as the secret to a successful essay in Montaigne's original and defining strategy—the marriage of alluring detail with instructive generality, all told with the stamp of an author's personal involvement.

I am not a modest man, but I do know my great weaknesses amidst one lucky strength. I am not innumerate, but how I wish for the mathematical creativity, a pure blank for me, that drives so many scientists to fine accomplishment. I am not illogical, but how I yearn for the awesome ability I note in many colleagues to identify, develop, and test the linear implications of an argument.

All people have oddly hypertrophied skills, but some folks never identify their uniqueness properly; for others, the specialness doesn't integrate into professional life and becomes, at best, a recreation or a party trick. I did receive one great gift from nature's preeminent goddess, Fortuna—a happy conjunction of my own hypertrophy with maximal utility in a central professional activity. I cannot forget or expunge any item that enters my head, and I can always find legitimate and unforced connections among the disparate details. In this sense, I am an essay machine; cite me a generality, and I will give you six tidbits of genuine illustration. A detail, by itself, is blind; a concept without a concrete illustration is empty. The conjunction defines the essay as a genre, and I draw connections in a manner that feels automatic to me.

But Fortuna's gifts languish unless reinforced by her sister Diligencia under the patronage of Amor. I am past fifty, but I retain every atom of the enthusiasm in my five-year-old self for any item of knowledge. I love to learn the details and the reasons of people's lives and nature's ways. I yearn to encounter these items in their original languages and first presentations, not through someone else's distillation. I do not believe in vicarious pleasure and will go to ridiculous lengths to stand on the right spot, or hold the object itself. *Ich kann nicht anders* by internal necessity, but I think the passion has paid off as well. I could identify King Gustav because I knew the story of Verdi's *A Masked Ball* (essay 32). I could read Fischer's Latin dedication to Razumovsky; I knew what had happened to Moscow in 1812; and I had to wonder whether Fischer's Razumovsky could be linked with Beethoven's patron (essay 20).

Such self-imposed moral pressure can be burdensome, but think of the boons in personal understanding. I had to read Burke on the sublime and the beautiful because my youthful arrows had injured my grandfather's book (essay 15)—but Burke's wise argument therefore lay in storage inside my head, all ready to illuminate Mary Roberts's genre of beautiful (not sublime) books on natural history by nineteenth-century women. I felt guilty for bypassing a youthful opportunity to learn the Papiamentu creole of Curaçao, but I found expiation twenty-five years later and tumbled upon the theme for an essay as well (essay 27).

I might therefore epitomize, in four statements, my own approach to the little, but old and honorable, literary genre of the scientific essay. First, I may wander through byzantine and arcane tidbits of collateral information—from baseball to beeswax to yo-yo crazes among New York schoolkids—but I will always return home to evolution and the great themes of time, change, and history.

Second, even my most general essays on maximally abstract subjects begin (in my mind at least, if not always and ultimately on paper) with an intriguing detail that piqued my interest: a reference to natural selection in the operative paragraph of history's most infamous document, the Wannsee Protocol (essay 24); snails printed backwards in seventeenth-century treatises on natural history (essay 16); the poignancy of a receipt documenting a few guilders for the bargeman's beer and sausage as key evidence for the provenance of an extinct antelope that can no longer speak for itself (essay 21).

Third, I try to link, in admittedly idiosyncratic, but not forced or artificial, ways, these lovely details into their own string, and then to the essay's generality as a truly helpful illustration, not a frill or an indulgence—the dinosaur to the haystack. I knew that King Gustav and Linnaeus must have some deep connection if they faced each other through the maximal thinness of a banknote (essay 32). Self-deprecation in Idaho does represent the same lamentable habit as self-hate among natural historians facing their molecular colleagues—both parochialisms must cede to mutual reinforcement in equality (essay 31). A humongous fungus does illustrate the most difficult and abstract problem in Darwinian theory, the nature of individuality and the identification of evolutionary agents (essay 26). Recent discoveries in the evolution of whales do both zing the creationists and illuminate the difficult evolutionary principle of multiple adaptive peaks, and the constraints of historical legacies (essay 28). Haldane's celebrated quip about God's fondness for beetles should be joined to the empirical issue of just how many species actually inhabit our earth (essay 29). *Jurassic Park*, just a movie perhaps, does provide a ground for discussing the nature of fads, the pitfalls of reductionism, and the essential "rightness" of museums centered on objects (essay 17). Edgar Allan Poe's book on shells is not the embarrassment that all literary sources proclaim (though the volume is substantially plagiarized), if you know the technical traditions of malacological writing and therefore recognize Poe's interesting originality (essay 14). My personal favorite in this volume will never be a popular hit because the connections are too byzantine and cascading, and the personae too anonymous, but how I loved moving from a dull and sycophantic dedication to a local count, to the torching of Moscow after Napoleon's conquest in 1812, to the opposite conduct of the Razumovsky brothers, one who did nothing for Fischer, the other who did everything for Beethoven, to the touching similarities in Mozart's and Fischer's middle names, to the old-fashioned virtue of self-sufficiency—and I had to find out absolutely all of this by myself; for no secondary sources exist (essay 20). What do Greek bus stops tell us about Erasmus Darwin's heroic couplets on the sex lives of plants? Find out in the closing essay 34.

Fourth, you and I must walk together. Most "popular" writing in science simplifies concepts (usually trivializing them as well, if unintentionally) in the belief, often false, that understanding will thereby be enhanced. Perhaps, sometimes—but for me, the essay is then not worth writing. I will, of course, clarify language, mainly to remove the jargon that does impede public access. But I will not make concepts either more simple or more unambiguous than nature's own complexity dictates. I intend my essays for professionals and lay readers alike—an old tradition, by the way, in scientific writing from Galileo to Darwin, though effectively lost today. I would not write these essays any differently if I intended them for my immediate colleagues alone. Thus, while I hope that you will appreciate my respect, our bargain may require a bit more from you than the usual item of American journalism demands. If I discuss Linnaeus's sexual system for the classification of plants, I will give you his names and rationales for all twenty-four categories (see essay 33). But isn't this better than the usual journalistic one-liner—"Linnaeus classified plants by differences in their sexual parts"—for I give you the opportunity to grasp the details of Linnaean order (and I then reward you with Erasmus Darwin's heroic couplets on Linnaeus's categories).

I love doing this monthly work, but all good things must end—and the imminent millennium provides a natural termination (some would say for the whole earth—see essay 2, my other personal favorite for the virtue of details, from Samuel Sewall's trumpeters on Boston Common in 1701 to poor Carry Nation holed up in jail on January 1, 1901). I will therefore try to write every month until January 2001 (despite my defense of 2000 in essay 2, but like Peary near the North Pole, you can't come this close and not attempt to satisfy all interpretations)—and this series should therefore run for two more volumes.

I must run out my skein, but nature never will—and we should all take the greatest pleasure in this, her infinite bounty. So, let me launch this volume seven with Lewis Carroll's most prescient recognition that seven of human design cannot dent the wondrous superfluity of nature. The Walrus and the Carpenter, walking on the beach, weep because they would love to clear away nature's presence in "such quantities of sand." But they recognize the hopelessness:

> If seven maids, with seven mops
> Swept it for half a year,
> "Do you suppose," the Walrus said,
> "That they could get it clear?"
> "I doubt it," said the Carpenter,
> And shed a bitter tear.

I am weeping tears of joy—as did Wordsworth contemplating "The Sparrow's Nest":

> She gave me eyes, she gave me ears;
> And humble cares, and delicate fears;
> A heart, the fountain of sweet tears;
> And love, and thought, and joy.

PART ONE

Heaven and Earth

Happy Thoughts on a Sunny Day in New York City

GALILEO DESCRIBED THE UNIVERSE in his most famous line: "This grand book is written in the language of mathematics, and its characters are triangles, circles, and other geometrical figures." Why should the laws of nature be subject to statement in such elegantly basic algebra? Why does gravity work by the principle of inverse squares? Why do simple geometries pervade nature—from the hexagons of the honeycomb, to the complex architecture of crystals? D'Arcy Thompson, author of *Growth and Form* and my earliest intellectual hero (along with my father and Charles Darwin), wrote that "the harmony of the world is made manifest in Form and Number, and the heart and soul and all the poetry of Natural Philosophy are embodied in the concept of mathematical beauty." Many scientists, if only to coin a striking metaphor, depict a creating God as a mathematician from the realm of Plato or Pythagoras. The physicist James Jeans wrote: "From the intrinsic evidence of his creation, the Great Architect of the Universe now begins to appear as a pure mathematician."

But much of nature is messy and multifarious, markedly resistant to simple mathematical expression (at least before fractals gave us a way to formulate the complexities of a mountaintop, a coastline, or a leaf). And other scientists have developed equally striking metaphors about a creator who revels in the unquantifiable details—as in J. B. S. Haldane's famous quip (see essay 29) that God must have an inordinate fondness for beetles.

We have, in many respects, been oversold on the mathematical precision of nature. Even the preeminent field for abstract, quantified beauty—a domain whose very name, celestial mechanics, seems to evoke ethereal harmony—includes ever so many awfully messy and downright inconvenient irregularities. Why, for example, couldn't God have arranged some simple and decent proportionality between the earth's axial rotation and solar revolution? Why didn't he give the year a nice even number of days, without elaborate fractions that demand complex seat-of-the-pants corrections in our calendars? Why 365 days and almost (but not quite) an extra quarter?—so that we have to add a leap day every fourth time round, but remove it again every hundred years (because God ordained a little less than an extra quarter after 365), except for every four hundred years when we put it back again. (And thus, if you penetrated that sentence, you will grasp why 2000 will be a leap year, even if, among some purists, it will not mark a millennium—see next essay.)

Nature also mocks our attempt to encase her in a Platonic straitjacket by establishing an almost laughably fortuitous reason for some apparent, highly visible regularities that have played a major role in human history. In my favorite example, much discussed by many commentators, solar and lunar eclipses produce a gorgeously precise and tight fit (as the moon's shadow snugly covers the sun and vice versa). Must not such exactitude be explicitly arranged, or at least arise as a predictable consequence from one of those mathematically elegant laws of nature? But the effect is only a happenstance of history. The sun's diameter is about four hundred times larger than the moon's. But the sun is also about four hundred times more distant—so their discs appear the same size to an observer on earth. (Now consider how much of human mythology rests upon an image of two guardians, intimately related by their common size: "And God made two great lights: the greater light to rule the day, and the lesser light to rule the night.")

When nature so mocks us, she often comes clean every once in a while, as if to offer confession for such a sublime joke. On May 10, 1994, a rare form of solar eclipse, far less spectacular than the conventional lid of darkness, but immensely fascinating for its own more subtle strangeness, enveloped much of North America. The moon's distance from the earth varies quite a bit during its revolution (planetary orbits are also not so regular as the charts in our high-school textbooks implied). If a solar eclipse occurs when the moon lies at maximal distance from the earth, then the lunar shadow does not fully cover the sun's disc. At totality, therefore, a ring of bright light remains at the sun's periphery. Such eclipses are called annular, from a Latin word for "ring." (Annular eclipses are much less spectacular than total eclipses at normal lunar distances, for a ring of bright sunshine still yields substantial light—as much or more than on an ordinary cloudy day—while the sky turns off as if God threw a light switch when the moon's larger disc fully covers the sun.)

I was angry with myself on May 10. The eclipse was eighty-eight percent complete at my Boston home, while totality beckoned only an hour or two north from Concord, New Hampshire, and many other pleasant sites throughout New England. The next annular eclipse in New England will occur on July 23, 2093, long past my watch—so May 10 or never (at least without substantial travel) for me. I ordered all my students to drive into the region of totality on pain of instant expulsion. (Professors—enjoying these odd moments of surcease from Shaw's observation that those who can, do; while those who can't, teach—really do revel in such application of limited power. I so ordered, and not a one of them went—to their eternal shame, but otherwise without consequence.) Meanwhile, duty-bound to honor a commitment made before I heard about the eclipse, I went south to New York City, toward less solar coverage by an already compromised lunar shadow.

Many things keep us going in this vale of tears—a baby's smile, Bach's B-minor Mass, a decent bagel. Every once in a while, as if to grant us the courage to go on, the powers that be turn one of life's little disasters into a bit of joy or an episode of instruction. The Lord of the (Partial) Ring must have been smiling on me this May 10, for he brought me in a sour mood to my natal city of New York and then rewarded me with a better experience than totality in Concord could ever have provided.

I love pristine nature, but I am a humanist at heart, and I revel more in complex interactions between fellow members of *Homo sapiens* and the great external world. Now think of every stereotype you hold about New Yorkers. (They are untrue, of course, but culturally powerful as a recognized type or icon nonetheless.) New Yorkers are harried, self-centered, cynical, rushed, acquisitive, uncurious, uncommunicative, and downright nasty to all humans who cannot be wheedled or manipulated for material gain. Right? Of course, as all Americans know, even those who have never been east of the Mississippi! A solar eclipse must therefore rank as the last thing that could ever intrigue a real New Yorker. I mean, gimme a break mister. You want me to stop what I'm doing and look into the sky—at a partial and annular eclipse? Get lost—and screw in your own light bulb.

Yet, as Joshua once stopped the sun over Gibeon, New York City returned the compliment on May 10. In midtown Manhattan, in the middle of a busy working day, New York stopped to watch the sun. Let me not exaggerate. Many folks just went on about their business, as the human tide of midday swept down Seventh Avenue. But large knots of eclipse watchers also stood their ground on every street. What features in this less spectacular version of the general phenomenon—partial and annular, rather than total and completely covered—could have inspired the interest of New Yorkers? Consider two aspects of this remarkable event.

First, in this age of artificially induced full-body shake-me-ups, from rollercoasters to all the electronic powers of film, video games, and amplified sound, we hardly think that anything so subtle, albeit pervasive, as the character of surrounding sunlight could move our passions, or even invite our notice (but then the Impressionist painters did have some insights about the power of light's quality). It does not get very dark when the sun is occluded by eighty percent on a bright day; an ordinary cloud cover reduces visibility by more than this. Thus, the sky did not darken precipitously over New York on May 10. But we are exquisitely sensitive to the usual character of light, even though we may not explicitly credit our awareness, and may not be able even to state what feels so odd.

It did not, I repeat, get very dark over New York, but the cloudless sky implied brightness, and the day turned eerily somber, while sunlight continued to reign—and people noticed, and trembled ever so slightly. "Moses and the children of Israel" sang a song to God to praise the stunning power of altered celestial events (Exodus, chapter 15): "The people shall hear and be afraid . . . the dukes of Edom shall be amazed; the mighty men of Moab, trembling shall take hold upon them . . . they shall be as still as a stone." And so New York, mightier by far and incomparably more sophisticated than these old kingdoms of the Middle East, noticed and stood still as a sky full of daylight darkened to the level of a clearly nonexistent thunderstorm. A woman said to her friend, "Holy shit, either the world is about to end, or it's going to rain—and it sure as hell ain't gonna rain."

Second, the sight of a crescent sun is so unusual, so outside our daily experience, that people do pause to notice, and wonder. If the first phenomenon, eerie (if slight) darkness, impelled a kind of visceral attention, the crescent sun, by contrast, provoked a more intellectual response.

At every eclipse, official purveyors of the news deluge us with warnings about grave dangers to our sight should we stare at the eclipsing sun. Don't look up, even for a moment. The sun will burn a painless hole in your retina faster than masturbating boys went blind in the bad old days of dire warnings. I do understand why such exaggerations must be presented. Staring at the sun for minutes on end is a very bad idea and can have all the consequences stated in eclipse warnings—so news sources must say, "Don't look at all," in order to impart sufficient fear for preventing these longer starings. So strident are these warnings that many people actually come to believe in a special power of eclipse light to do such harm. But one can, of course, stare directly at the sun for a moment without danger on all days, both regular and eclipsed. After all, we do glance inadvertently into the sun's disc every once in a while, and we do not go blind.

But most people, and quite rightly, did not look directly at the sun, and took official advice for observation by a clever set of devices for filtering or projecting images. And I became grateful for this panoply of strategies during my humanistic "field trip" for science through the streets of New York, for the viewing devices provoke discussion and encourage sharing, thus helping to forge the eclipse community.

Some people look up through filters. A young man had prepared several strips of overdeveloped film, and he passed them around, a double layer for each observer (as the newspapers had advised), to all interested parties. A welder on 53rd Street spent his work break sharing his goggles with the gathering crowd.

Others took advantage of a wonderful phenomenon in optics, using the principle that almost any small hole or space will act as a pinhole camera to project the image of the crescent sun. Here New York City even holds an advantage over the country—for an image projects badly upon rumpled ground, but ever so well against a smooth white sidewalk. New York is such a wondrous mixture of colors, classes, dress, and activity (I have seen many cities more beautiful and more exotic, but none more diverse). But we so rarely come together, for what can transcend our differences and forge common concern? And what answer to this query could be more elegant or literal than the pervasive sun itself?

On 58th Street, a West Indian janitor in his work clothes stood in front of an apartment building, where a ripped awning contained several small holes, each projecting a beautiful image of the crescent sun upon the sidewalk. The janitor, acting the part of a carnival barker, gathered in the passersby under his awning to see the grand sight, gratis of course. At the next building, like the proprietor of the adjacent stall at the same carnival, an Asian man pierced holes in envelopes, sheets of paper, and manila folders, showing people how to project the sun's image upon the ground—again for free and for the sheer pleasure of sharing.

People gathered on every street to show off their newly discovered devices for projecting images. Trees attracted the greatest crowds, for the spaces between leaves act as little cameras, and hundreds of dancing crescent suns appeared on the sidewalk amid the shadows of branches and leaves. One woman, elegantly dressed and dangling a cigarette from her lips, held up her hand into the path of sunlight at the eclipse's acme, and a crescent image appeared at the bottom of the space between each pair of adjacent fingers. She squealed with delight, and the people around her cheered. A boy then took off his adjustable baseball cap, unsnapped the connecting band, and projected a sun through each tiny hole of the band. And again, the people cheered.

I have watched eclipses with relish for all my conscious life. Like all devotees, I have my favorite stories and main events. I remember my best lunar eclipse, seen when I was a teenager from the twenty-fifth floor of a friend's apartment, high over Manhattan. The fully covered moon often turns dark, but may also glow with a variety of colors. On this night the entire disc of the eclipsed moon turned red, a deep, dark red that I had never seen in the heavens, or perhaps even on earth. And I understood that two verses from "The Saints" are descriptions of solar and lunar eclipses, not abstract scare stories of eschatology (I played washtub bass in a folk group at the time, and we often performed this song): "When the sun refuse to shine . . . when the moon turns red with blood; oh Lord, I want to be in that number, when the saints go marching in"—a description, after all, of the Last Judgment, when eclipses will accompany the panoply of awful events. Did not the prophet Joel (2:31) also speak as an astronomer in citing the same image for the same purpose: "The sun shall be turned into darkness, and the moon into blood, before the great and the terrible day of the Lord come."

And I remember, for how can one blessed with an opportunity to witness this most spectacular of all celestial events ever forget, the total solar eclipse of early 1970. Our department rented a fishing boat to sail off Nantucket, the only bit of New England real estate privileged with a view of totality. I longed to see the moon's shadow fully cover the sun; I thrilled at a chance to observe the sun's corona. But I had not understood the most awesome phenomenon of all. We live in a natural world of shadings. Even catastrophes have foreshadowings; clouds precede thunderstorms, and tornadoes can be seen in the distance. But when the sun enters total eclipse, the sky turns off as if a celestial janitor threw a switch. For the sun is powerful, and a fraction of one percent of sunlight is daytime, while totality is nighttime—and the transition is a moment, a twinkling of an eye. The sky turned off, and my infant son cried in my arms.

We hear so many dire warnings about the poor quality of science teaching in our schools, so many lamentations over the profound ignorance of most Americans about nearly any phenomenon of the natural world. Perhaps these jeremiads have validity; half of my own students could not explain to me why our planet has seasons. Surely we should be struggling to increase literacy in science, for no issue of education could be more important.

But I am convinced that the problem does not arise from lack of interest. Such a false charge is often made amid the litany of correct accusations mentioned in the last paragraph. Interest is immense, but not always expressed as activity traditionally called science or ranked among its pursuits (and our misattribution therefore arises from our inadequate taxonomies of intellectual enterprise). My colleague Phil Morrison is fond of cataloguing the large number of common activities requiring a good deal of scientific understanding, but not usually so classified: the astronomical knowledge of people who build and maintain telescopes; the deep botanical experience of members in gardening clubs (a fine example of power concentrated in older women); or even the people who frequent racetracks and bet intelligently on horses, for misunderstanding of probability may be the greatest of all general impediments to scientific literacy.

May I now add to this list the aggregate intellectual power (how I wish we could quantify it) of all the dinosaur names accurately memorized (and spelled) by millions of five-year-old kids in America today. And also the accumulated joy and pleasure of millions upon millions of Americans who paused to watch the sun and to wonder on May 10, 1994. New York City was the best place to be on that date; my faith in raw interest is fully affirmed—and raw interest is the substrate and *sine qua non* of any real reform in education and larger understanding.

We often argue that only misfortune can bring us together. We do help each other during snowstorms; we do open our hearts and our houses to victims of an immediate disaster in our vicinity; we will search all night in the woods for a lost child we do not know. All these observations properly give us hope about common humanity in a world more often characterized by thoughtlessness, self-serving action, and even downright cruelty. But we also suppose that only disaster can provoke this effect, never pleasure, and certainly not intellectual as opposed to purely visceral delight. But interest and curiosity can also bring us together—and my observations of New Yorkers reveling in nature and spontaneously talking about the sun somehow give me more hope than our joint courage in times of crisis can provide, even though unity in disaster may make me cry in sublime appreciation, while the bonding of eclipses only makes me smile.

And so I end this essay by quoting the greatest of all tributes to the sun. I have often stated my personal theory about popular writing in science. I divide this genre into two modes, which I call Galilean, for intellectual essays about nature's puzzles, and Franciscan, for lyrical pieces about nature's beauty. I honor Galileo for writing his two major works as dialogues in Italian, and therefore addressed to all thinking people in his orbit, and not in the formal Latin of churches and universities. And I honor Saint Francis of Assisi for his tributes to nature's loveliness.

I am an unrepentant Galilean. I work in a tradition extending from the master himself, to Thomas Henry Huxley in the last century, down to J. B. S. Haldane and Peter Medawar in our own. I greatly admire Franciscan lyricism, but I don't know how to write in that mode. I began this essay with a quotation from the eponymous hero of my literary bloodline, Galileo himself. But my essay talks about the power of the sun to unify our diverse cultures and concerns, so I must end with a man I have never quoted before in these columns, the eponym of the other style—Saint Francis of Assisi. Saint Francis composed his beautiful "Canticle of Brother Sun" in 1225. He wrote in the Umbrian dialect of his local people, and his poem is often regarded as the first preserved in any modern language:

Brother Sun, who brings the day . . .
How beautiful is he, how radiant in all his splendor!

Dousing Diminutive Dennis's Debate

(or DDDD = 2000)

In 1697, on the day appointed for repenting mistakes in judgment at Salem, Samuel Sewall of Boston stood silently in old South Church, Boston, while his confession of error was read aloud. He alone among judges of the falsely accused (and truly executed) "witches" of Salem had the courage to undergo such public chastisement. Four years later, the same Samuel Sewall made a most joyful noise unto the Lord—and at a particularly auspicious moment. He hired four trumpeters to herald, as he wrote, the "entrance of the eighteenth century" by sounding a blast on Boston Common right at daybreak. He also paid the town crier to read out his "verses upon the New Century." The opening stanzas seem especially poignant today, the first for its relevance (I am writing this essay on a bleak January day in Boston, and the temperature outside is –2° Fahrenheit), and the second for a superannuated paternalism that highlights both the admirable and the dubious in our history:

> Once more! Our God vouchsafe to shine:
> Correct the coldness of our clime.
> Make haste with thy impartial light,
> and terminate this long dark night.
> Give the Indians eyes to see

The light of life, and set them free.
So men shall God in Christ adore,
And worship idols vain, no more.

I do not raise this issue either to embarrass the good judge for his tragic error, or to praise his commendable courage, but for an aspect of the tale that may seem peripheral to Sewall's intent, but that nevertheless looms large as we approach the millennium destined to climax our current decade. Sewall hired his trumpeters for January 1, 1701, not January 1, 1700—and he therefore made an explicit decision in a debate that the cusp of his new century had kindled, and that has increased mightily at every similar transition since (see my main source for much of this essay, the marvelously meticulous history of *fins de siècle—Century's End* by Hillel Schwartz). When do centuries end?—at the termination of years marked '99 (as common sensibility suggests), or at the close of years marked '00 (as the narrow logic of a peculiar system dictates)?

The debate is already more intense than ever, though we are still several years from our own forthcoming transition, and for two obvious reasons. First—o cursèd spite—our disjointed times, and our burgeoning press, provide greatly enhanced opportunity for rehearsal of such *narrishkeit* ad nauseam; do we not feast upon trivialities to divert attention from the truly portentous issues that engulf us? Second, this time around really does count as the ultimate blockbuster: for this is the millennium,[1] the great and indubitable unicum of any human observer (though a few trees, and maybe a fungus or two, but not a single animal, have been through it before—see essay 26).

On December 26, 1993, *The New York Times* ran a piece to bury the Christmas buying orgy and welcome the new year. This article, on commercial gear-up for the century's end, began by noting: "There is money to be made on the millennium . . . in 999 feelings of gloom ran rampant. What the doomsayers may have lacked was an instinct for mass marketing." The commercial cascade of this millennium is already in full swing: in journals, datebooks, the inevitable coffee mugs and

1. In this essay's spirit of dispelling a standard set of confusions that have already surrounded the forthcoming millennium, may I at least devote a footnote to the most trivial, but also the most unambiguously resolvable. *Millennium* has two *n*'s—honest to God, it really does, despite all the misspellings, even in most of the books and product names already dedicated to the event. The adjective *millennial* also has two, but the alternative *millenarian* has only one. (The etymologies are slightly different. *Millennium* is from Latin *mille*, one thousand, and *annus*, year—hence the two *n*'s. *Millenarian* is from the Latin *millenarius*, "containing a thousand [of anything]," hence no *annus*, and no two *n*'s.)

T-shirts, and a thousand other products being flogged by the full gamut, from New Age fruitcakes of the counterculture, to hard-line apocalyptic visionaries at the Christian fringe, to a thicket of ordinary guys out to make an honest buck. The article even tells of a consulting firm explicitly established to help others market the millennium—so we are already witnessing the fractal recursion that might be called metaprofiteering, or growing clams of advice in the clam beds of your advisees' potential profits.

I am truly sorry that I cannot, in current parlance, "get with the program." I feel compelled to mention two tiny little difficulties that could act as dampers upon the universal ballyhoo. First—though I will not make a big deal of this technicality—millennia are not transitions at the ends of thousand-year periods, but particular periods *lasting* one thousand years; so I'm not convinced that we even have the name right. Second, if we insist on a celebration (as we should), no matter what name be given, we had better decide when to celebrate. I devote this essay to explaining why the second issue cannot be resolved—a situation that should be viewed as enlightening, not depressing. For just as Tennyson taught us to prefer love lost over love unexperienced, it is better to not know and know why one can't know, than to be clueless about why the hell so many people are so agitated about 1999 vs. 2000 for the great divide. At least when you grasp the conflicting, legitimate, and unresolvable claims of both sides, you can then celebrate both alternatives with equanimity—or neither (with informed self-righteousness) if your persona be sour, or smug.

RIGHTFUL NAMES. *Millennium* does mean, by etymology, a period of one thousand years. This concept, however, did not arise within the field of practical calendrics, or the measurement of time, but in the domain of eschatology, or futuristic visions about a blessed *end* of time. Millennial thinking is embedded in the two apocalyptic books of the Bible—Daniel in the Old Testament and Revelation in the New. In particular the traditional Christian millennium is a future epoch that will last for one thousand years, and end with a final battle and Last Judgment of all the dead. As described by Saint John in one of his oracular visions (Revelation 20), Satan shall be bound for one thousand years and cast into the bottomless pit; Christ shall return and reign for this millennium with resurrected Christian martyrs. Satan shall then be loosed; he shall team up with Gog, Magog, and a host of other baddies, for a final battle; Christ and the good guys win, the devils end up in "the lake of fire and brimstone"; all the dead are now resurrected and, in a Last

Judgment at this end of time, either rise to live with Jesus, or end up in that other, unpleasant place along with most of history's interesting characters.

> And I saw an angel come down from heaven . . . And he laid hold on . . . Satan, and bound him a thousand years, and cast him into the bottomless pit, and shut him up, and set a seal upon him, . . . and I saw the souls of them that were beheaded for the witness of Jesus . . . and they lived and reigned with Christ a thousand years . . . And when the thousand years are expired, Satan shall be loosed out of his prison, and shall go out to deceive the nations which are in the four quarters of the earth, Gog and Magog, to gather them together to battle . . . and fire came down from God out of heaven, and devoured them. And the devil that deceived them was cast into the lake of fire and brimstone . . . And I saw the dead, small and great, stand before God; and the books were opened . . . And whosoever was not found written in the book of life was cast into the lake of fire [Revelation 20:1–15].

How, then, did this original concept of a forthcoming reign of Christ become transmogrified in popular speech into a word for calendric transitions at multiples of one thousand? The main reason must be simple confusion, and loss of knowledge about the original meaning, as apocalyptic versions of Christianity, not to mention Bible reading in general, decline in popularity (despite, to say the least, vigorous continuing support in some circles)! But a rationale of sorts for the transfer of meaning does exist within the history of eschatology, particularly in its intersection with my profession of geology in attempts to ascertain the age of the earth.

Many biblical passages state that God's day may be compared with a thousand human years: "Be not ignorant of this one thing, that one day is with the Lord as a thousand years, and a thousand years as one day" (2 Peter 3:8; see also Psalm 90). This comparison, read literally, led many interpreters to conclude that the seven days of creation must correspond with a maximal duration of seven thousand years for the earth from creation to final destruction at the Last Judgment. In this scheme, the seventh or last cosmic epoch, corresponding to God's day of rest after six days of furiously creative activity, would be a one-thousand-year period of bliss, the grand sabbath of the traditional millennium. If either science or hermeneutics could then determine the time of the earth's origin, we might know the moment of inception for this last happy age.[2]

Most calculations of the earth's antiquity, if done literally from biblical life spans and other ancient sources, place the creation somewhere between 3761 B.C. (the Jewish calendar) and more than 5500 B.C. (the Septuagint, or Greek Bible). Therefore, a transition into the millennial age might well be on the horizon—or should have occurred just a while ago, according to your favored calculation. True, none of the suggested times of creation give any reason to redefine a millennium as a transition around a date with three zeros in its written form, but at least we may understand why people might conflate a future period of millennial bliss with some system for counting historical time in periods of one thousand years.

RIGHTFUL TIMES. As a man of below-average stature myself, I am delighted to report that the source of all our infernal trouble about the ends of centuries may be laid on the doorstep of a sixth-century monk named Dionysius Exiguus, or (literally) Dennis the Short. Instructed to prepare a chronology for Pope Saint John I, Little Dennis decided to begin countable years with the foundation of Rome. But, neatly balancing his secular and sacred allegiances, Dionysius then divided time again at Christ's appearance. He reckoned Jesus' birth at December 25, near the end of year 753 A.U.C. (*ab urbe condita*, or "from the foundation of the city," that is, of Rome). Dionysius then restarted time just a few days later, on January 1, 754 A.U.C.—not Christ's birth, but the Feast of the Circumcision on his eighth day of life, and also, not coincidentally, New Year's Day in Roman and Latin Christian calendars.

Dionysius's legacy has provided little but trouble. First of all, he

2. In this sense, the great seventeenth-century revival of millenarian thought should not only be viewed, though we usually do so, as a rearguard and visionary attack against developing science, but in part as an outgrowth of the contemporary scientific revolution. The Catholic Church, at least since Augustine, had suppressed literal millenarianism with an allegorical argument (see Augustine's *City of God*) that the millennium must be viewed as a spiritual state collectively entered by the Church at Pentecost—the descent of the Holy Ghost to the apostles soon after Christ's resurrection—and fully subject to contemporary personal experience by mystical communion with God. This argument, needless to say, serves a social purpose for a powerful and conservative institution wishing to maintain a status quo of daily influence, and not encourage wild theories about actual and imminent ends of the world. But when science and other developing forms of scholarship, including history, philosophy, and textual analysis, began to devise methods for arguing about such subjects as the age of the earth, hopes for calculation did encourage active searches for actual beginnings and ends of time, while the developing myth of scientific progress also fueled hopes for gradual transition into an earthly millennium (see *Encyclopedia Britannica* on Millennialism, particularly on the contrast of early Christian apocalyptic versions—sudden divine overturn of an exhausted and sinful world—with this seventeenth-century account of "progressive millennialism." Since these threads intertwine in such modern apocalyptic groups as Jehovah's Witnesses and Seventh Day Adventists, this history is scarcely of antiquarian interest alone!)

didn't even get the date right, for Herod died in 750 A.U.C. Therefore, if Jesus and Herod overlapped (and the gospels will have to be drastically revised if they did not), then Jesus must have been born in 4 B.C. or earlier—thus granting the bearer of time's title several years of life before the inception of his own era!

But Dennis's misdate of Jesus counts as a mere peccadillo compared with the consequences of his second bad decision. He started time again on January 1, 754 A.U.C.—and he called this date January 1 of year one A.D. (*Anno Domini*, or "year of the Lord")—not the year zero (which would, in retrospect, have spared us from ever so much trouble!).

In short, Dennis neglected to begin time with year zero, thus discombobulating all our usual notions of counting. During the year that Jesus was one year old, the time system that supposedly started with his birth was two years old. (Babies are zero years old until their first birthday; modern time was already one year old at its inception.) The absence of a year zero also means that we cannot calculate algebraically (without making a correction) through the B.C.–A.D. transition. The time from 1.5 B.C. to A.D. 1.5 is one year, not three years.

The problem of centuries arises from Dennis's unfortunate decision to start with year one, rather than year zero—and for no other reason. If we insist that all decades must have ten years, and all centuries one hundred years, then year 10 belongs to the first decade—and, sad to say, year 100 must remain in the first century. Thenceforward, the issue never goes away. Every year with a '00 must count as the hundredth and final year of its century—no matter what common sensibility might prefer: 1900 went with all the other 1800 years to form the nineteenth century; and 2000 must be the completing year of the twentieth century, and not the inception of the next millennium. Or so the pure logic of Dennis's system dictates. If our shortsighted monk had only begun with a year zero, then logic and sensibility would coincide, and the wild millennial bells could ring forth but once and resoundingly at the beginning of January 1, 2000. But he didn't.

Since logic and sensibility do not coincide, and since both have legitimate claims upon our decision, the great and recurring debate about century boundaries simply cannot be resolved. Some questions have answers because obtainable information decrees a particular conclusion. The earth does revolve around the sun and evolution does regulate the history of life. Some questions have no answers because we cannot get the required information (I doubt that we will ever identify Jack the Ripper with certainty). Many of our most intense debates, however, are not resolvable by information of any kind, but arise from conflicts in values or modes of analysis. (Shall we permit abortion, and in what circumstances? Does God exist?) A subset of these unresolvable debates—ultimately trivial, but capable of provoking great agitation, and thus the most frustrating of all—have no answers because they are about words and systems, rather than things, and phenomena of the world (that is, "things") therefore have no bearing upon potential solutions. The century debate lies within this vexatious category.

The logic of Dionysius's arbitrary system dictates one result—that centuries change between '00 and '01 years. Common sensibility leads us to the opposite conclusion: we want to match transitions with the extent or intensity of apparent sensual change, and 1999 to 2000 just looks more definitive than 2000 to 2001, so we set our millennial boundary at the change in all four positions, not at the mere increment of 1 to the last position. (I refer to this side as "common sensibility" rather than "common sense" because support invokes issues of aesthetics and feeling, rather than logical reasoning.)

One might argue that humans, as creatures of reason, should be willing to subjugate sensibility for logic; but we are, just as much, creatures of feeling. And so the debate has progressed at every go-round. Hillel Schwartz, for example, cites two letters to newspapers, written from the camp of common sensibility in 1900. "I defy the most bigoted precisian to work up an enthusiasm over the year 1901, when we will already have had twelve months' experience of the 1900s." "The centurial figures are the symbol, and the only symbol, of the centuries. Once every hundred years there is a change in the symbol, and this great secular event is of startling prominence. What more natural than to bring the century into harmony with its only visible mark?"

I do so love human foibles; what else can keep us laughing (as we must) in this tough world of ours? The more trivial an issue, and the more unresolvable, so does the heat of debate, and the assurance of absolute righteousness, intensify on each side (just consider professorial debates over parking places at university lots). The same clamor arises every hundred years. An English participant in the debate of 1800 vs. 1801 wrote of "the idle controversy, which has of late convulsed so many brains, respecting the commencement of the current century." On January 1, 1801, a poem in the *Connecticut Courant* pronounced a plague on both houses (but sided with Dionysius):

Precisely twelve o'clock last night,
The Eighteenth Century took its flight.
Full many a calculating head

Has rack'd its brain, its ink has shed,
To prove by metaphysics fine
A hundred means but ninety-nine;
While at their wisdom others wonder'd
But took one more to make a hundred.

The same smugness reappeared a century later. *The New York Times*, with anticipatory diplomacy, wrote in 1896: "As the present century draws to its close we see looming not very far ahead the venerable dispute which reappears every hundred years—*viz:* When does the next century begin? . . . There can be no doubt that one person may hold that the next century begins on the 1st of January, 1900, and another that it begins on the 1st of January, 1901, and yet both of them be in full possession of their faculties." But a German commentator remarked: "In my life I have seen many people do battle over many things, but over few things with such fanaticism as over the academic question of when the century would end . . . Each of the two parties produced for its side the trickiest of calculations and maintained at the same time that it was the simplest matter in the world, one that any child should understand."

You ask where I stand? Well, publicly of course I take no position because, as I have just stated, the issue is unresolvable: for each side has a fully consistent argument within the confines of different but equally defensible systems. But privately, just between you and me, well, let's put it this way: I know a young man with severe cognitive limits as a result of inborn mental handicaps, but who happens to be a prodigy in day-date calculation (he can, instantaneously, give the day of the week for any date, thousands of years, past or future; we used to call such people idiot savants, a term now happily fading from use, though I have no love for the euphemistic substitute, "savant syndrome"). He is fully aware of the great century debate, for nothing could interest him more. I asked him recently whether the millennium comes in 2000 or 2001—and he responded unhesitatingly, "In 2000. The first decade had only nine years."

What an elegant solution, and why not? After all, no one then living had any idea whether they were toiling in year zero or year one—or whether their first decade had nine or ten years, their first century ninety-nine or one hundred. The B.C./A.D. system wasn't invented until the sixth century, and wasn't generally accepted in Europe until the eleventh century. So why don't we just proclaim that the first century had ninety-nine years—since not a soul then living either knew or cared about the anachronism that would later be heaped upon all the years of their lives? Centuries can then turn when common sensibility desires, and we underscore Dionysius's blessed arbitrariness with a caprice, a device of our own that marries the warring camps. Neat, except that I think people want to argue passionately about trivial unresolvabilities—lest they be compelled to invest such rambunctious energy in real battles that might kill somebody.

What else might we salvage from rehearsing the history of a debate without an answer? Ironically, such arguments contain the possibility for a precious sociological insight: since no answer can arise from the "externalities" of nature or logic, changing viewpoints provide "pure" trajectories of evolving human attitudes—and we can therefore map societal trends without impediments of such confusing factors as discovered truth.

I had intended to spend only a few hours in research for this essay, but as I looked up documents from century transitions, I noticed something interesting in this sociological realm. The two positions—I have called them "logical" and "common sensible" so far in this essay—also have clear social correlations that I would not have anticipated. The logical position—that centuries must have one hundred years and transitions must therefore occur, because Dionysius included no year zero, between '00 and '01 years—has always been overwhelmingly favored by scholars, and by people in power (press and business in particular), representing what we may call "high culture." The common sensible position—that we must honor the appearance of maximal change between '99 and '00 years, and not fret overly about Dionysius's unfortunate lack of foresight—has been the perpetual favorite of that mythical composite once designated as John Q. Public, or "the man in the street," and now usually called vernacular or pop culture.

The distinction goes back to the very beginning of this perpetually recurring debate about century transitions. Hillel Schwartz traces the first major hassle to the 1699–1701 passage (place the moment where you wish), the incarnation that prompted Samuel Sewall's trumpeting in Boston. Interestingly, part of the discussion then focused upon an issue that has been persistently vexatious ever since: *viz.*, did the first millennial transition of 999–1001 induce a period of fear about imminent apocalyptical endings of the world—called "the great terror" by supporters of this position. Opinions range from luridly supportive (see the remarkably uncritical book by Richard Erdoes, who elevates every hint of rumor into a dramatic assertion), to the fully debunking (see Hillel

Schwartz, previously cited, and scores of references cited in chapter one therein). I will, in my ignorance, take refuge in the balanced position of the French historian Henri Focillon (in his book *The Year 1000*).

Focillon allows that apocalyptic stirring certainly occurred, at least locally in France, Lorraine, and Thuringia, toward the middle of the tenth century. But he finds strikingly little evidence for any general fear surrounding the year 1000 itself—nothing in any papal bull, nothing from any pope, ruler, or king.

On the plus side, one prolific monk named Raoul Glaber certainly spoke of millennial terrors, stating that "Satan will soon be unleashed because the thousand years have been completed." He also claimed, though no documentary or archaeological support has been forthcoming, that a wave of new church-building began just a few years after 1000, when folks finally realized that Armageddon had been postponed: "About three years after the year 1000," wrote Glaber, "the world put on the pure white robe of churches."

Glaber's tale provides a striking lesson in the dangers of an *idée fixe*. He was still alive in 1033, still trumpeting the forthcoming millennium—though he admitted that he must have been wrong about Christ's nativity for the beginning of a countdown, and now proclaimed that the apocalypse would surely arrive instead at the millennium of Christ's Passion, in 1033. He read a famine of that year as a sure sign: "Men believed that the orderly procession of the seasons and the laws of nature, which until then had ruled the world, had relapsed into the eternal chaos; and they feared that mankind would end."

I doubt that we should grant much critical acclaim to Fra Glaber (who, according to other sources, was quite a wild character, having been expelled from several monasteries during his checkered career). I do tend to side with critics of the great terror. Why, after all, should 1000 have provoked any great reaction at the time—especially since Dionysius's system had not been generally accepted, and different cultures hadn't even agreed on a date for the inception of a new year. I suspect that the notion of a great terror must arise largely as an anachronistic back-reading, combined with clutching at a few legitimate straws.

As another reason for doubting a great terror in 999–1001, the legend of such an episode only begins with a brief mention in a late-sixteenth-century work by Cardinal Cesare Baronio. Once the debate on century endings got started in the 1690s, however, back-reading into the first millennium became inevitable. Did the legendary terror occur at the end of 999 or 1000? Interestingly, the high-culture-vs.-pop-culture distinction can be traced even to this anachronistic reconstruction, with scholars favoring 1000, and popular legends 999. Hillel Schwartz writes:

> Sarcastic, bitter, sometimes passionate debates *in re* a terminus on New Year's Eve '99 vis-à-vis New Year's Eve '00, have been prosecuted since the 1690's and confusion has spread to the mathematics of the millennial year. For Baronio and his (sparse) medieval sources, the excitements of the millennium were centered upon the end of the year 1000, while the end of 999 has figured more prominently in the legend of the panic terror.

The pattern has held ever since, as the debate bloomed in the 1690s, spread in the 1790s with major centers in newspapers of Philadelphia and London (with added poignancy as America mourned the death of George Washington at the very end of 1799), and burst out all over the world in a frenzy of discussion during the 1890s.

The 1890s version displays the clearest division of high vs. vernacular culture. A few high-culture sources did line up behind the pop favorite of 1899–1900. Kaiser Wilhelm II of Germany officially stated that the twentieth century had begun on January 1, 1900. A few barons of scholarship, including such unlikely bedfellows as Sigmund Freud and Lord Kelvin, agreed. But high culture overwhelmingly preferred the Dionysian imperative of 1900–1901. An assiduous survey showed that the presidents of Harvard, Yale, Princeton, Cornell, Columbia, Dartmouth, Brown, and the University of Pennsylvania all favored 1900–1901—and with the entire Ivy League so firmly behind Dionysius, why worry about a mere kaiser (even though the King of Sweden rallied to Wilhelm's defense).

In any case, 1900–1901 won decisively, in the two forums that really matter. Virtually every important public celebration for the new century, throughout the world (and even in Germany), occurred from December 31, 1900, into January 1, 1901. Moreover, essentially every major newspaper and magazine officially welcomed the new century with their first issue of January 1901. I made a survey of principal sources and could find no exceptions. *The Nineteenth Century*, a leading British periodical, changed its name to *The Nineteenth Century and After*, but only with the January 1901 issue, which also featured a new logo of bi-faced Janus, with an old bearded man looking down and left into the nineteenth century, and a bright youth looking right up into the twentieth. Such reliable standards as *The Farmer's Almanack* and *The Tribune Al-*

manac declared their volumes for 1901 the "first number of the twentieth century." On December 31, 1899, *The New York Times* began a story on *The Nineteenth Century* by noting, "Tomorrow we enter upon the last year of a century that is marked by greater progress in all that pertains to the material well-being and enlightenment of mankind than all the previous history of the race." On January 1, 1901, the lead headline proclaimed "Twentieth Century's Triumphant Entry," and described the festivities in New York City: "The lights flashed, the crowds sang, the sirens of craft in the harbor screeched and roared, bells pealed, bombs thundered, rockets blasted skyward, and the new century made its triumphant entry." Meanwhile, poor Carry Nation never got to watch the fireworks, or even to raise a glass, for a small story on the same first page announced, "Mrs. Nation Quarantined—Smallpox in jail where Kansas saloon wrecker is held—says she can stand it."

So high culture still held the reins of opinion last time around—even in such organs of pop culture as *The Farmer's Almanack*, no doubt published by men who considered themselves among the elite. But consider the difference as we approach this millennium—for who can doubt that pop culture will win decisively on this most important replay. Arthur C. Clarke and Stanley Kubrick stood by Dionysius in book and film versions of *2001*, but I can hardly think of another source that does not specify the inception of 2000 as the great moment of transition. All book titles of our burgeoning literature honor pop culture's version of maximal numerical shift—including Ben Bova's *Millennium: A Novel about People and Politics in the Year 1999*; J. G. de Beus's *Shall We Make the Year 2000*; Raymond Williams's *The Year 2000*; and even Richard Nixon's *1999: Victory Without War*. Prince's album and lead song *1999* cites the same date from this *ne plus ultra* of pop sources.

Cultural historians have often remarked that expansion of pop culture, including both respect for its ways and diffusion of its influence, marks a major trend of the twentieth century. Musicians from Benny Goodman to Wynton Marsalis play their instruments in jazz bands and classical orchestras. The Metropolitan Opera has finally performed *Porgy and Bess*—and bravo for them. Scholars write the most damnedly learned articles about Mickey Mouse.

This remarkable change has been well documented and much discussed, but commentary has so far missed this important example from the great century debate. The distinction still mattered in 1900, and high culture won decisively by imposing January 1, 1901, as the inception of the twentieth century. Pop culture (or the amalgam of its diffusion into courts of decision-makers) may already declare clear victory for the millennium, which will occur at the beginning of the year 2000 because most people so feel it in their bones, Dionysius notwithstanding—and again I say bravo. My young friend wanted to resolve the debate by granting the first century only ninety-nine years; now ordinary humanity has spoken for the other end—and the transition from high-culture dominance to pop-culture diffusion will resolve this issue of the ages by granting the twentieth century but ninety-nine years!

How lovely—for eternal debates about the unresolvable really do waste a great deal of time, put us in bad humor, and sap our energy from truly important pursuits. Let us, instead, save our mental fight—not to establish the blessed millennium (for I doubt that humans are capable of such perfection), but at least to build Jerusalem upon our planet's green and pleasant land.

The Celestial Mechanic and the Earthly Naturalist

DURING THE SAN FRANCISCO EARTHQUAKE of 1906, a statue of Louis Agassiz fell off the front of a building at Stanford University and landed just as neatly as could be, but upside down—feet in the air and head buried in the pavement. Agassiz had been both the greatest ichthyologist (student of fishes) of his day, and the last serious creationist holdout against evolution when he died in 1873. David Starr Jordan, the president of Stanford, was the greatest ichthyologist of the generation after Agassiz, and an early and fervent Darwinian as well. Thus the two men shared a similar passion for the same group of organisms, but couldn't have disagreed more on theoretical issues.

According to legend, Jordan delivered one of history's cleverest quips when he went out to survey the damage and saw the inverted statue: "Oh, well, I always thought better of Agassiz in the concrete than in the abstract." A lovely story that surely deserves to be true. But, alas, it is not. In his own autobiography, *The Days of a Man,* written in 1922, David Starr Jordan felt duty-bound to debunk this tale and admit that he had never uttered the famous line, while the originator had used a less quotable and opposite version. Jordan wrote:

> About the quadrangle the only touch of humor was furnished by the large marble statue of Agassiz, which had plunged from

its place head-first and waist-deep into the concrete pavement. Somebody—Dr. Argyll, perhaps—remarked that "Agassiz was great in the abstract but not in the concrete."

People are clever, but almost no one ever devises an optimal quip precisely at the needed moment. Therefore, virtually all celebrated one-liners are later inventions—words that people wished they had spouted, but failed to manufacture at the truly opportune instant. Thus the most famous of all scientific epithets is also, and alas, surely embellished if not completely fictional.

We have all heard the story of Napoleon's meeting with the great astronomer Pierre-Simon Laplace (1749–1827), identified by the *Dictionary of Scientific Biography* as "among the most influential scientists in all history." Laplace, or so the story goes, gave Napoleon a copy of his multivolume *Mécanique céleste (Celestial Mechanics)*. Napoleon perused the tomes and asked Laplace how he could write so much about the workings of the heavens without once mentioning God, the author of the universe. Laplace replied: "Sire, I have no need of that hypothesis."

The actual quip, well attested in a surviving letter, is mildly clever, but pretty insipid compared with the legend, and made by the general rather than the scientist. Laplace had first met Napoleon in 1785 when he examined the future emperor, then an artillery cadet, in mathematics at the Ecole Militaire in Paris. In October 1799, three weeks before the coup d'état that brought Napoleon to power, Laplace did present the very weighty first two volumes of his work to his former student. Napoleon hefted the books and promised to read them "in the first six months I have free." He then invited Laplace to dinner the next day, "if you have nothing better to do."

I suspect that this legend attached itself to Laplace because he does represent the best candidate for such a tale. Laplace is science's chief apostle of strict determinism and heavenly stability based on obedience of all bodies to laws of nature that damp out any perturbation to restore regularity of motion and position (Laplace coined the term "celestial mechanics").

Even Isaac Newton, so often cited as the apostle of such a view, happily invoked a little help from divine intervention either to get things going, or to restore regularity at any time in subsequent celestial history when nature's usual laws could not correct a perturbation. Newton, for example, sought to reconcile geological evidence for the earth's antiquity with the Genesis story of creation in six days by arguing that the earth then rotated very slowly, thus producing "days" of any desired length.

But Newton could not then fathom how an acceleration of rotation to days of twenty-four hours could be accomplished under nature's laws, so he invoked a positive spin from God himself. He wrote to Thomas Burnet (a colleague who upheld universal constancy and sufficiency of nature's laws and who therefore favored an allegorical interpretation of biblical language about "days"):

> Where natural causes are at hand God uses them as instruments in his works, but I do not think them alone sufficient for the creation and therefore may be allowed to suppose that amongst other things God gave the earth its motion by such degrees and at such times as was most suitable.

By contrast, the most famous quote genuinely attributed to Laplace vigorously defends a strict determinism that does make a conventional view of God's continuous role both irrelevant and unnecessary (God may still be a clock-winder, an instigator of nature's immutable laws at the outset, but he has no need ever to intervene in subsequent history; after all, a truly omnipotent God can surely establish optimal laws right at the start, thus avoiding any necessity for direct miraculous correction of a solar system gone awry). In an epigram that has defined strict determinism ever since, Laplace boasted that if anyone could provide a complete account of the position and motion of every particle in the universe at any single moment, then total knowledge of nature's laws would permit a full determination of all future history. Laplace's boast is usually cited from the introduction to his *Analytical Theory of Probabilities* (1812), but the *Dictionary of Scientific Biography* cites a much earlier and crisper version from a youthful article written in the seminal year of 1776:

> The present state of the system of nature is evidently a consequence of what it was in the preceding moment, and if we conceive of an intelligence which at a given instant comprehends all the relations of the entities of this universe, it could state the respective position, motions, and general affects of all these entities at any time in the past or future.

Beyond his work in celestial mechanics, Laplace won most renown for his pioneering studies of probability. One might ask why the prophet of determinism and heavenly constancy should have focused upon probability, now so strongly associated with opposing ideas of randomness, but the solution is not far to seek. Laplace firmly believed that, in reality, every event is fully determined by general laws of the universe. But nature is complex and we are woefully ignorant of her ways; we must therefore calculate probabilities to compensate for our limitations. Events, in other words, are probable only relative to our meager knowledge.

Celestial mechanics is the most triumphant realm of deterministic predictability because our instruments are precise and the laws relatively simple (primarily Newton's principle of universal gravitation). But more-complex terrestrial events are just as determined, if only we knew the laws and conditions as well—as one day, perhaps, we will. Laplace wrote in his popular book of 1796, the one that shall be the focus of this essay:

> Everything in nature obeys these general laws; everything derives from them by necessity and with as much regularity as the cycle of seasons. The path followed by a light atom [*atôme léger*] that the winds seem to transport at random, is ruled in as certain a manner as the planetary orbits. [My translation—by "atom," Laplace only means a tiny particle, not the invisible and chemically indivisible building block that later theory would identify.]

Later in the book, he explicitly states that we will eventually learn the more complex laws for smaller terrestrial objects, and that earthly physics will then be as deterministic as celestial mechanics:

> Several experiments already made give us reason to hope that, one day, these laws will be perfectly known; then, by applying mathematics, we will be able to raise the physics of terrestrial bodies to the same degree of perfection that the discovery of universal gravitation has given to celestial physics.

In his 1776 article, cited previously, Laplace directly links the need for a theory of probability to human ignorance of nature's deterministic ways, and he makes the same comparison between a simpler and well-known celestial mechanics and a much more difficult earthly physics:

> Man owes that advantage [in celestial mechanics] to the power of the instrument he employs, and to the small number of re-

> lations that [this field] embraces in its calculations. But ignorance of the different causes involved in the production of events, as well as their complexity, taken together with the imperfection of analysis, prevents our reaching the same certainty about the vast majority of phenomena. Thus there are things that are uncertain for us, things more or less probable, and we seek to compensate for the impossibility of knowing them by determining their different degrees of likelihood. So it is that we owe to the weakness of the human mind one of the most delicate and ingenious of mathematical theories, the science of chance or probability.

(I think that Laplace's view of probability is still commonly held both by some scientists and, more widely, by well-educated people in general. Such is the allure of physical determinism, and our hope for a simple order of things—though I suspect that nature contains much truly intrinsic randomness at all levels.)

In celestial mechanics, the primary focus of his career, Laplace emphasized one theme above all others: the laws of nature, with Newton's principle of universal gravitation in the lead, decree a permanent stability that could only be disturbed by foreign causes (like God's miraculous hand—the unneeded hypothesis!). Laplace attacked this issue by studying all the classical and apparent exceptions that studies of planetary motion had accumulated over the centuries. These exceptions all took the same form: measurement of planetary orbits had detected a slight but accumulating irregularity, which, if continued over aeons, would destabilize the solar system. In each case, Laplace devised the same style of solution: these irregularities are not accumulative, but self-correcting. They are cycling oscillations that maintain the broader and permanent stability of the solar system. For this brilliant work, Laplace justly earned his common epithet as the Newton of France.

In 1773, Laplace took up the troubling problem of why Jupiter's orbit seemed to be shrinking while Saturn's expanded (a situation that, if continued, would destroy the regularity of planetary motion; the great Newton, in fact, had thrown up his hands and invoked occasional divine intervention to safeguard equilibrium). Laplace showed that these inequalities are periodic (with a cycle of nearly one thousand years), and not accumulating. In the next phase of the cycle, Jupiter's orbit will expand, and Saturn's shrink. Then, in 1786, Laplace developed a general proof that eccentricities and inclinations of planetary orbits must remain small and be fully self-correcting, thus maintaining the stability of the solar system.

Finally, in 1787, Laplace resolved the last major anomaly in planetary motion by relating the moon's orbit to changes in eccentricity of the earth's revolution about the sun. The moon's orbit had been expanding and our satellite would eventually escape, should the trend continue. Laplace showed that the moon's mean motion is accelerated when the earth's orbit becomes more circular, but will be retarded when the earth's eccentricity increases. He then argued that the earth's orbital eccentricity cycles with a period measured in millions of years; the lunar orbit will therefore be self-correcting, and the moon will not escape.

In 1788, with the fall of the Bastille and the great secular revolution just one year away, Laplace summarized his views on the fact and meaning of celestial stability:

> Thus the system of the world only oscillates around a mean state from which it never departs except by a very small quantity. By virtue of its constitution and the law of gravity, it enjoys a stability that can be destroyed only by foreign causes, and we are certain that their action is undetectable from the time of the most ancient observations until our own day. This stability in the system of the world, which assures its duration, is one of the most notable among all phenomena, in that it exhibits in the heavens the same intention to maintain order in the universe that nature has so admirably observed on earth for the sake of preserving individuals and perpetuating species.

All the foregoing properly leads us to view Laplace as the archetypal defender of a certain view of science, all too commonly equated with the entire varied enterprise: stability in the heavens, determinism of all events under the aegis of natural laws with clean, mathematical formulation—an almost antihistorical view that we might contrast with alternative models of complex unpredictability and dynamic change, often in accumulating and directional modes.

Fair enough, but we now encounter the anomaly that inspired this essay. Laplace is also the author of the first widely credited historical theory for the origin of the solar system—the so-called nebular hypothesis of Kant and Laplace, first enunciated in 1796. (The great philosopher Immanuel Kant published a similar theory in the same year as Laplace;

the two men were not in contact and surely developed their ideas independently.) How could the apostle of nonchange and antihistory also devise a theory that, according to the *Dictionary of Scientific Biography*, "has conventionally been cited as an early instance, perhaps as marking the introduction, of a historical dimension into physical science. That attribution, indeed, has been its chief attraction."

In 1796, Laplace published a wonderful book, honored and regarded as a prototype ever since, in a tradition that the French call *haute vulgarisation* (or high-class popularization—not at all an oxymoron, but the worthiest of all goals for scientific writers). The work, titled *Exposition du système du monde (Exposition of the System of the World)*, is suffused with the rationalistic spirit of a revolutionary France that had thrown off the shackles of past history. The title page, in fact, does not say 1796, but only "*L'an IV de la République Française,*" since the revolutionary government had started time all over again on September 22, 1792, the founding day of the French republic.

In the opening "*avertissement,*" Laplace states that he will divide the circle into four hundred degrees (one hundred for each quadrant), the day into ten hours, the hour into one hundred minutes, the minute into one hundred seconds, and temperature into one hundred degrees from freezing to boiling of water—the only survivor, as the Centigrade scale, of these attempts to rationalize old measures. (Do not infer that Laplace was a revolutionary zealot. Quite the contrary. He was shrewd and basically unpolitical. His major accomplishment, as the old quip says of Talleyrand as well, was to serve every government from revolution to restoration, and die in bed. Laplace flourished by supporting any group in power, while not alienating the probable successors. His written dedication to Napoleon in his *Théorie analytique des probabilités* [1812] seemed so embarrassingly sycophantic to later editors that the semiofficial *Oeuvres complètes [Complete Works]*, published after Laplace's death, left it out.)

The *Exposition* is a two-volume work in five books—the first, on what may be seen by observing the heavens on a clear night; the second, on the "real" motion of planets, moons, and comets; the third, on laws of motion; the fourth, on Laplace's own work in celestial mechanics and gravity; and the fifth, on the history of astronomy. Laplace shows his distrust and discomfort for real history with all its messiness, its backings and forthings, by stating that he will not discuss astronomy as people actually developed the ideas, but will instead provide a rationally ordered chronological account of successes:

> The order in which I am about to discuss the principal results of the system of the world is not that which the human mind followed in its research. The march of the mind has been encumbered and uncertain; often, it only reached the true cause of phenomena after having exhausted all the false hypotheses that imagination had suggested; and discovered truths have almost always been allied to errors that time and observation finally separated out. I will offer in a few words the tableau of attempts, and of successes.

The nebular hypothesis is undoubtedly the most famous legacy of Laplace's *Exposition*, but this theory appears only as an afterthought in a few pages of a final chapter appended to the end of book five—*Considerations sur le système du monde, et sur les progrès futurs de l'astronomie (Considerations on the System of the World, and on the Future Progress of Astronomy).* This remarkable chapter also features a correct hypothesis that many "nebulae" (resolved in the best telescopes of the time as diffuse clouds) are actually distant galaxies of stars (with the Milky Way as an arm of our own galaxy), and that the universe is therefore even vaster than we had ever conceived. In this section, Laplace even recognizes that some stars may be so dense that gravity precludes the escape of their own light—the phenomenon now recognized (in different form) as black holes. Thus, Laplace argues (incorrectly this time), much apparent darkness in the night sky may really be occupied by enormous, dense stars. His figures and sizes are wrong by irrelevant modern standards, but his conjecture is fascinating:

> A luminous star of the same density as the earth, but with a diameter 250 times greater than that of the sun, would not allow any of its light to reach us, by virtue of its own gravitational attraction. It is therefore possible that the largest luminous bodies in the universe are, for this reason, invisible.

According to the nebular hypothesis, the sun, early in its history, was surrounded by an atmosphere that extended well beyond the current planetary orbits. This atmosphere rotated with the sun. At successive intervals, large segments broke off and coalesced in an equatorial plane at the periphery of this contracting mass; these segments also began to rotate on their own behalf, and formed the planets at their centers. Moons formed by a similar process of atmospheric rotation and spalling off around planetary cores. Laplace argued that no other mechanism could account for all the primary regularities of motion in the solar system—particularly the revolution of all planets in the same direction and virtually in the same plane, the revolution of satellites in the same direction, and the rotation in the same direction of all planets and satellites (not true, but Laplace didn't know).

How, then, can we resolve the paradox that the scientific apostle of stability, a man who seemed to distrust and reject any real history either for celestial objects or for his own profession, should also be godfather to the first important theory for the origin of the solar system? Part of the answer may simply lie in the fact that Laplace dedicated only a few pages to the nebular hypothesis—and anyone might allow himself an uncharacteristic speculation, or a fanciful sally into a field usually considered alien, for such little dedicated space. (Until I read the *Exposition* after buying a copy recently, I had never realized how few pages the nebular hypothesis occupied, so the anomaly seemed greater to me. We so often make the silly mistake of equating later importance with length of original effort. Many of the most famous ideas in science began as paragraphs or footnotes in weighty tomes otherwise entirely forgotten. Have we not all been surprised and amused to find that some of the best known biblical stories occupy only a line or two among pages of begats and other dull lists?)

But the main reason for Laplace's short excursion into history is far more interesting, and entirely conceptual rather than practical. Most intellectuals never abandon their motivating belief; if they seem to write about something contrary, more-careful reading usually reveals the passage as a form of support for the familiar central doctrine. Of course the nebular hypothesis is a historical statement about the origin of planets, but as I read Laplace's conjecture and came to the last paragraph, I saw the evident solution and chuckled. Laplace invoked the nebular hypothesis in his usual interest of bolstering stability in the solar system! Planets must have some origin, after all, and Laplace argues that this particular style of formation best guarantees permanence thereafter. The striking last paragraph, virtually cribbed from his 1788 article previously cited, triumphantly proclaims:

> Whatever one makes of this origin for the planetary system . . . it is certain these elements are ordered in such a manner that they must enjoy the greatest stability, if foreign causes never trouble them. Only by this means [formation by the nebular hypothesis] are the movements of planets and satellites almost cir-

> cular, and directed in the same sense and almost in the same plane. This system can only oscillate about a mean state, from which it can only deviate by very tiny amounts. The average movements of rotation and revolution of these different bodies are uniform . . . It seems that nature arranged all bodies in the heavens in order to assure the duration of the system, and by means similar to those so admirably followed on earth for the conservation of individuals and the perpetuation of species.

I was arrogant enough to think that I had made some sort of discovery when I read the *Exposition* and recognized the antihistorical basis for the hypothesis that made Laplace so famous as the first historian for the universe. But I soon discovered that others had followed the same path of argument. C. C. Gillipie, perhaps America's finest senior historian of science, put the point most forcefully in his long article on Laplace in the *Dictionary of Scientific Biography*:

> If the text is allowed to speak for Laplace, it will be altogether evident that evolutionary considerations in the 19th-century sense formed no part of his mentality. The conclusions that he had reached concerned stability; the evidence for that he had calculated, many and many a time . . . He again referred to it as a warranty for the care that nature had taken to ensure the duration of the physical universe, just as it has the conservation of organic species . . . Clearly, it was not about the development of the solar system that he was thinking. It was about the birth.

We can, I think, best grasp the contrast between Laplace's antihistorical thinking and a truly developmental approach by comparing the nebular hypothesis with the only serious contemporary competitor as a theory for planetary origins—the hypothesis of cometary collision devised by the greatest of all eighteenth-century French naturalists, Georges Buffon (1707–1778). Laplace himself admitted Buffon as his only competition, writing in the *Exposition:* "Buffon is the only one I know who, since the discovery of the true system of the world, has tried to go back to the origin of planets and satellites."

Buffon argued that a comet had struck the sun, knocking out a large plume of solar material that then broke up to form the planets and satellites. Laplace rejected this idea because Buffon's theory could not, in his view, explain all the regularities of planetary motion. Cometary impact would account for the common direction of planetary revolution, with all planets in virtually the same plane (a result of the motion and orientation imparted to the plume knocked from the sun). But Laplace argued that Buffon's theory could not explain the common direction of planetary rotation or the origin of satellites.

Buffon and Laplace seem so different at first glance. Their generation or two of separation spans a world of change from Buffon's service to the last two King Louis before the revolution, to Laplace's work with various revolutionary governments and Napoleon. But their lives and studies include some striking similarities relevant to their joint interest in theories of planetary origin. Buffon was also a fine mathematician with two special interests that matched Laplace with uncanny precision. He was, first of all, a committed Newtonian who translated *The Method of Fluxions* into French from an English version of Newton's original Latin. Second, his greatest interest lay in probability, and he made a major contribution in first applying integral and differential calculus by extending the theory of probability to surfaces. (Interestingly, both Buffon and Laplace won admission to the French Academy of Sciences for monographs on probability, Buffon in 1734, Laplace in 1773.)

But the two men, in their scientific maturity, occupied opposite ends on the spectrum of professional activity, and the contrasting ethos of these termini set their profoundly differing attitudes to history, making Laplace indifferent and Buffon intrinsically committed. Laplace stuck to the mathematical bent of his youth and became the greatest celestial mechanic of his time. Buffon, on the other hand, changed course and devoted his career to botany and zoology; he became, in short, the greatest earthly naturalist of his day (only Linnaeus himself might have been granted higher rank—see essay 32).

Buffon's magnificent and multivolume *Histoire naturelle* took a lifetime (Buffon died before its completion) and fills a large library shelf. Students of the heavens may revel in constancy and precision. Students of earthly organisms also search for general patterns, and frequently succeed; but naturalists must also take delight in the uniqueness of each creature, and they must be sensitive to the developmental histories of organisms, both in the courses of their lifetimes and (if they study the fossil record, as Buffon did) in the vastly larger domain of geological time. Good naturalists must be historians.

In 1749, Buffon introduced his cometary theory of planetary formation in his first work on geology, *Histoire et théorie de la terre (History and Theory of the Earth)*. Much later, in 1778, the year of his death, Buffon published a much expanded and altered version entitled

Epoques de la nature (Epochs of Nature). Most biologists and historians consider the *Epoques* as Buffon's masterpiece and as one of the finest examples of scientific prose ever written. The *Epoques* also includes an explicit defense and exposition of historical methodology, thus providing a striking contrast with Laplace and helping us to grasp the criteria of proper history. In particular, two differences between Buffon and Laplace sharpen our understanding of the nature of historical inquiry.

CRITERIA OF INFERENCE. Historians use and cherish the narrative methods of explanation by antecedent events and situations; current results are outcomes of the unique and contingent web of all that came before, and all that holds continuity with a present world in need of explanation. Historians also know that records of the past must be imperfect, for many kinds of data are not recorded as material remains, and much that could be preserved in principle has not survived in actuality. We always mourn lost data and hope for greater completion, but we do not apologize for the necessarily fragmentary record of our past, and we may treat spotty information as a delicious puzzle and a challenge. Antihistorians, like Laplace, get very nervous when they must use narrative data; they often become downright apologetic when they base a claim on anything other than a calculation or a direct observation of a present event.

Laplace ended his discussion of the nebular hypothesis with just such an apology, speaking of "this planetary system, which I present with the mistrust which must accompany everything that is not the result of an observation or a calculation." Buffon, on the other hand, begins the *Epoques* with a paean of praise to the excitement and efficacy of digging into the archives of the past with narrative methods. Consider Buffon's opening words:

> In civil history, we consult titles, we research medals, we decipher ancient inscriptions in order to determine the time of human revolutions and to fix the dates of events in the moral order. Similarly, in natural history, it is necessary to excavate the archives of the world, to draw old monuments from the entrails of the earth, to collect their debris, and to reassemble into a single body of proof all the indices of physical changes which enable us to go back to the different ages of nature. This is the only way to fix points in the immensity of space, and to place a certain number of milestones on the eternal route of time. [My translation.]

CHARACTER OF EVENTS. History must respect (and even love) the last two syllables of its name. Narratives must tell a story, a tale that captures our interest as a series of unique events with interesting causal connections. There is no history in Laplace's heavens, but only a suite of bodies going nowhere as they cycle endlessly in obedience to simple laws; any promising hope for directionality or accumulating instability is soon dashed by the self-correcting cyclicity of all perturbations. His nebular hypothesis is history, but only for the geological instant of the solar system's birth; ahistorical timelessness rules forever after. The thing that hath been, it is that which shall be; and that which is done, is that which shall be done; and there is no new thing under the sun.

By contrast, Buffon's *Epochs of Nature* rests upon an opposite conviction that the time of our planet tells an engrossing story of accumulating change through several stages (called "epochs" by Buffon to set his title). He divided the history of the earth into seven directional epochs: first, the origin of the earth and planets by cometary impact; second, the formation of the solid earth and its mineral deposits; third, the covering of continents by water and the production of marine life; fourth, the retreat of waters and the emergence of new continents by volcanic action; fifth, the appearance of animal life on land; sixth, the fragmentation of continents and formation of the earth's current topography; and, seventh, the appearance of humans and our accession to power. Could any contrast with Laplace's ever-cycling heavens be more profound?

Buffon explicitly challenged the idea of constancy by noting that the narrative record of geology and paleontology proclaims a story of directional change:

> Although it may appear at first sight that the great works [of nature] do not alter and never change, and that its productions, even the most fragile and most evanescent, must be always and constantly the same . . . nevertheless, in observing more closely, we note that [nature's] course is not absolutely uniform, that it undergoes successive alterations giving rise to new combinations and to mutations of matter and form; and that, finally, however fixed nature may appear in its ensemble, so is it variable in each one of its parts; and if we embrace nature in its full extent, we cannot doubt that it is very different today from what it was at the beginning and from what it has become in the succession of time: it is these changes that we are calling epochs.

Lawful timelessness is awesome, but the pageant of history thrills us too, and in a different way that makes time sensible. Everyone needs a good mechanic, including the heavens, but give me an earthly naturalist any day, for humans are storytellers. In the nearly 250 essays of this series, I have tried to avoid repetition (if only to honor the principles of history cited above). But, like a broken record (a metaphor from the last epoch of history, soon to be rendered unintelligible, I fear), one quotation keeps recurring. I have used it to end nearly half a dozen essays (shameful in a way, but we all have our Laplacean side). This quotation also includes the masthead that I use for the entire series, "this view of life" (I guess we all need our constancies). I love this quotation because it affirms the power of life and history by making the same contrast between Laplace's ever-cycling heavens, always moving yet always the same, and the glorious tale of life, always different, always going somewhere, always telling a story. It is the last paragraph of Darwin's *Origin of Species:*

> There is grandeur in this view of life, with its several powers, having been originally breathed into a few forms or into one; and that, whilst this planet has gone cycling on according to the fixed law of gravity, from so simple a beginning endless forms most beautiful and most wonderful have been, and are being, evolved.

4

The Late Birth of a Flat Earth

THE MORTAL REMAINS OF the Venerable Bede (673–735) lie in Durham Cathedral, under a tombstone with an epitaph that must win all prizes for a "no nonsense" approach to death. In rhyming Latin doggerel, the vault proclaims: *Hac sunt in fossa, Baedae venerabilis ossa*—"The bones of the Venerable Bede lie in this grave." (*Fossa* is, literally, a ditch or a trough, but we will let this gentler reading stand.)

In the taxonomy of Western history that I learned as a child, Bede shone as a rare light in the "Dark Ages" between Roman grandeur and a slow medieval recovery culminating in the renewed glory of the Renaissance. Bede's fame rests upon his scriptural commentaries and his *Historia ecclesiastica gentis Anglorum (Ecclesiastical History of the English People)*, completed in 732. Chronology sets the basis of good history, and Bede preceded his great work with two treatises on the reckoning and sequencing of time: *De temporibus (On Times)* in 703, and *De temporum ratione (On the Measurement of Times)* in 725.

Bede's chronologies had their greatest influence in popularizing our inconvenient system of dividing recent time into B.C. and A.D. (see essay 2) on opposite sides of Christ's supposed nativity (almost surely incorrectly determined, as Herod had died by this time of transition, and could not have seen the Wise Men or slaughtered the innocent at the onset of year one). In his chronologies, Bede sought to order the events of Christian history, but the primary motive and purpose of his calculations centered on a different, and persistently vexatious, problem in ecclesiastical timing—the reckoning of Easter. The complex definition of this holiday—the first Sunday following the first full moon occurring on or after the vernal equinox—requires considerable astronomical sophistication, for lunar and seasonal cycles must both be known with precision.

Such computations entail a theory of the heavens, and Bede clearly presented his classical conception of the earth as a sphere at the hub of the cosmos—*orbis in medio totius mundi positus* (an orb placed in the center of the universe). Lest anyone misconstrue his intent, Bede then explicitly stated that he meant a three-dimensional sphere, not a flat plate. Moreover, he added, our planetary sphere may be considered as perfect because even the highest mountains produce no more than an imperceptible ripple on a globe of such great diameter.

I also once learned that most other ecclesiastical scholars of the benighted Dark Ages had refuted Aristotle's notion of a spherical earth, and had depicted our home as a flat, or at most a gently curved, plate. Didn't we all hear the legend of Columbus at Salamanca, trying to convince the learned clerics that he would reach the Indies and not fall off the ultimate edge?

The human mind seems to work as a categorizing device (perhaps even, as many French structuralists argue, as a dichotomizing machine, constantly partitioning the world into dualities of raw and cooked [nature vs. culture], male and female, material and spiritual, and so forth). This deeply (perhaps innately) ingrained habit of thought causes us particular trouble when we need to analyze the many continua that form so conspicuous a part of our surrounding world. Continua are rarely so smooth and gradual in their flux that we cannot specify certain points or episodes as decidedly more interesting, or more tumultuous in their rates of change, than the vast majority of moments along the sequence. We therefore falsely choose these crucial episodes as boundaries for fixed categories, and we veil nature's continuity in the wrappings of our mental habits.

We must also remember another insidious aspect of our tendency to divide continua into fixed categories. These divisions are not neutral; they are established for definite purposes by partisans of particular viewpoints. Moreover, since many continua are temporal, and since we have a lamentable tendency to view our own age as best, these divisions often saddle the past with pejorative names, while designating successively more modern epochs with words of light and progress. As an obvious example, many people (including yours truly) view the great medieval cathedrals of Europe as the most awesome of all human constructions. (For me—and I say this as a humanist and non-theist—Chartres is off scale, a place of mystery and magic, not truly of this world.) Yet we designate the style of these buildings "Gothic"—originally a pejorative term (traced to seventeenth-century origin in the *Oxford English Dictionary*) applied by self-styled sophisticates who viewed medieval times as a barbaric interlude between the classical forms of Greece and Rome, and their revival in Renaissance and later times. These cathedrals, after all, were not built by German tribes who had their heyday in the third to fifth centuries! The names of several peoples who conquered the waning classical world—Goths and Vandals in particular—became pejorative terms for anything considered rude or mean. For that matter, the word *barbarian* comes from the Latin term for "foreigner."

Our conventional divisions of Western history are mired in these twinned errors of false categorization and pejorative designation. I know that professional historians no longer use such a taxonomy, but popular impression still supports a division into classical times (glory of Greece and grandeur of Rome), followed by the pall of the Dark Ages, some improvement in the Middle Ages, and an éclat of culture's rediscovery in the Renaissance. But consider the origin of the two pejorative terms in this sequence—and the relationship of taxonomy to prejudiced theories of progress becomes clear.

According to the historian J. B. Russell, Petrarch devised the term "Dark Ages" in about 1340 to designate a period between classical times and his own form of modernism. The term "Middle Ages" for the interval between classical fall and Renaissance revival originated in the fifteenth century, but gained popularity only in the seventeenth century. Some people consider everything from the fall of Rome to the Renaissance as Dark, others as Middle. Still others make a sequential division into an earlier Dark and later Middle, separated by Charlemagne or by the arbitrary millennial transition of 1000. Such uncertainty only shows the foolishness of attempting to define fixed categories within continua. In any case, the intent of Darks and Middles could not be more clear—to view Western history as possessing a Greek and Roman acme, with supposed loss as tragic, followed by the beginning of salvation in Renaissance rediscovery.

Such prejudicial tales of redemption require a set of stories to support their narrative. Most of these legends feature art, literature, or architecture, but science has also contributed. I write this essay to point out that the most prominent of all scientific stories in this mode—the supposed Dark and Medieval consensus for a flat earth—is entirely mythological. Moreover, when we trace the invention of this fable in the nineteenth century, we receive a double lesson in the dangers of false taxonomies—the second and larger purpose of this essay. For the myth itself only makes sense under a prejudicial view of Western history as an era of darkness between lighted beacons of classical learning and Renaissance revival—while the nineteenth-century invention of the flat earth, as we shall see, occurred to support another dubious and harmful separation wedded to another legend of historical progress—the supposed warfare between science and religion.

Classical scholars, of course, had no doubt about the earth's sphericity. Our planet's roundness was central to Aristotle's cosmology and was assumed in Eratosthenes' measurement of the earth's circumference in the third century B.C. The flat-earth myth argues that this knowledge was then lost when ecclesiastical darkness settled over Europe. For a thousand years of middle time, almost all scholars held that the earth must be flat—like the floor of a tent, held up by the canopy of the sky, to cite a biblical metaphor read literally. The Renaissance rediscovered classical notions of sphericity, but proof required the bravery of Columbus and other great explorers who should have sailed off the edge, but (beginning with Magellan's expedition) returned home from the opposite direction after going all the way round.

The inspirational, schoolchild version of the myth centers upon Columbus, who supposedly overcame the calumny of assembled clerics at Salamanca to win a chance from Ferdinand and Isabella. Consider this version of the legend, cited by Russell from a book for primary-school children written in 1887, soon after the myth's invention (but little different from accounts that I read as a child in the 1950s):

> "But if the world is round," said Columbus, "it is not hell that lies beyond that stormy sea. Over there must lie the eastern strand of Asia, the Cathay of Marco Polo" . . . In the hall of the convent there was assembled the imposing company—shaved monks in gowns . . . cardinals in scarlet robes . . . "You think the earth is round . . . Are you not aware that the holy fathers of the church have condemned this belief . . . This theory of yours looks heretical." Columbus might well quake in his boots at the mention of heresy; for there was that new Inquisition just in fine running order, with its elaborate bone-breaking, flesh-pinching, thumb-screwing, hanging, burning, mangling system for heretics.

Dramatic to be sure, but entirely fictitious. There never was a period of "flat earth darkness" among scholars (regardless of how many uneducated people may have conceptualized our planet both then and now). Greek knowledge of sphericity never faded, and all major medieval scholars accepted the earth's roundness as an established fact of cosmology. Ferdinand and Isabella did refer Columbus's plans to a royal commission headed by Hernando de Talavera, Isabella's confessor and, following defeat of the Moors, Archbishop of Granada. This commission, composed of both clerical and lay advisers, did meet, at Salamanca among other places. They did pose some sharp intellectual objections to Columbus, but all assumed the earth's roundness. As a major critique, they argued that Columbus could not reach the Indies in his own allotted time, because the earth's circumference was too great. Moreover, his critics were entirely right. Columbus had "cooked" his figures to favor a much smaller earth, and an attainable Indies. Needless to say, he did not and could not reach Asia, and Native Americans are still called Indians as a legacy of his error.

Virtually all major medieval scholars affirmed the earth's roundness. I introduced this essay with the eighth-century view of the Venerable Bede. The twelfth-century translations into Latin of many Greek and Arabic works greatly expanded general appreciation of natural sciences, particularly astronomy, among scholars—and convictions about the earth's sphericity both spread and strengthened. Roger Bacon (1220–1292) and Thomas Aquinas (1225–1274) affirmed roundness via Aristotle and his Arabic commentators, as did the greatest scientists of later medieval times, including John Buriden (1300–1358) and Nicholas Oresme (1320–1382).

So who, then, was arguing for a flat earth, if all the chief honchos believed in roundness? Villains must be found for any malfeasance, and Russell shows that the great English philosopher of science William Whewell first identified major culprits in his *History of the Inductive Sciences*, published in 1837—two minimally significant characters named Lactantius (245–325) and Cosmas Indicopleustes, who wrote his "Christian Topography" in 547–549. Russell comments: "Whewell pointed to the culprits . . . as evidence of a medieval belief in a flat earth, and virtually every subsequent historian imitated him—they could find few other examples."

Lactantius did raise the old saw of absurdity in believing that people at the antipodes might walk with their feet above their heads in a land where crops grow down and rain falls up. And Cosmas did champion a literal view of a biblical metaphor—the earth as a flat floor for the rectangular, vaulted arch of the heavens above. But both men played minor roles in medieval scholarship. Only three reasonably complete medieval manuscripts of Cosmas are known (with five or six additional fragments), and all in Greek. The first Latin translation dates from 1706—so Cosmas remained invisible to medieval readers in their own lingua franca.

Purveyors of the flat-earth myth could never deny this plain testimony of Bede, Bacon, Aquinas, and others—so they argued that these men acted as rare beacons of brave light in pervasive darkness. But consider the absurdity of such a position. Who formed the orthodoxy representing this consensus of ignorance? Two pipsqueaks named Lactantius and Cosmas Indicopleustes? Bede, Bacon, Aquinas, and their ilk were not brave iconoclasts. They formed the establishment, and their convictions about the earth's roundness stood as canonical, while Lactantius and colleagues remained entirely marginal. To call Aquinas a courageous revolutionary because he promoted a spherical earth would be akin to labeling Fisher, Haldane, Wright, Dobzhansky, Mayr, Simpson, and all the other great twentieth-century evolutionists as radical reformers because a peripheral creationist named Duane Gish wrote a pitiful little book during the same years called *Evolution, the Fossils Say No*!

Where then, and why, did the myth of medieval belief in a flat earth arise? Russell's historiographic work gives us a good fix on both times and people. None of the great eighteenth-century anticlerical rationalists—not Condillac, Condorcet, Diderot, Gibbon, Hume, or our own Benjamin Franklin—accused the scholastics of believing in a flat earth, though these men were all unsparing in their contempt for medieval versions of Christianity. Washington Irving gave the flat-earth story a good boost in his largely fictional history of Columbus, published in 1828—but his version did not take hold. The legend grew during the nineteenth century, but did not enter the crucial domains of schoolboy pap or tour-guide lingo. Russell did an interesting survey of nineteenth-century history texts for secondary schools, and found that very few mentioned the flat-earth myth before 1870, but that almost all texts after 1880 featured the legend. We can therefore pinpoint the invasion of general culture by the flat-earth myth to the period between 1860 and 1890.

Those years also featured the spread of an intellectual movement based on the second error of taxonomic categories explored in this essay—the portrayal of Western history as a perpetual struggle, if not an outright "war," between science and religion, with progress linked to the victory of science and the consequent retreat of theology. Such movements always need whipping boys and legends to advance their claims. Russell argues that the flat-earth myth achieved its canonical status as a primary homily for the triumph of science under this false dichotomization of Western history. How could a better story for the army of science ever be concocted? Religious darkness destroys Greek knowledge and weaves us into a web of fears, based on dogma and opposed both to rationality and experience. Our ancestors therefore lived in anxiety, restricted by official irrationality, afraid that any challenge could only lead to a fall off the edge of the earth into eternal damnation. A fit tale for an intended purpose, but entirely false because few medieval scholars ever doubted the earth's sphericity.

I was especially drawn to this topic because the myth of dichotomy and warfare between science and religion—an important nineteenth-century theme with major and largely unfortunate repercussions extending to our times—received its greatest boost in two books that I own and treasure for their firm commitment to rationality (however wrong and ultimately harmful their dichotomizing model of history), and for an interesting Darwinian connection with each author. (I have often said that I write these essays as a tradesman, not a polymath, and that my business is evolutionary theory.) Russell identifies these same two books as the primary codifiers of the flat-earth myth: John W. Draper's *History of the Conflict between Religion and Science*, first published in 1874; and Andrew Dickson White's *A History of the Warfare of Science with Theology in Christendom*, published in 1896 (a great expansion of a small book first written in 1876 and called *The Warfare of Science*).

Draper (1811–1882) was born in England, but emigrated to the United States in 1832, where he eventually became head of the medical school at New York University. His 1874 book ranks among the great publishing successes of the nineteenth century—fifty printings in fifty years as the best-selling volume of the *International Scientific Series*, the most successful of nineteenth-century publishing projects in popular science. Draper states his thesis in the preface to his volume:

> The history of Science is not a mere record of isolated discoveries; it is a narrative of the conflict of two contending powers, the expansive force of the human intellect on one side, and the compressing arising from traditionary faith and human interests on the other . . . Faith is in its nature unchangeable, stationary; Science is in its nature progressive; and eventually a divergence between them, impossible to conceal, must take place.

Draper extolled the flat-earth myth as a primary example of religion's constraint and science's progressive power:

> The circular visible horizon and its dip at sea, the gradual appearance and disappearance of ships in the offing, cannot fail to incline intelligent sailors to a belief in the globular figure of the earth. The writings of the Mohammedan astronomers and philosophers had given currency to that doctrine throughout Western Europe, but, as might be expected, it was received with disfavor by theologians . . . Traditions and policy forbade [the Papal Government] to admit any other than the flat figure of the earth, as revealed in the Scriptures.

Russell comments on the success of Draper's work:

> *The History of the Conflict* is of immense importance, because it was the first instance that an influential figure had explicitly declared that science and religion were at war, and it succeeded as few books ever do. It fixed in the educated mind the idea that "science" stood for freedom and progress against the superstition and repression of "religion." Its viewpoint became conventional wisdom.

Andrew Dickson White (1832–1918) grew up in Syracuse, New York, and founded Cornell University in 1865 as one of the first avowedly secular institutions of higher learning in America. He wrote of the goals he shared with his main benefactor, Ezra Cornell:

> Our purpose was to establish in the State of New York an institution for advanced instruction and research, in which science, pure and applied, should have an equal place with literature; in which the study of literature, ancient and modern, should be emancipated as much as possible from pedantry . . . We had especially determined that the institution should be under the control of no political party and of no single religious sect.

White avowed that his decision to found a secular university reflected no hostility to theology, but only recorded his desire to foster an ecumenical religious spirit:

> It had certainly never entered into the mind of either of us that in all this we were doing anything irreligious or unchristian . . . I had been bred a churchman, and had recently been elected a trustee of one church college, and a professor in another . . . my greatest sources of enjoyment were ecclesiastical architecture, religious music, and the more devout forms of poetry. So far from wishing to injure Christianity, we both hoped to promote it; but we did not confound religion with sectarianism.

But the calumnies of conservative clergymen dismayed him profoundly and energized his fighting spirit:

> Opposition began at once . . . from the good protestant bishop who proclaimed that all professors should be in holy orders, since to the Church alone was given the command "Go, teach all the nations," to the zealous priest who published a charge that . . . a profoundly Christian scholar had come to Cornell in order to inculcate infidelity . . . from the eminent divine who went from city to city denouncing the "atheistic and pantheistic tendencies" of the proposed education, to the perfervid minister who informed a denominational synod that Agassiz, the last great opponent of Darwin, and a devout theist, was "preaching Darwinism and atheism" in the new institution.

These searing personal experiences led White to a different interpretation of the "warfare of science with theology." Draper was a genuine anti-theist, but he confined his hostility almost entirely to the Catholic Church, as he felt that science could coexist with more liberal forms of Protestantism. White, on the other hand, professed no hostility to religion, but only to dogmatism of any stripe—while his own struggles had taught him that Protestants could be as obstructionist as anyone else. He wrote: "Much as I admired Draper's treatment of the questions involved, his point of view and mode of looking at history were different from mine. He regarded the struggle as one between Science and Religion. I believed then, and am convinced now, that it was a struggle between Science and Dogmatic Theology." White therefore argued that the triumph of science in its warfare with dogmatism would benefit true religion as much as science. He expressed his credo as a paragraph in italics in the introduction to his book:

> In all modern history, interference with science in the supposed interest of religion, no matter how conscientious such interference may have been, has resulted in the direst evils both to religion and to science, and invariably; and, on the other hand, all untrammelled scientific investigation, no matter how dangerous to religion some of its stages may have seemed for the time to be, has invariably resulted in the highest good both of religion and of science.

Despite these stated disagreements, White's and Draper's accounts of the actual interaction between science and religion in Western history do not differ greatly. Both tell a tale of bright progress continually sparked by science. And both develop and utilize the same myths to support their narrative, the flat-earth legend prominently among them. Of Cosmas Indicopleustes's flat-earth theory, for example, White wrote, "Some of the foremost men in the Church devoted themselves to buttressing it with new texts and throwing about it new outworks of theological reasoning; the great body of the faithful considered it a direct gift from the Almighty."

As another interesting similarity, both men developed their basic model of science vs. theology in the context of a seminal and contemporary struggle all too easily viewed in this light—the battle for evolution, specifically for Darwin's secular version based on natural selection. No issue, certainly since Galileo, had so challenged traditional views of the deepest meaning of human life, and therefore so contacted a domain of religious inquiry as well (see essay 25). It would not be an exaggeration to say that the Darwinian revolution directly triggered this influential nineteenth-century conceptualization of Western history as a war between two taxonomic categories labeled science and religion. White made an explicit connection in his statement about Agassiz (the founder of the museum where I now work, and a visiting lecturer at Cornell). Moreover, the first chapter of his book treats the battle over evolution, while the second begins with the flat-earth myth.

Draper wraps himself even more fully in a Darwinian mantle. The end of his preface designates five great episodes in the history of science's battle with religion: the debasement of classical knowledge and the descent of the Dark Ages; the flowering of science under early Islam; the battle of Galileo with the Catholic Church; the Reformation (a plus for an anti-Catholic like Draper); and the struggle for Darwinism. No one in the world had a more compelling personal license for such a view, for Draper had been an unwilling witness—one might even say an instigator—of the single most celebrated incident in overt struggle between Darwin and divinity. We all have heard the famous story of Bishop Wilberforce and T. H. Huxley duking it out at the British Association meeting in 1860 (for more on this incident, see essay 26 in my earlier book *Bully for Brontosaurus*). But how many people know that their verbal pyrotechnics did not form the stated agenda of this meeting, but only arose during free discussion following the formal paper officially set for this session—an address by the same Dr. Draper on the "intellectual development of Europe considered with reference to the views of Mr. Darwin." (I do love coincidences of this sort. Sociologists tell us that we can touch anyone through no more than six degrees of separation, given the density of networks in human contact. But to think of Draper, taking the first degree just inches from Hooker, Huxley, and Wilberforce, can only be viewed as God's gift to an essayist who traffics in connections.)

This essay has discussed a double myth in the annals of our bad habits in false categorization: (1) the flat-earth legend as support for a biased ordering of Western history as a story in redemption from classical to Dark to Medieval to Renaissance; and (2) the invention of the flat-earth myth to support a false dichotomization of Western history as another story of progress, a war of victorious science over religion.

I would not be agitated by these errors if they led only to an inadequate view of the past without practical consequence for our modern world. But the myth of a war between science and religion remains all too current, and continues to impede a proper bonding and conciliation between these two utterly different and powerfully important institutions of human life. How can a war exist between two vital subjects with such different appropriate turfs—science as an enterprise dedicated to discovering and explaining the factual basis of the empirical world, and religion as an examination of ethics and values?

I do understand, of course, that this territorial separation is a modern decision—and that differing past divisions did entail conflict in subsequent adjustment of boundaries. After all, when science was weak to nonexistent, religion did extend its umbrella into regions now properly viewed as domains of natural knowledge. But shall we blame religion for these overextensions? As thinking beings, we are internally compelled to ponder the great issues of human origins and our relationship with the earth and other creatures; we have no other option but ignorance. If science once had no clue about these subjects, then they fell, albeit uncomfortably and inappropriately, into the domain of religion by default. No one gives up turf voluntarily, and the later expansion of science into rightful territory temporarily occupied by religion did evoke some lively skirmishes and portentous battles. These tensions were also exacerbated by particular circumstances of contingent history—including the resolute and courageous materialism of Darwin's personal theory, and the occupation (at the same time) of the Holy See by one of the most fascinating and enigmatic figures of the nineteenth century: the strong, embittered, and increasingly conservative pope Pio Nono (Pius IX).

But these adjustments, however painful, do not justify a simplistic picture of history as continual warfare between science and theology. Exposure of the flat-earth myth should teach us the fallacy of such a view and help us to recognize the complexity of interaction between these institutions. Irrationality and dogmatism are always the enemies of science, but they are no true friends of religion either. Scientific knowledge has always been helpful to more generous views of religion—as preservation, by ecclesiastical scholars, of classical knowledge about the earth's shape aided religion's need for accurate calendars, for example.

I began this essay with a story about the Venerable Bede's use of cosmology to set a chronology for the determination of Easter. Let me end with another story in the same mold—and another illustration of science's interesting and complex potential bond with religion. Two days before my visit to the Venerable Bede's tomb in Durham, I marveled at an intricate astronomical device prominently displayed in the Church of St. Sulpice in Paris. Precisely at noon each day, the sun's light shines through a tiny hole in a window high in the south transept, and illuminates a copper meridian laid into the floor of the transept and ending at an obelisk surmounted by a globe at the north wall.

The line and obelisk are appropriately marked so that the days of solstices and equinoxes can be determined with precision by the position of noon light. Why should such a scientific instrument be contained within a church? The inscription on the obelisk gives the answer—*ad certam paschalis* (for the determination of Easter), a calculation that requires precise reckoning of the vernal equinox. Interestingly, as a further illustration of complexities in the relationship between science and religion, St. Sulpice became a temple to humanism during the French Revolution, and most of the religious glass and statuary was smashed. The names of kings and princes, once carved on the obelisk, were thoroughly obliterated, but these fervid revolutionaries spared the beautiful blue marble balustrade of the choir because the copper meridian passes right through, and they did not wish to disrupt a scientific instrument.

I would not choose to live in any age but my own; advances in medicine alone, and the consequent survival of children with access to these benefits, should preclude any temptation to trade for the past. But we cannot understand history if we saddle the past with pejorative categories based on our bad habits for dividing continua into compartments of increasing worth toward the present. These errors apply to the vast paleontological history of life, as much as to the temporally trivial chronicle of human beings. I cringe every time I read that this failed business, or that defeated team, has become a dinosaur in succumbing to progress. *Dinosaur* should be a term of praise, not opprobrium. Dinosaurs reigned for more than 100 million years and died through no fault of their own; *Homo sapiens* is nowhere near a million years old, and has limited prospects, entirely self-imposed, for extended geological longevity.

Honor the past at its face value. The city of York houses the next great cathedral south of Durham. As Durham displays some amusing Latin doggerel to honor the Venerable Bede, so does York feature a verse to illustrate this principle of respect for the past in the service of understanding. On the wall of the chapter house, we read,

Ut rosa flos florum
Sic est domus ista domorum

As the rose is the flower of flowers, so is this the house of houses.

PART TWO

Literature and Science

The Monster's Human Nature

An old Latin proverb tells us to "beware the man of one book"—*cave ab homine unius libri*. Yet Hollywood knows only one theme in making monster movies, from the archetypal *Frankenstein* of 1931 to the recent mega-hit *Jurassic Park* (see essay 17). Human technology must not go beyond an intended order decreed by God or set by nature's laws. No matter how benevolent the purposes of the transgressor, such cosmic arrogance can only lead to killer tomatoes, very large rabbits with sharp teeth, giant ants in the Los Angeles sewers, or even larger blobs that swallow entire cities as they grow. Yet these films often use far more subtle books as their sources and, in so doing, distort the originals beyond all thematic recognition.

The trend began in 1931 with *Frankenstein*, Hollywood's first great monster "talkie" (though Mr. Karloff only grunted, while Colin Clive, as Henry Frankenstein, emoted). Hollywood decreed its chosen theme by the most "up front" of all conceivable strategies. The film begins with a prologue (even before the titles roll) featuring a well-dressed man standing on stage before a curtain, both to issue a warning about potential fright, and to announce the film's deeper theme as the story of "a man of science who sought to create a man after his own image without reckoning upon God."

In the movie, Dr. Waldman, Henry's old medical school professor, speaks of his pupil's "insane ambition to create life," a diagnosis supported by Frankenstein's own feverish words of enthusiasm: "I created it. I made it with my own hands from the bodies I took from graves, from the gallows, from anywhere."

The best of a cartload of sequels, *The Bride of Frankenstein* (1935), makes the favored theme even more explicit in a prologue featuring Mary Wollstonecraft Shelley, who published *Frankenstein* in 1818 when she was only nineteen years old, in conversation with her husband Percy and their buddy Lord Byron. She states: "My purpose was to write a moral lesson of the punishment that befell a mortal man who dared to emulate God."

Shelley's original *Frankenstein* is a rich book of many themes, but I can find little therein to support the Hollywood reading. The text is neither a diatribe on dangers of technology nor a warning about overextended ambition against a natural order. We find no passages about disobeying God—an unlikely subject for Mary Shelley and her freethinking friends (Percy had been expelled from Oxford in 1811 for publishing a defense of atheism). Victor Frankenstein (I do not know why Hollywood changed him to Henry) is guilty of a great moral failing, as we shall see later, but his crime is not technological transgression against a natural or divine order.

We can find a few passages about the awesome power of science, but these words are not negative. Professor Waldman, a sympathetic character in the book, states, for example, "They [scientists] penetrate into the recesses of nature, and show how she works in her hiding places. They ascend into the heavens; they have discovered how the blood circulates, and the nature of the air we breathe. They have acquired new and almost unlimited powers." We do learn that ardor without compassion or moral consideration can lead to trouble, but Shelley applies this argument to any endeavor, not especially to scientific discovery (her examples are, in fact, all political). Victor Frankenstein says:

> A human being in perfection ought always to preserve a calm and peaceful mind, and never to allow passion or a transitory desire to disturb his tranquility. I do not think that the pursuit of knowledge is an exception to this rule. If the study to which you apply yourself has a tendency to weaken your affections . . . then that study is certainly unlawful, that is to say, not befitting the human mind. If this rule were always observed . . . Greece

> had not been enslaved; Caesar would have spared his country; America would have been discovered more gradually, and the empires of Mexico and Peru had not been destroyed.

Victor's own motivations are entirely idealistic: "I thought, that if I could bestow animation upon lifeless matter, I might in process of time (although I now found it impossible) renew life where death had apparently devoted the body to corruption." Finally, as Victor lies dying in the Arctic, he makes his most forceful statement on the dangers of scientific ambition, but he only berates himself and his own failures, while stating that others might well succeed. Victor says his dying words to the ship's captain who found him on the polar ice: "Farewell, Walton! Seek happiness in tranquility, and avoid ambition, even if it be only the apparently innocent one of distinguishing yourself in science and discoveries. Yet why do I say this? I have myself been blasted in these hopes, yet another may succeed."

But Hollywood dumbed these subtleties down to the easy formula—"man must not go beyond what God and nature intended" (you almost have to use the old gender-biased language for such a simplistic archaicism)—and has been treading in its own footsteps ever since. The latest incarnation, *Jurassic Park,* substitutes a *Velociraptor* re-created from old DNA for Karloff cobbled together from bits and pieces of corpses, but hardly alters the argument an iota.

Karloff's *Frankenstein* contains an even more serious and equally prominent distortion of a theme that I regard as the primary lesson of Mary Shelley's book—another lamentable example of Hollywood's sense that the American public cannot tolerate even the slightest exercise in intellectual complexity. Why is the monster evil? Shelley provides a nuanced and subtle answer that, to me, sets the central theme of her book. But Hollywood opted for a simplistic solution, so precisely opposite to Shelley's intent that the movie can no longer claim to be telling a moral fable (despite protestations of the man in front of the curtain, or Mary Shelley herself in the sequel), and becomes instead, as I suppose the makers intended all along, a pure horror film.

James Whale, director of the 1931 *Frankenstein,* devoted the movie's long and striking opening scenes to this inversion of Shelley's intent—so the filmmakers obviously viewed this alteration as crucial. The movie opens with a burial at a graveyard. The mourners depart, and Henry with his obedient servant, the evil hunchbacked Fritz, dig up the body and cart it away. They then cut down another dead man from a gallows, but

Henry exclaims, "The neck's broken. The brain is useless; we must find another brain."

The scene now switches to Goldstadt Medical College, where Professor Waldman is lecturing on cranial anatomy and comparing "one of the most perfect specimens of the normal brain" with "the abnormal brain of a typical criminal." Waldman firmly locates the criminal's depravity in the inherited malformations of his brain; anatomy is destiny. Note, Waldman says, "the scarcity of convolutions on the frontal lobes and the distinct degeneration of the middle frontal lobes. All of these degenerate characteristics check amazingly with the case history of the dead man before us, whose life was one of brutality, of violence, and of murder."

Fritz breaks in after the students leave and steals the normal brain, but the sound of a gong startles him and he drops the precious object, shattering its container. Fritz then has to take the criminal brain instead, but he never tells Henry. The monster is evil because Henry unwittingly makes him of evil stuff. Later in the film, Henry expresses his puzzlement at the monster's nasty temperament, for he made his creature of the best materials. But Waldman, finally realizing the source of the monster's behavior, tells Henry, "The brain that was stolen from my laboratory was a criminal brain." Henry then counters with one of cinema's greatest double takes, and finally manages a feeble retort: "Oh, well, after all, it's only a piece of dead tissue." "Only evil will come from it," Waldman replies. "You have created a monster and it will destroy you." True enough, at least until the sequel.

Karloff's intrinsically evil monster stands condemned by the same biological determinism that has so tragically, and falsely, restricted the lives of millions who committed no transgression besides membership in a despised race, sex, or social class. Karloff's actions record his internal state. He manages a few grunts and, in *The Bride of Frankenstein*, even learns some words from a blind man who cannot perceive his ugliness, though the monster never gets much beyond "eat," "smoke," "friend," and "good." Shelley's monster, by contrast, is a most remarkably literate fellow. He learns French by assimilation after hiding, for several months, in the hovel of a noble family temporarily in straitened circumstances. His three favorite books would bring joy to the heart of any college English professor who could persuade students to read and enjoy even one: Plutarch's *Lives*, Goethe's *Sorrows of Young Werther*, and Milton's *Paradise Lost* (of which Shelley's novel is an evident parody). The original monster's thundering threat certainly packs more oomph than Karloff's pitiable grunts: "I will glut the maw of death, until it be satiated with the blood of your remaining friends."

Shelley's monster is not evil by inherent constitution. He is born unformed—carrying the predispositions of human nature, but without the specific behaviors that can only be set by upbringing and education. He is the Enlightenment's man of hope, whom learning and compassion might mold to goodness and wisdom. But he is also a victim of post-Enlightenment pessimism as the cruel rejection of his natural fellows drives him to fury and revenge. (Even as a murderer, the monster remains fastidious and purposive. Victor Frankenstein is the source of his anger, and he kills only the friends and lovers whose deaths will bring Victor most grief; he does not, like Godzilla or the Blob, rampage through cities.)

Mary Shelley chose her words carefully to take a properly nuanced position at a fruitfully intermediate point between nature and nurture—whereas Hollywood opted for nature alone to explain the monster's evil deeds. Frankenstein's creature is not inherently good by internal construction—a benevolent theory of "nature alone," but no different in mode of explanation from Hollywood's opposite version. He is, rather, born *capable* of goodness, even with an *inclination* toward kindness, should circumstances of his upbringing call forth this favored response. In his final confession to Captain Walton, before heading north to immolate himself at the Pole, the monster says:

> My heart was fashioned to be *susceptible of love and sympathy;* and, when wrenched by misery to vice and hatred, it did not endure the violence of the change without torture, such as you cannot even imagine. [My italics to note Shelley's careful phrasing in terms of potentiality or inclination, rather than determinism.]

He then adds:

> Once my fancy was soothed with dreams of virtue, of fame, and of enjoyment. Once I falsely hoped to meet with beings who, pardoning my outward form, would love me for the excellent qualities which I was *capable of bringing forth.* I was nourished with high thoughts of honor and devotion. But now vice has degraded me beneath the meanest animal . . . When I call over the frightful catalogue of my deeds, I cannot believe that I am he whose thoughts were once filled with sublime and tran-

> scendent visions of the beauty and the majesty of goodness. But it is even so; the fallen angel becomes a malignant devil.

Why, then, does the monster turn to evil against an inherent inclination to goodness? Shelley gives us an interesting answer that seems almost trivial in invoking such a superficial reason, but that emerges as profound when we grasp her general theory of human nature. He becomes evil, of course, because humans reject him so violently and so unjustly. His resulting loneliness becomes unbearable. He states:

> And what was I? Of my creation and creator I was absolutely ignorant; but I knew that I possessed no money, no friends, no kind of property. I was, besides, endowed with a figure hideously deformed and loathsome . . . When I looked around, I saw and heard none like me. Was I then a monster, a blot upon the earth, from which all men fled, and whom all men disowned?

But why is the monster so rejected, if his feelings incline toward benevolence, and his acts to evident goodness? He certainly tries to act kindly, in helping (albeit secretly) the family in the hovel that serves as his hiding place:

> I had been accustomed, during the night, to steal a part of their store for my own consumption; but when I found that in doing this I inflicted pain on the cottagers, I abstained, and satisfied myself with berries, nuts, and roots, which I gathered from a neighboring wood. I discovered also another means through which I was enabled to assist their labors. I found that the youth spent a great part of each day in collecting wood for the family fire; and, during the night, I often took his tools, the use of which I quickly discovered, and brought home firing sufficient for the consumption of several days.

Shelley tells us that all humans reject and even loathe the monster for a visceral reason of literal superficiality: his truly terrifying ugliness—a reason both heartrending in its deep injustice, and profound in its biological accuracy and philosophical insight about the meaning of human nature.

The monster, by Shelley's description, could scarcely have been less attractive in appearance. Victor Frankenstein describes the first sight of his creature alive:

> How can I describe my emotions at this catastrophe, or how delineate the wretch whom with such infinite pains and care I had endeavored to form? His limbs were in proportion, and I had selected his features as beautiful. Beautiful!—Great God! His yellow skin scarcely covered the work of muscles and arteries beneath; his hair was a lustrous black, and flowing; his teeth of a pearly whiteness; but these luxuriances only formed a more horrid contrast with his watery eyes, that seemed almost of the same color as the dun white sockets in which they were set, his shriveled complexion, and straight black lips.

Moreover, at his hyper-NBA height of eight feet, the monster scares the bejeezus out of all who cast eyes upon him.

The monster quickly grasps this unfair source of human fear and plans a strategy to overcome initial reactions, and to prevail by goodness of soul. He presents himself first to the blind old father in the hovel above his hiding place and makes a good impression. He hopes to win the man's confidence, and thus gain a favorable introduction to the world of sighted people. But, in his joy at acceptance, he stays too long. The man's son returns and drives the monster away—as fear and loathing overwhelm any inclination to hear about inner decency.

The monster finally acknowledges his inability to overcome visceral fear at his ugliness; his resulting despair and loneliness drive him to evil deeds:

> I am malicious because I am miserable; am I not shunned and hated by all mankind? . . . Shall I respect man when he contemns me? Let him live with me in the interchange of kindness, and, instead of injury, I would bestow every benefit upon him with tears of gratitude at his acceptance. But that cannot be; the human senses are insurmountable barriers to our union.

Our struggle to formulate a humane and accurate idea of human nature focuses on proper positions between the false and sterile poles of nature and nurture. Pure nativism—as in the Hollywood version of the monster's depravity—leads to a cruel and inaccurate theory of biological determinism, the source of so much misery and such pervasive suppression of hope in millions belonging to unfavored races, sexes, or social classes. But pure "nurturism" can be just as cruel, and just as wrong—as in the blame, once heaped upon loving parents in bygone days of rampant Freudianism, for failures in rearing as putative source of mental illness or retardation that we can now identify as genetically based—for all organs, including brains, are subject to inborn illness.

The solution, as all thoughtful people recognize, must lie in properly melding the themes of inborn predisposition and shaping through life's experiences. This fruitful joining cannot take the false form of percentages adding to 100—as in "intelligence is 80 percent nature and 20 percent nurture," or "homosexuality is 50 percent inborn and 50 percent learned," and a hundred other harmful statements in this foolish format. When two ends of such a spectrum are commingled, the result is not a separable amalgam (like shuffling two decks of cards with different backs), but an entirely new and higher entity that cannot be decomposed (just as adults cannot be separated into maternal and paternal contributions to their totality).

The best guide to a proper integration lies in recognizing that nature supplies general ordering rules and predispositions—often strong, to be sure—while nurture shapes specific manifestations over a wide range of potential outcomes. We make classical "category mistakes" when we attribute too much specificity to nature—as in the pop sociobiology of supposed genes for complexly social phenomena like rape and racism; or when we view deep structures as purely social constructs—as in earlier claims that even the most general rules of grammar must be learned contingencies without any universality across cultures. Noam Chomsky's linguistic theories represent the paradigm for modern concepts of proper integration between nature and nurture—principles of universal grammar as inborn learning rules, with peculiarities of any particular language as a product of cultural circumstance and place of upbringing.

Frankenstein's creature becomes a monster because he is cruelly ensnared by one of the deepest predispositions of our biological inheritance—our instinctive aversion toward seriously malformed individuals. (Konrad Lorenz, the most famous ethologist of the last generation, based much of his theory on the primacy of this inborn rule.) We are now appalled by the injustice of such a predisposition, but this proper moral feeling is an evolutionary latecomer, imposed by human consciousness upon a much older mammalian pattern.

We almost surely inherit such an instinctive aversion to serious malformation, but remember that nature can only supply a predisposition, while culture shapes specific results. And now we can grasp—for Mary Shelley presented the issue to us so wisely—the true tragedy of Frankenstein's monster, and the moral dereliction of Victor himself. The predisposition for aversion toward ugliness can be overcome by learning and understanding. I trust that we have all trained ourselves in this essential form of compassion, and that we all work hard to suppress that frisson of rejection (which in honest moments we all admit we feel), and to judge people by their qualities of soul, not by their external appearances.

Frankenstein's monster was a good man in an appallingly ugly body. His countrymen could have been educated to accept him, but the person responsible for that instruction—his creator, Victor Frankenstein—ran away from his foremost duty, and abandoned his creation at first sight. Victor's sin does not lie in misuse of technology, or hubris in emulating God; we cannot find these themes in Mary Shelley's account. Victor failed because he followed a predisposition of human nature—visceral disgust at the monster's appearance—and did not undertake the duty of any creator or parent: to teach his own charge and to educate others in acceptability.

He could have schooled his creature (and not left the monster to learn language by eavesdropping and by scrounging for books in a hiding place under a hovel). He could have told the world what he had done. He could have introduced his benevolent and educated monster to people prepared to judge him on merit. But he took one look at his handiwork, and ran away forever. In other words, he bowed to a base aspect of our common nature, and did not accept the particular moral duty of our potential nurture:

> I had worked hard for nearly two years, for the sole purpose of infusing life into an inanimate body. For this I had deprived myself of rest and health. I had desired it with an ardor that far exceeded moderation; but now that I had finished, the beauty of the dream vanished, and breathless horror and disgust filled my heart. Unable to endure the aspect of the being I had created, I rushed out of the room . . . A mummy again endued with animation could not be so hideous as that wretch. I had gazed on him while unfinished; he was ugly then; but when those muscles and joints were rendered capable of motion, it became a thing such as even Dante could not have conceived.

The very first line of the preface to *Frankenstein* has often been misinterpreted: "The event on which this fiction is founded has been supposed, by Dr. Darwin, and some of the physiological writers of Germany, as not of impossible occurrence." People suppose that "Dr. Darwin" must be Charles of evolutionary fame. But Charles Darwin was born on Lincoln's birthday in 1809, and wasn't even ten years old when Mary Shelley wrote her novel. "Dr. Darwin" is Charles's grandfather Erasmus, one of England's most famous physicians, and an atheist who believed in the material basis of life (see essays 32–34). (Shelley is referring to his idea that such physical forces as electricity might be harnessed to quicken inanimate matter—for life has no inherently spiritual component, and might therefore emerge from nonliving substances infused with enough energy.)

I will, however, close with my favorite moral statement from Charles Darwin, who, like Mary Shelley, also emphasized our duty to foster the favorable specificities that nurture and education can control. Mary Shelley wrote a moral tale, not about hubris or technology, but about responsibility to all creatures of feeling and to the products of one's own hand. The monster's misery arose from the moral failure of other humans, not from his own inherent and unchangeable constitution. Charles Darwin later invoked the same theory of human nature to remind us of duties to all people in universal bonds of brotherhood: "If the misery of our poor be caused not by the laws of nature, but by our institutions, great is our sin."

The Tooth and Claw Centennial

IF BUTTERCUPS BUZZED after the bee
If boats were on land, churches at sea
If ponies rode men and if grass ate the cows
If cats were chased into holes by the mouse . . . .

"Then all the world would be upside down"—an apt description of his plight, or so Charles Cornwallis undoubtedly thought when he instructed his pipers and drummers to play this ditty during his surrender to Washington at Yorktown. (The Americans responded with "Yankee Doodle.")

Such reversals of established order intrigue us for their challenge to our "safe" assumptions. I keep a file for biological examples—carnivorous plants, worms that eat frogs, marine phytoplankton (single-celled, photosynthetic forms) that release toxins, poison fish, and then digest flecks of tissue dislodged from the dying vertebrates. I am writing this essay on the one hundredth anniversary of another curious reversal, sociological this time but from the heart of British science. The November 1892 edition of *The Nineteenth Century*, perhaps the leading British review of the time, published a series of tributes to Alfred, Lord Tennyson, the poet laureate who had died the month before. A memorial from Thomas Henry Huxley leads the parade—in verse. This tribute has never won any prizes for rhyme or meter, but I still take delight in the thought that Britain's leading scientist chose to honor Tennyson in the poet laureate's own medium. Huxley spoke of Tennyson's company in Westminster Abbey, undoubtedly evoking his old friend Darwin as exemplar of the last couplet:

> And lay him gently down among
> The men of state, the men of song:
> The men who would not suffer wrong:
> The thought-worn chieftains of the mind:
> Head servants of the human kind.

But why did Huxley choose to memorialize Tennyson? They knew each other only slightly. Both belonged to the Metaphysical Society, an elite club of Victorian intellectuals; but Tennyson almost always remained silent at meetings. Tennyson liked Huxley, but recorded only two visits of the scientist to his home. Huxley resolves this riddle for us in a letter to the Secretary of the Royal Society (Britain's leading association of scientists), urging that an official representative be sent to Tennyson's funeral. Huxley honored Tennyson from general respect, not personal friendship:

> He was the only modern poet, in fact I think the only poet since the time of Lucretius, who has taken the trouble to understand the work and tendency of the men of science.

Even so, why should this series of essays, devoted to evolutionary subjects, choose the old device of a funerary centennial to honor Tennyson? His general interest in science will not suffice, especially in the light of other monumental events in 1892, equally worth memorializing: the election of Grover Cleveland, the birth of Haile Selassie, Monet's beginning of the Rouen Cathedral paintings, the pugilistic victory of Gentleman Jim Corbett over John L. Sullivan, and the composition and first performance of "Ta-ra-ra boom-de-ay."

I choose Tennyson (and have, in fact, been looking forward to this excuse for several years) for a definite and parochial reason. Many subjects have canonical descriptors, snippets of phrase that, in knee-jerk fashion, identify the item with all the immediacy of a psychiatist's test in word association. If I say "The Georgia Peach," you will reply "Ty Cobb" (if you know anything about baseball). If I say "the Big Apple," you will reply "New York City" (if you know anything about anything). Darwinian evolution also has a canonical descriptor—this time, a snatch of poetry: "Nature red in tooth and claw."

Every evolutionist knows the line. It slips out lecture after lecture, article after article—even following that New Year's pledge never to quote the cliché again. Its parodies are also legion. My colleague Michael Ruse, for example, once subtitled a book about Darwin's intellectual struggles with his contemporaries: "Science red in tooth and claw."

Every evolutionist can cite the line (we would draw and quarter any impostor who couldn't); we all think that it describes a biological world reconfigured by evolutionary theory; nearly all of us know that it began as a line of poetry; most of us could cite Tennyson as the source; I suspect half of us even know that the phrase comes from *In Memoriam;* I'll bet a thousand bucks that fewer than one in a hundred of us have ever read the poem (and I stood among the ninety-nine until last week). Don't be too quick with your opprobrium. *In Memoriam,* after all, is not a haiku with seventeen syllables, or a sonnet with fourteen lines. *In Memoriam* runs to 131 sections, and more lines than I care to count (filling eighty pages in my edition). And long Victorian poems are not high on the hit parade, even of most serious intellectuals, these days. So, initially, I decided to write this essay in the year of Tennyson's centenary in order to discover for myself—and to convey to my colleagues and readers—the actual context of a line that we have all repeated too often and without any background. As so often happens, the story became broader and more interesting as I probed.

Tennyson was born in 1809, the same year as Darwin. As an undergraduate at Trinity College Cambridge, he met Arthur Hallam, handsome and brilliant son of the historian Henry Hallam. Their ardent friendship was clearly the key emotional experience of Tennyson's life. (I will not speculate on its nature, and the literature continues to maintain a discreet silence, largely for lack of evidence—for Hallam's father destroyed all of Tennyson's letters to his son, while Tennyson's son later burned all of Hallam's letters to his father. Arthur Hallam was engaged to Tennyson's sister at the time of his death, so complexity probably reigned, but if the intense bond between Arthur Hallam and Alfred Tennyson didn't have at least a repressed sexual basis, then . . . well, I'll be a monkey's uncle.)

On October 1, 1833, Tennyson received a letter from Henry Olden, Hallam's uncle, and his world collapsed: "Your friend, Sir, and my much loved nephew, Arthur Hallam, is no more—it has pleased God to remove him from this his first scene of Existence, to a better world, for which he was created. He died at Vienna on his return from Buda, by apoplexy—and I believe his Remains come by sea from Trieste." Arthur Hallam was twenty-two years old when he died.

In Memoriam, published in 1850, is Tennyson's extensive tribute to this extraordinary friendship, and to the emotional, religious, and philosophical meaning of such loss. (Tennyson originally published anonymously—though his distinctive authorship didn't elude a soul in the know—under the full title *In Memoriam A.H.H. Obit* [died] *MDCCCXXXIII.*) The poem was an instant success, and surely played a large role in Tennyson's appointment (following the death of Wordsworth) as poet laureate later in 1850. Queen Victoria and her husband, Prince Albert, especially liked the poem. After Albert's death in 1861, Victoria regarded *In Memoriam* as important solace in her extended grief. "Next to the Bible," she stated, "*In Memoriam* is my comfort." She even altered one of Tennyson's verses in her private copy, substituting "widow" for Tennyson's "widower," and "she" for "he"—so that the lines could be recast as mourning for Albert:

> Tears of the widow, when she sees
> A late-lost form that sleep reveals—
> And moves her doubtful arms, and feels
> His place is empty, fall like these.

Victoria requested (I think one says "commanded") a visit from Tennyson in 1862, and later wrote in her diary:

> I went down to see Tennyson who is very peculiar looking, tall, dark, with a fine head, long black flowing hair and a beard—oddly dressed, but there is no affectation about him. I told him how much I admired his glorious lines to my precious Albert and how much comfort I found in his *In Memoriam*.

With this background, we can grasp the setting for Tennyson's famous image of "nature red in tooth and claw." And we can understand why evolutionists have so misinterpreted the phrase as either a harbinger or a description of Darwin's world. First of all—and sorry to sound so defensive—the error is not entirely our fault. (The most basic of all facts can make evolutionists seem mighty stupid when they argue that Darwin's new formulation inspired Tennyson's line—for *In Memoriam* appeared in 1850, while Darwin kept his views close to his chest before publishing *The Origin of Species* in 1859.) A long tradition of literary criticism has read evolution into the biological passages of *In Memoriam*, and the uniformitarian geology of Charles Lyell into Tennyson's lines about the earth and its historical changes. The Dutch historian of science Nicolaas A. Rupke presents an extensive list of such literary citations and writes in his important book *The Great Chain of History*:

> It has become customary to read in . . . *In Memoriam* not only organic evolution, but also the geology of . . . Lyell. Because the relevant sections were written . . . long before Darwin's *Origin of Species* . . . some literary critics have interpreted these passages as an anticipation of the theory of organic evolution by the intuitive genius of a poet, before the analytical mind of Darwin dared to arrive at the same conclusion. "How did the poet come to forestall the scientists in their own game?" one critic asks.

If *In Memoriam* is a grieving man's quest for peace, transcendence, renewed faith, resolution, acceptance, or whatever (all and more have been proposed), then what role does science play in this search—and remember that Tennyson was a champion of science, not an embodiment of the unjust (and probably nonexistent) stereotype of an affected, antitechnological, romantic poet. The scientific verses of *In Memoriam* are among the most famous, and critical commentary has always viewed them as essential to the narrator's quest in the poem.

Tennyson first dismisses a silly argument about science, the kind of sophistry woven by such tormentors as Job's "comforters." How can anyone grieve so deeply in a world made so exciting by scientific advance?

> A time to sicken and to swoon,
>   When Science reaches forth her arms
>   To feel from world to world, and charms
> Her secret from the latest moon?

Tennyson replies with two affecting verses, also invoking an image from nature. How can you compare a happy generality with my private desolation?

> Behold, ye speak an idle thing:
>   Ye never knew the sacred dust:

I do but sing because I must,
And pipe but as the linnets sing:

And one is glad; her note is gay,
For now her little ones have ranged;
And one is sad; her note is changed,
Because her brood is stol'n away.

Tennyson's serious examination of nature as a possible source of solace occupies a crucial place in three consecutive sections (54–56), just before the poem's midpoint. In section 54 (which I shall quote in full), Tennyson uses the first four verses to express a standard argument of the "natural theology" so popular in the generation just before his—that good must lie behind nature's apparent evil:

Oh yet we trust that somehow good
Will be the final goal of ill,
To pangs of nature, sins of will,
Defects of doubt, and taints of blood;

That nothing walks with aimless feet;
That not one life shall be destroy'd,
Or cast as rubbish to the void,
When God hath made the pile complete;

That not a worm is cloven in vain;
That not a moth with vain desire
Is shrivell'd in a fruitless fire,
Or but subserves another's gain.

Behold, we know not anything;
I can but trust that good shall fall
At last—far off—at last, to all,
And every winter change to spring.

Nicely said; but then, as a stunning reversal in the section's last verse, Tennyson labels this conventional belief as a vain reverie:

So runs my dream: but what am I?
An infant crying in the night:

An infant crying for the light:
And with no language but a cry.

The narrator must now examine nature more honestly—as Tennyson does in section 55. In some of the poem's most famous lines, Tennyson expresses a theme that would be important to Darwin (though scarcely original with him), and that strikes the narrator as such a mockery in his grief: Why does nature, while preserving stability of species, permit such a hecatomb of individual and untimely deaths?

Are God and Nature then at strife,
That Nature lends such evil dreams?
So careful of the type she seems,
So careless of the single life;

That I, considering everywhere
Her secret meaning in her deeds,
And finding that of fifty seeds
She often brings but one to bear.

Tennyson then looks to large scales for an answer. Perhaps the carnage of individuals (like Hallam) subserves a larger good over aeons:

I stretch lame hands of faith, and grope,
And gather dust and chaff, and call
To What I feel is Lord of all,
And faintly trust the larger hope.

But Tennyson is not optimistic; he has already labeled this solution as a "faint" hope. He then opens section 56 with my favorite lines of all, the geological verses of *In Memoriam.* Nature mocks his own observation in section 55 by showing that, in the fullness of time, even species ("types") must die. "All shall go," and momentary suffering supports no permanent stability:

"So careful of the type?" but no.
From scarpèd cliff and quarried stone
She cries, "A thousand types are gone:
I care for nothing, all shall go.

"Thou makest thine appeal to me:
I bring to life, I bring to death:
The spirit does but mean the breath:
I know no more." . . .

And so we finally come to "nature red in tooth and claw." One hope still remains: nature may savage individuals and eventually remove species, but does this carnage (however paradoxically) eventually lead to human nobility and the soul's immortality? Tennyson, in a long question spread through four verses, answers "no" and, invoking the famous image, even chides the narrator for imagining such a solution in the light of nature's factual rapacity:

. . . And he, shall he,
Man, her last work, who seem'd so fair,
Such splendid purpose in his eyes,
Who roll'd the psalm to wintry skies,
Who built him fanes of fruitless prayer,

Who trusted God was love indeed
And love Creation's final law—
Tho' Nature, red in tooth and claw
With ravine, shriek'd against his creed—

Who loved, who suffer'd countless ills,
Who battled for the True, the Just,
Be blown about the desert dust,
Or seal'd within the iron hills?

Later in the poem, as the narrator resolves his grief, Tennyson does take comfort from the progressive pathway that he infers from geological history. "Contemplate all this work of time," he states in beginning section 118. Perhaps the ancient dead are harbingers of better things to come: "But trust that those we call the dead / Are breathers of an ampler day / For ever nobler ends." He then describes geological history—an earth that "in tracts of fluent heat began . . . Till at the last arose the man . . . The herald of a higher race." The section ends with a plea for human betterment: "Move upward, working out the beast, / And let the ape and tiger die." (I assume he means the ape and tiger within us, and that these lines are not pro-hunting propaganda!)

Tennyson ends *In Memoriam* (section 131) with a happy epithalamium (a fancy name for a marriage ode). He returns to the theme of historical progress and compares the growth of the child that will issue from this marriage with advance of the race: "And, moved thro' life of lower phase, / Result in man, be born and think." This ray of comfort, drawn from nature at the very end of the poem, may strike us as hokey today, but I am inclined to respect any rationale invoked (however tenuously) to describe a gain of emotional peace by someone who has grieved so long and deeply. Tennyson argues that modern humans are in transition to some higher stage, that our current sufferings aid this progress, and that Hallam was a premature representative of these nobler beings:

> No longer half-akin to brute,
> For all we thought and loved and did,
> And hoped, and suffer'd is but seed
> Of what in them is flower and fruit;
>
> Whereof the man, that with me trod
> This planet, was a noble type
> Appearing ere the times were ripe,
> That friend of mine who lives in God.

The sources of these "nature passages" have almost always been misconstrued because we remember and honor supposed "winners" and forget the scientists now branded as wrong. Tennyson's biology is almost always viewed as evolutionary, and his geology as following Lyell's uniformity of slow and steady change. In fact, as Rupke shows so well (and as should be clear to anyone who reads *In Memoriam* with adequate knowledge of British nineteenth-century geology), both facets of Tennyson's natural history derive from a single and different source—the progressivist and catastrophist geology that represented the main line of early-nineteenth-century thought (and that Tennyson had studied at Cambridge under his tutor, the great philosopher of science William Whewell, who knew and supported the catastrophists, and who even coined their name—see essay 13).

The catastrophists—Buckland, Sedgwick, Conybeare, and others—though not generally remembered today, were the geological giants of Tennyson's youth. They argued for a *nonevolutionary*, directional history based on successive creations of increasing excellence, separated by catastrophic episodes of extinction. Tennyson often cites them directly. Section 118 describes the origin of the earth by the nebular hypothesis (coalescence from rings of hot gases spun off from the sun—see essay 3), a central idea for catastrophists (as a starting point for life's progressive history, keyed to the earth's cooling), but denied by Lyell, who advocated a climatic steady state—"in tracts of fluent heat began," Tennyson writes. The famous lines about extinction ("From scarpèd cliff . . . all shall go") are descriptions of catastrophic episodes. Later in section 56, Tennyson even cites the favorite case-study of catastrophists, the Mesozoic "sea-monsters" (ichthyosaurs and plesiosaurs): "Dragons of the prime, / That tare each other in their slime."

In this light, we have no reason to regard Tennyson as an incipient evolutionist because he speaks so often about progress in life's history, for advance by successive creation was a hallmark of catastrophist geology. Tennyson may have been favorable to some form of evolutionary thinking—a subject widely discussed in the years before Darwin's *Origin*. He presumably imagined human spiritual progress as gradual and uninterrupted. But the biological and geological passages of *In Memoriam* record the progressivist catastrophism of his time, not the evolutionary theory of Darwin's world to come.

An old cliché proclaims that each generation reads great works of literature in a different and distinctive way, and that a primary sign of greatness lies in the intrinsic richness that permits so many changing interpretations. Tennyson's contemporaries read *In Memoriam* as a great religious poem, an odyssey in the rediscovery of faith, following deep grief and doubt born of a senseless and untimely death. The great liberal theologian Charles Kingsley, good friend of Huxley and author of *The Water Babies* and *Westward Ho!*, wrote a major review of *In Memoriam* (*Fraser's Magazine*, September 1850). He called Tennyson a "willing and deliberate champion of vital Christianity," and labeled *In Memoriam* "the noblest Christian poem which England has produced for two centuries . . . [expressing] an orthodoxy the more sincere because it has worked upward through the abyss of doubt."

In a famous essay on *In Memoriam*, written in 1936, T. S. Eliot takes an almost opposite position. Tennyson, he says, "had the finest ear of any English poet since Milton." Eliot acknowledges the usual Victorian reading: "Tennyson's contemporaries . . . regarded it as a message of hope and reassurance to their rather fading Christian faith." But Eliot then demurs, and supports the religious character of *In Memoriam* from an inverted, modern perspective:

> *In Memoriam* can, I think, justly be called a religious poem, but for another reason than that which made it seem religious to his

> contemporaries. It is not religious because of the quality of its faith, but because of the quality of its doubt. Its faith is a poor thing, but its doubt is a very intense experience . . . Tennyson seems to have reached the end of his spiritual development with *In Memoriam;* there followed no reconciliation, no resolution.

As a totally naive reader, I experienced the poem differently again, and in an unoriginal manner that will probably be viewed as stereotypical for this generation. I find it hard to discern any consistent intellectual or philosophical answer to the key issue of Hallam's death and its meaning. The poem is full of contradictions and nonresolutions—as in Tennyson's shifting use of historical progress (dismissal as solace early in the poem, acceptance as subsidiary comfort at the end, when he views Hallam as a harbinger of higher stages).

I read the poem instead, and with an intensity that brought me to tears in places, as a wonderful and deeply truthful account of the psychology of mourning. You are devastated by an event that cannot be explained or reconciled: the love of your life dies at age twenty-two. Fundamentally, you can do little more than wait—so that the long process of emotional healing, something deeply constitutional within us, I suspect, can play out its long course of years. If you succeed, and do not sink permanently into despair, you eventually reconstruct your life. You find no answers, but you do accept because you must, and you move on. To me, *In Memoriam* is an odyssey in the working out of extended grief. I am awestruck all the more because Tennyson composed the verses in haphazard fashion over seventeen years, yet the sequence of 131 sections rings so true as a chronological account of grieving. How could Tennyson remember and capture the sequence so beautifully? How could he integrate the swirling and swinging moods: the anger, the despair, the emptiness, the search for answers, the exultation of temporary resolution ("Ring out, wild bells, to the wild sky" of section 106), the renewed despondency, the final acceptance without real answers.

Above all, I admire Tennyson's treatment of the relationship between science and human values—for I believe that his answer is entirely right, and just as important (if not more so) in our times. The narrator of *In Memoriam* probes several sources for answers to his quest for meaning—science prominently among them. He presents several characterizations of nature, some contradictory—red in tooth and claw, a domain of death for all species, a realm of steady progress through history. And he rejects them all as potential resolutions of his ethical and emotional quest.

On one obvious level, the narrator must reject science or any source of objective information—for how can any exterior knowledge extinguish the primal pain of personal grief? But Tennyson goes further and argues that, in principle, science cannot provide answers for moral questions about life's meaning. As a champion of science, not a detractor sniping from another profession, Tennyson lauds its power to build a global network of railroads, feed nations, answer empirical riddles of the universe—but he knows that science cannot tell us why a man should die so young, or how a grieving lover should resolve his suffering.

Tennyson consistently held this position on the separateness of scientific and moral knowledge. He stated (as reported by his son, whom, to add a footnote to the poignancy of this story, he named Hallam):

> We do not get this faith from Nature or the world. If we look at Nature alone, full of perfection and imperfection, she tells us that God is disease, murder and rapine. We get this faith from ourselves, from what is highest within us.

In Memoriam features the same sentiments. Early in the poem, in section 3, Tennyson considers and rejects nature as a source of moral instruction ("my natural good"):

> And shall I take a thing so blind,
>   Embrace her as my natural good;
>   Or crush her, like a vice of blood,
> Upon the threshold of the mind?

Later on, in section 120, he refutes the idea that our essence is nothing but our material self: "I think we are not wholly brain, / Magnetic mockeries." In a wonderful couplet, Tennyson allows that science might establish a material basis, but still would not speak to our moral struggles: "Let Science prove we are, and then / What matters Science unto men."

Setting forth the proper logic of a question does not guarantee an answer. I accept Tennyson's separation of scientific from moral and ethical quests; but few of us these days would be satisfied with Tennyson's particular answer, especially for resolving his grief at Hallam's death. Tennyson, by his own statements and all his friends' memories, was obsessed with the issue of personal identity for the soul after death. Following one of his few long talks with Tennyson, Huxley remarked that "immortality was the one dogma to which Tennyson was passionately devoted." And Tennyson himself stated, "The cardinal point of Christianity is the Life after Death." Thus, insofar as *In Memoriam* reaches a moral conclusion at all, Tennyson celebrates his voyage from religious doubt to confidence that he will meet Hallam again in heaven—a lame resolution after so much struggle, at least for most modern readers.

> My own dim life should teach me this,
> That life shall live for evermore
> Else earth is darkness at the core,
> And dust and ashes all that is.

Immortality, moreover, must be personal. The fusion of Hallam's soul into a glob of general good is not enough: "And I shall know him when we meet: / And we shall sit at endless feast, / Enjoying each the other's good."

But the particular character of Tennyson's personal solution doesn't vitiate the principle that answers to questions about ethical meaning cannot come from science. Tennyson, in fact, revered both sources and knew that "the good life"—a clichéd phrase, perhaps, but do we not all seek it?—required their successful integration. Two separate sources—Huxley's world and Tennyson's. (Huxley, by the way, took the same position on the separate and equally necessary contributions of science and ethics to a reasoned life—see his famous essay, *Evolution and Ethics*.) Tennyson called these two sources knowledge and reverence, personified as mind and soul. And he spoke of their union with a metaphor from the discipline that owns "harmony" as a technical concept:

> Let knowledge grow from more to more,
> But more of reverence in us dwell;
> That mind and soul, according well,
> May make one music, as before.

7

Sweetness and Light

"SWEET IS PLEASURE AFTER PAIN." This motto will be familiar to all, but I venture that few readers will know the source. The line is a paean to Bacchus, as voiced by soldiers to celebrate the joy that can follow a fight:

> Bacchus, ever fair and young,
> Drinking joys did first ordain.
> Bacchus's blessings are a treasure,
> Drinking is the soldier's pleasure . . .
> Sweet is pleasure after pain.

John Dryden's *Alexander's Feast*, source of these lines, tells the story of music's power over the emotions. (Dryden wrote the poem as an ode to Saint Cecilia, patron of music. Several composers set the verses for chorus and orchestra; Handel's version of 1736, written thirty-six years after Dryden's death and quoted above, remains a staple of the choral literature.) Alexander the Great, with a woman at his side and a glass from Bacchus in his hand, listens to Timotheus playing his lyre—and succumbs to every emotion that the great musician evokes:

> Sooth'd with the sound, the king grew vain,
> Fought all his battles o'er again,

And thrice he routed all his foes,
And thrice he slew the slain.

The phenomenon of replaying old tunes is pervasive in our lives and cultures. And how could it be otherwise, for talented folks are many, but the good tunes few. Alfred North Whitehead commented that the entire European philosophical tradition "consists of a series of footnotes to Plato"—by which he did not accuse his colleagues (and, by implication, himself) of stupidity or plagiarism, but merely noted that truly great problems are evident and finite—and therefore properly delineated (if not solved) by the first comprehensive thinker of extensive record. Alexander, at least, vaguely knew what he was doing—even if he was drunk, emotionally manipulated, and playing the oldest of all games in embellishing war stories.

Better stories, more ripe for satire in Dryden's mode, arise from *unconscious* rediscovery—especially when people, intoxicated with the excitement of a novel personal insight, think that they have just divined an ancient truth for the very first time. A lovely example of this phenomenon recently circulated through the "Letters to the Editor" section of the leading British scientific journal, *Nature.*

On January 16, 1992, Zakaria Erzinclioglu of the zoology department of Cambridge University wrote to complain most vociferously about a false and harmful use of language:

> One of the most absurd and persistent misuses of words in science is the use of the word "ancient" to describe species that flourished millions of years ago . . . The truth is that organisms that lived long ago are "young" relative to those alive today . . . The animals themselves are not ancient in any sense at all. They died out a long time ago when the Earth was young. Why do we persist in misusing the term in this way?

Ralph Estling then responded on February 20 (British periodicals do have a tendency to go on and on with disputes of such quirky interest but limited import. Some regard the tradition as cranky; I find it charming):

> From our point of view which is generally the coordinate system we favor, trilobites and Aristotle are both; they were here early (earlier than we were), younger than we are, but from our habitual standpoint of looking back through time from where

> we happen to be at the moment, they are ancient and therefore older than we are.

Now, I have no quarrel whatever with Estling's pluralistic judgment, and I also agree with Erzinclioglu's closing point that choice of words is neither trivial nor arbitrary: "What does it matter whether we call early life 'ancient' or not? It can be said that it is merely a convention and that it is sheer pedantry to split hairs over this matter. I do not believe so. Words can play subtle tricks in the mind and any scientific word that might lead us astray in our thinking must surely be replaced with one that is more suitable."

I enter this fray to make a historical point, not a judgment. This dispute may be a goodie, but it is also a real oldie. Our protagonists, in fact, have just rediscovered one of our most ancient (or youngest?) linguistic wrangles, a classic paradox of our literary traditions. Francis Bacon made the canonical statement, in Latin epitome, in his *Advancement of Learning* (1605): *Antiquitas saeculi, juventus mundi* (or, roughly, "the good old days were the world's youth"). He later expanded this theme as aphorism 84 of his *Novum Organum.* (*The Organon* is a collective title for Aristotle's treatises on logic. In claiming to start things again with a New Organon, Bacon displayed his characteristic flair for self-promotion.)

> The old age of the world . . . is the attribute of our own times, not of that earlier age in which the ancients lived; and which, though in respect of us it was the elder, yet in respect of the world it was the younger.

This puzzle became a major issue in British seventeenth-century thought and polemic—for an interesting reason, centered on the origins of modern science. Scholars referred to this problem as "the Baconian Paradox." Robert K. Merton, our leading sociologist of science, has traced the history and uses of this debate in his wonderful book *On the Shoulders of Giants* (first published in 1965, and continuously in print ever since). Merton grants Bacon the classical formulation, but finds earlier hints and partial statements ranging from Giordano Bruno right back to the apocryphal book II Esdras of the Vulgate, or Latin Bible. Merton also documents a history of later rediscovery, with credit claimed for originality each time—a continuing tale with a new chapter now added from *Nature.* Jeremy Bentham, for example, left this aphorism among his unfinished papers (published posthumously in 1824): "What is the wisdom of the times called old? Is it the wisdom of gray hairs? No. —It is the wisdom of the cradle."

I can offer no resolution of this issue to accompany my documentation of its venerability. The seventeenth-century debaters were right; Bacon's observation is a true paradox—that is, a seemingly self-contradictory or absurd statement that happens to be true. We get different answers from two equally proper and justifiable vantage points. Trilobites are both young (looking up from the origin of multicellular life) and old (looking back from 1992). Both perspectives, from both ends, are "correct"—and they do contradict. As with all the classical paradoxes that so fascinate and frustrate us, we both wince and revel in Bacon's dictum because it embodies one of the inherent ambiguities of our complex lives. Just a week after Estling's response to Erzinclioglu, Rossini reached both his two hundredth and his forty-eighth birthday on February 29, 1992 (yes, forty-eight and not fifty; 1800 and 1900 were not leap years in our Gregorian calendar). Frederick, the pirate apprentice, subject to the same ambiguity (and faced with the prospect of bondage until his twenty-first birthday at age eighty-eight—not eighty-four because 1900 wasn't a leap year), gently touched the heart of this predicament in W. S. Gilbert's words, "How quaint the ways of paradox! / At common sense she gaily mocks!"

So why bother with this story at all? Is my correction and commentary any more than worthless pedantry and antiquarian "gotcha"? So might we judge this entire tale if Bacon's paradox did not illustrate such a central episode in the history of science and continue to prompt our attention to such an important issue today. Why did R. K. Merton bother to trace Bacon's paradox, and what are the "shoulders of giants" (in his title) anyway? Merton's book is a delightfully whimsical search through history—going all the way back to Bernard of Chartres in 1126—for the origin of a quotation usually attributed to Isaac Newton (from a letter he wrote to Robert Hooke): "If I have seen farther, it is by standing on the shoulders of giants." As Merton shows, Newton was staking no claim to literary novelty, but merely repeating a phrase so widely regarded as common property that it demanded no quotation marks. Merton had two larger purposes beyond the sheer intellectual fun of romping through centuries in search of a phrase. First, he has spent a good slug of his career studying the phenomenon of "multiples" in scientific discovery. Our mythology venerates the lone genius, but most great innovations arise several times, often in virtual simultaneity (the calculus by Newton and Leibniz, natural selection by Darwin and Wallace). What better illustration for the social context of knowledge than a documentation of thoughts so pervasively "in the air" that several smart people pull them down at the same time? And what gentler illustration than a great quotation, supposedly the unique property of a singular genius, but actually a common phrase for a millennium?

Second, Merton is an expert on that most fascinating period of the seventeenth century when the concept of progress, prompted by the beginnings of modern science, first became a dominant motif and driving force in Western culture. Dwarfs on the shoulders of giants may have been kicking around as a metaphor of modesty (or false modesty) for centuries, but Newton and his seventeenth-century colleagues elevated this image to new prestige and familiarity because it epitomized the ambivalence that any intellectual must feel in a world that sees itself as continually improving—namely, how can we pay proper homage to the past while still acknowledging our superiority? Dwarfs on the shoulders of giants is nearly ideal for such a purpose: we can praise our forebears as our intellectual superiors, while still maintaining that we now stand on higher ground.

Bacon's paradox, of course, features another facet of the same imagery—and Merton treats this phrase as a subsidiary theme to dwarfs and giants. In some ways, Bacon's metaphor is even more biting, particularly if you wish to defend the rights of dwarfs (moderns) against the supposed wisdom of the giants (ancients). If those giants were just little kids in the world's youth, perhaps they weren't so transcendently brilliant—and perhaps we moderns, with our gray beards in the earth's old age, are the true repositories of accumulated wisdom.

As a scientist, I feel the tug and relevance of these twinned metaphors (dwarfs vs. giants, and ancient times as the earth's youth) because their potent imagery is a product of the world that my profession helped to create. Debate on the comparative values of past and present did not dominate our discourse until historicism and the notion of progress became paramount in Western traditions—and these seminal ideas first arose with force during the seventeenth century, when the beginnings of modern science and commerce conjured up visions of steady and inexorable improvement. Why worry about whether the past was old or young, if time runs in the repetitive cycles of Plato's Great Year and does not either advance or regress? Bacon's paradox only has meaning in the world constructed by science.

In the seventeenth century, all these swirling issues came to a sharp focus in a dispute that evoked great passion then, and, in only slightly modified form, now engulfs the academy again—the "battle of the books," or the relative merit of ancient and modern learning and literature. (The seventeenth century pitted Aristotle against Descartes; now we counterpose the canon of "great books" by DWEMs—dead, white, European males—against a more diverse contemporary literature.) Consider the seventeenth-century context of the original battle: Latin and Greek formed the cornerstone of the curriculum; the learning of ancient Greece and Rome set an unsurpassable standard for all that would ever arise afterward (remember that the Renaissance received its name, meaning "rebirth," for attempting to recover, not to exceed, the glories of classical civilization).

Thus the original "battle of the books" contrasted a curriculum of Greek and Roman classics with an attempt by modernists to grant equal (or greater) status to contemporary works of literature, philosophy, and science. Bacon himself supported the moderns and used his paradox to good effect, invoking the obvious simile of time's passage with the accumulation of knowledge in human aging:

> And truly as we look for greater knowledge of human things and a riper judgment in the old man than in the young . . . so in like manner from our age . . . much more might fairly be expected than from the ancient times, inasmuch as it is a more advanced age of the world.

For contrast, let us turn to the most famous defense ever penned by the other side—the brief for the ancients embodied in Jonathan Swift's celebrated satire of 1704, "A Full and True Account of the Battle Fought Last Friday Between the Ancient and the Modern Books in St. James's Library," usually called, for short, the "Battle of the Books." All might have been well had the two parties made a concordat, and kept to their own proper spaces. But the librarian had fostered discord by intemperate mixtures in shelving. (As with nearly everything in Swift, the satirical reference is to something contemporary and specific, not universal and obvious. Swift comments here on an attempt by one of his colleagues to brand many of the fables attributed to Aesop as modern interpolations—hence meriting a reshelving of this ancient on the other side): "In replacing his books, he was apt to mistake, and clap Descartes next to Aristotle; poor Plato had got [with] Hobbes . . . and Virgil was hemmed in with Dryden." (Note how the source of my opening line, Swift's unloved colleague, becomes a primary symbol of modernism in this passage.)

Early in the text, both sides use Bacon's paradox to advance their respective arguments (Swift, in a marginal note inserted at this point, explicitly places Bacon's argument in the context of disputes between early and recent learning by writing, "according to the Modern Paradox"):

> Discord grew extremely high, hot words passed on both sides, and ill blood was plentifully bred. Here a solitary ancient, squeezed up against a whole shelf of moderns, offered fairly to dispute the case, and to prove by manifest reasons, that the priority was due to them, from long possession . . . But these [the moderns] denied the premises, and seemed very much to wonder, how the ancients could pretend to insist upon their antiquity, when it was so plain (if they went to that) that the moderns were much the more ancient of the two.

The bulk of Swift's text describes the actual battle, with his own sympathies for the ancients scarcely hidden—as in this passage, where Aristotle misses the inventor of "the Modern Paradox," but kills Descartes instead (as the greatest French modern falls into the vortex of his own theory):

> Then Aristotle observing Bacon advance with a furious mien, drew his bow to the head, and let fly his arrow, which missed the valiant modern, and went hizzing over his head; but Descartes it hit: the steel point quickly found a defect in his head-piece; it pierced the leather and the pasteboard, and went in at his right eye. The torture of the pain, whirled the valiant bowman round, till death, like a star of superior influence, drew him into his own vortex.

But Swift introduces the actual battle with a verbal curtain-raiser—a three-page gem that forms one of the greatest extended metaphors in all Western literature: the dispute of the spider (representing the moderns) and the bee (the ancients). In the library, a spider dwells "upon the highest corner of a large window." He is fat and satisfied, "swollen up to the first magnitude, by the destruction of infinite numbers of flies, whose spoils lay scattered before the gates of his palace, like human bones before the cave of some giant." (Swift, I assume, did not know that males of most orb-weaving spiders are small and do not build webs—and that his protagonist was undoubtedly a "she." So, for that matter, come to think of it, is the industrious bee, also called "he" in this text.)

Swift leaves us in no doubt about the intended comparison; the spider, spinning such a mathematically sophisticated web from his own innards (and not relying on an external source of succor), is a scientific modern:

> The avenues to his castle were guarded with turnpikes, and palissadoes, all after the *modern* way of fortification [Swift's own italics]. After you had passed several courts, you came to the center, wherein you might behold the constable himself in his own lodgings, which had windows fronting to each avenue, and ports to sally out upon all occasions of prey or defense. In this mansion, he had for some time dwelt in peace and plenty.

A bee then flies in through a broken pane and happens "to alight upon one of the outward walls of the spider's citadel." His weight breaks the spider's web, and the convulsions of the resulting tumult awaken the spider, causing him to run out in fear "that Beelzebub with all his legions, was come to revenge the death of many thousands of his subjects, whom the enemy had slain and devoured." (A nice touch. Beelzebub, a popular name for the devil, is literally "lord of the flies".) Instead, he finds only the bee, and curses in a style that has been called Swiftian ever since: "A plague split you . . . giddy son of a whore . . . Could you not look before you, and be damned? Do you think I have nothing else to do (in the Devil's name) but to mend and repair after your arse?"

The spider, calming down, now takes up his intellectual role as a modern and excoriates the bee with the crucial argument from his side: You advocates of the ancients are just pitiful and unoriginal drones who can create nothing yourselves, but can only forage among other people's antique insights (the flowers in the field, including nettles as well as objects of admitted beauty). We moderns build new intellectual structures from the heart of our own genius and discovery:

> What art thou, but a vagabond without house or home, without stock or inheritance? Born to no possessions of your own, but a pair of wings, and a drone-pipe. Your livelihood is an universal plunder upon nature; a freebooter over fields and gardens; and for the sake of stealing, will rob the nettle as readily as a violet. Whereas I am a domestic animal, furnished with a native stock within myself. This large castle (to show my improvements in the mathematics) is all built with my own hands, and the materials extracted altogether out of my own person.

The bee then responds for all devotees of ancient learning: I borrow but cause no harm in so doing, and I transmute what I borrow into new objects of great beauty and utility—honey and wax. But you, while claiming to build only from your own innards, must still destroy a hecatomb of flies for the raw material. Moreover, your vaunted web is weak, temporary and ephemeral, whatever its supposed mathematical beauty (while the distillation of ancient knowledge endures forever). Finally, how can you claim virtue for a product of your own spinning if the material be poison based on your own gall, and the effect thereof destruction?

> I visit, indeed, all the flowers and blossoms of the field and the garden, but whatever I collect from thence, enriches myself, without the least injury to their beauty, their smell, or their taste . . .
>
> You boast, indeed, of being obliged to no other creature, but of drawing and spinning out all from yourself; that is to say, if we may judge of the liquor in the vessel by what issues out, you possess a good plentiful store of dirt and poison in your breast; and, tho' I would by no means, lessen or disparage your genuine stock of either, yet, I doubt, you are somewhat obliged for an increase of both, to a little foreign assistance . . . In short, the question comes to this; whether is the nobler being of the two, that which by a lazy contemplation of four inches round; by an overweening pride, which feeding and engendering on itself, turns all into excrement and venom; produces nothing at last, but fly-bane and a cobweb: or that, which, by an universal range, with long search, much study, true judgment, and distinction of things, brings home honey and wax.

No one has ever set forth the issues better in nearly three hundred years of subsequent writing. Most thoughtful people come down somewhere between the bee and the spider, but extremists on both sides are still using the same arguments. Current partisans of the spider claim that the "great books" of traditional learning (now including such former moderns as Swift and his *Gulliver's Travels*) have become both unreadable and irrelevant for modern students—and might as well be dropped (or lightly retained as a few excerpts for a lick and a promise) in favor of direct engagement with modern literature and science. At worst, they may actively disparage the old mainstays as nothing but repositories of prejudice written by that biased subset of humans called white males.

Current partisans of the bee can dispense worthy platitudes about upholding standards and retaining a canon universally validated by endurance through so much time and turmoil. But these good arguments are often accompanied by blindness, or actual aversion, to the scientific and political complexities that permeate our daily lives and that all educated people must understand in order to be effective and thoughtful in their professions. Moreover, defense of the "great books" too often becomes a smokescreen for political conservatism and maintenance of old privileges (particularly among folks like me—white professors past fifty who don't wish to concede that other kinds of people might have something important, beautiful, or enduring to say).

How can we resolve this ancient debate from the youth of time? In one sense we can't, at least to anyone's clear victory—for both sides have good arguments, just as in Bacon's paradox that once epitomized the ongoing struggle. But an obvious solution stares us all in the face, if only we could overcome the narrowness and parochiality that leads any partisan to fortify his barricade. The answer has been with us since Aristotle—in the form of the "golden mean." The solution speaks to us whenever the still, small voice of reason reconciles two warring parties by compelling attention to good points on both sides. It lies embodied in a famous epigram of Edmund Burke (1729–1797), once a modern in the original battle, but now an arch-conservative among the DWEMs: "All government—indeed, every human benefit and enjoyment, every virtue and every prudent act—is founded on compromise and barter." We must hybridize the bee and the spider—and then, in good Darwinian fashion, select for the best traits of both parents in a rigorous program of good breeding (education). The spider is surely right in extolling the technical beauty of his web, and the absolute need for all contemporary people to understand both the mechanics and aesthetics of its structure. But the bee cannot be faulted for insisting that fields of well-distilled wisdom await our entirely benign exploitation for enjoyment and enlightenment—and that we would be utter fools to bypass such a rich store.

I can argue the virtues of both sides, but since I live in the world of science, and experience its parochialities on a more sustained and daily basis, I feel more impelled to advance the bee's cause. Distillation may be biased, but anything that endures for hundreds or thousands of years (at least in part by voluntary enjoyment rather than forced study) must contain something of value. No one celebrates diversity more than evolutionary biologists like myself; we love every one of those million beetle species, every variation of every scale count, every nuance in the coloration of feathers. But without some common mooring, we cannot talk to each other. And if we cannot talk, we cannot bargain, compromise, and understand. I am sad that I can no longer cite the most common lines from Shakespeare or the Bible in class, and hold any hope for majority recognition. I am troubled that the primary lingua franca of shared culture may now be rock music of the last decade—not because I regard the genre as inherently unworthy, but because I know that the language will soon change and therefore sow more barriers to intelligibility across generations. I am worried that people with inadequate knowledge of the history and literature of their culture will ultimately become entirely self-referential, like science fiction's most telling symbol—the happy fool who lives in the one-dimensional world of pointland, and thinks he knows everything because he forms his entire universe. In this sense, the bee criticizes the spider properly—an ephemeral cobweb "four inches round" is a paltry sample of our big and beautiful world. I can't do much with a student who doesn't know multivariate statistics and the logic of natural selection; but I cannot make a good scientist—though I can forge an adequate technocrat—from a person who never reads beyond the professional journals of his own field.

I give the last word to Swift. When the bee and the spider finish their argument, Aesop steps up and praises both parties, who have "admirably managed the dispute between them, have taken in the full strength of all that is to be said on both sides, and exhausted the substance of every argument pro and con." But he then, as befits his station and status, supports the bee. A person who ignores accumulated wisdom perishes in his own thin web:

> Erect your schemes with as much method and skill as you please; yet if the materials be nothing but dirt, spun out of your own entrails (the guts of modern brains) the edifice will conclude at last in a cobweb: the duration of which, like that of other spiders webs, may be imputed to their being forgotten, or neglected, or hid in a corner.

Aesop ends by praising the bee and inventing a proverb that forms one of the loveliest conjunctions in English. I began this essay with a rough saying about sweetness; I end with the most noble trope on the same word. Did you know that the phrase "sweetness and light"—meaning, literally, honey and wax—entered our lexicon of sayings as the culmination of Swift's defense, via Aesop, for the extended hive of our greatest intellectual traditions?

> As for us, the ancients; we are content with the bee, to pretend to nothing of our own, beyond our wings and our voice: that is to say, our flights and our language; for the rest, whatever we have got, has been by infinite labor, and search, and ranging through every corner of nature: the difference is, that instead of dirt and poison, we have rather chose to fill our hives with honey and wax, thus furnishing mankind with the two noblest of things, which are sweetness and light.

PART THREE

Origin, Stability, and Extinction

ORIGIN

8

In the Mind of the Beholder

A VARIETY OF ancient mottoes proclaims that no principle of aesthetics can specify the gorgeous and the ugly to everyone's satisfaction. "Beauty," we are told, "is in the eye of the beholder"; there is no accounting for tastes—an observation old enough to have a classical Latin original *(de gustibus non disputandum)*, and sufficiently universal to boast a trendier version in our current vernacular (different strokes for different folks).

Science, by contrast, is supposed to be an objective enterprise, with common criteria of procedure, and standards of evidence that should lead all people of good will to accept a documented conclusion. I do not, of course, deny a genuine difference between aesthetics and science on this score: we have truly discovered—as a fact of the external world, not a preference of our psyches—that the earth revolves around the sun and that evolution happens; but we will never reach consensus on whether Bach or Brahms was the greater composer (and professionals in the field of aesthetics would not ask so foolish a question).

But I would also reject any claim that personal preference, the root of aesthetic judgment, does not play a key role in science. True, the world is indifferent to our hopes—and fire burns whether we like it or not. But our ways of learning about the world are strongly influenced by the social preconceptions and biased modes of thinking that each scientist must apply to any problem. The stereotype of a fully rational and objective "scientific method," with individual scientists as logical (and interchangeable) robots, is self-serving mythology.

Historians and philosophers of science often make a distinction between the logic and psychologic of a scientific conclusion—or, in the jargon, "context of justification" and "context of discovery." After conclusions are firmly in place, a logical pathway can be traced from data through principles of reasoning, to results and new theories—context of justification. But scientists who make discoveries rarely follow this optimal pathway of subsequent logical reconstruction. Scientists reach their conclusions for the damnedest of reasons: intuitions, guesses, redirections after wild-goose chases, all combined with a dollop of rigorous observation and logical reasoning to be sure—context of discovery.

This messy and personal side of science should not be disparaged, or covered up, by scientists for two major reasons. First, scientists should proudly show this human face to display their kinship with all other modes of creative human thought. (The myth of a separate mode based on rigorous objectivity and arcane, largely mathematical, knowledge vouchsafed only to the initiated may provide some immediate benefits in bamboozling a public to regard us as a new priesthood, but must ultimately prove harmful in erecting barriers to truly friendly understanding, and in falsely persuading so many students that science lies beyond their capabilities.) Second, while biases and preferences often impede understanding, these mental idiosyncrasies may also serve as powerful, if quirky and personal, guides to solutions. C. S. Peirce (1839–1914), America's greatest philosopher of science, even coined a new word to express the imaginative mode of reasoning involved in such mental leaping: abduction, or "leading from" (one place to another), to contrast with the more sedate and classical modes of deduction, or logical sequencing, and induction, or generalization from accumulated particulars (all from the Latin *ducere*, "to lead").

This general theme leaped (or crept) into my mind as I contemplated the three hottest paleontological news items of 1993. In particular, I noted a discordance, common to all three items, between their coverage in the press and my personal reaction to the claim. All three were described as particularly surprising (they would not have ranked as "hot" items otherwise)—whereas I found each claim intensely interesting but entirely expected. This led me, naturally, to wonder why these (to me) perfectly reasonable claims seemed so unusual to others.

One might posit that my lack of surprise only recorded the professional knowledge of all practicing paleontologists—and that the discordance therefore lies between public and professional perception (thus reinforcing the myth of an arcane and enlightened priesthood of scientists). But many, probably most, of my professional colleagues were surprised as well—so the reasons for my expectations must be sought elsewhere.

I then recognized an abstract linkage among the three news items, and finally understood the coordinated source of my complacency and the surprise of others. On an overt level, the three items could not be more different—for they span a maximal range of time and subject in the evolutionary history of multicellular animals (and this disparity provides an added benefit in making their conjunction a good theme for an essay—so my literary thanks go out to them as well). The first item comes from the very beginning, the second from the middle, and the third from the latest moment in the history of animal life. The three seem just as different in subject—for the first examines evolutionary rate, the second interaction among organisms, and the third biogeography or place of origin for a key species.

But the three stories are linked at a level sufficiently abstract to evoke the underlying attitudes so basic to any person's individual being that popular culture speaks of a "philosophy of life" or "worldview." Scholars have also struggled with this notion of a personal or social model so pervasive that all particulars are judged in its light. Being scholars, they may use a fancy German term like *Weltanschauung,* which sounds complex, but only means "outlook upon the world." In the most celebrated use in a social sense, T. S. Kuhn referred to the shared worldview of scientists as a paradigm (see his classic 1962 book, *The Structure of Scientific Revolutions*). Such paradigms, in Kuhn's view, are so constraining, and so unbreakable in their own terms, that fundamentally new theories must be imported from elsewhere (insights of other disciplines, conscious radicalism of young rebels within a field), and must then triumph by rapid replacement (scientific revolution) rather than by incremental advance. But the most eloquent testimony to the power and pervasiveness of worldviews was surely provided by Gilbert and Sullivan's Private Willis (in *Iolanthe*), as he mused on guard duty outside the Victorian House of Commons:

> I often think it's comical
> How Nature always does contrive
> That every boy and every gal
> That's born into the world alive
> is either a little Liberal
> Or else a little Conservative!

Nothing is more dangerous than a dogmatic worldview—nothing more constraining, more blinding to innovation, more destructive of openness to novelty. On the other hand, a fruitful worldview is the greatest shortcut to insight, and the finest prod for making connections—in short, the best possible agent for a Peircean abduction. So much in our material culture is both alluring and dangerous at the same time—try fast cars and high-stakes poker for starters. Why shouldn't a fundamental issue in our intellectual lives possess the same properties?

In short, I realized that my linkage of the three issues, and my lack of surprise at claims reported in newspapers as startling, emanated from a worldview, or model of reality, different in some crucial respects from the expectations held by many scientific colleagues and by the general public. I do not know that my view is more correct; I do not even think that "right" and "wrong" are good categories for assessing complex mental models of external reality—for models in science are judged as useful or detrimental, not as true or false.

I do know that chosen models dictate our parsing of nature, and either channel our thoughts toward novel insight, or blind us to evident and important aspects of reality. Beauty must be in the eye of the beholder, but *access* to truth lies within the mind of the beholder—and our minds are as varied as our hairstyles. "Great is truth, and shall prevail"—but we only get there along pathways of our own mental construction. Science is as resolutely personal an enterprise as art, even if the chief prize be truth rather than beauty (though artists also seek truth, and good science is profoundly beautiful).

1. TIMING THE CAMBRIAN EXPLOSION: HOW FAST IS FAST? Paleontologists have long known, and puzzled over, the rapid appearance of nearly all major animal phyla during a short interval at the beginning of the Cambrian period (a subject frequently treated in these essays, and in my book *Wonderful Life*). The earth's fossil record extends back 3.5 billion years to the earliest rock sufficiently unaltered by later heat and pressure to preserve traces of ancient organisms. But, with the exception of some multicellular algae that play no role in the genealogy of animals, all life, including the ancestors of animals, remained unicellular for five-sixths of subsequent history—until about 550 million years ago, when an evolutionary explosion introduced all major groups of animals in just a few million years.

When geologists use the word *explosion*, please take this expression with a grain of salt and recognize that, in my professional world, explosions have very long fuses. No one has ever doubted that the Cambrian explosion must be measured in millions of years—a long time for anyone who has ever set a dynamite charge, but awfully quick relative to a history of life measured in billions (remember that one thousand millions make a billion). But how many millions?

Paleontologists have always hedged on this crucial question because we had no precise dates for the inception of the Cambrian period. The Cambrian ended some 505–510 million years ago, but we had no good fix on the beginning until last September, when several of my colleagues in the Cambridge mafia (Harvard plus MIT) joined with Russian geologists in finally nailing the early Cambrian, based on data "so beautiful you could cry" (to quote my grandmother, who would have understood).

Previous estimates for the Cambrian's beginning ranged from nearly 600 to 530 million years ago. The older dates (favored by most) permitted quite a good stretch for the Cambrian explosion, perhaps 30 million years or so (still a moment among billions, but at least a relaxed moment). My colleagues—see Bowring et al. in this book's bibliography—have now pinpointed the explosion by calibrating the radioactive decay of uranium to lead within zircon crystals obtained from volcanic rocks interbedded with Siberian sediments containing earliest Cambrian fossils.

The earliest Cambrian, like Caesar's Gaul, is divided into three parts, called, from oldest to youngest, Manakayan, Tommotian, and Atdabanian. (The names are all derived from Russian localities, where early Cambrian rocks are particularly well exposed.) The Manakayan contains many fossilized bits and pieces of cousins and precursors, but not the remains of major modern phyla. The Manakayan therefore predates the Cambrian explosion. By the end of the Atdabanian, virtually all modern phyla had made their appearance. The Cambrian explosion therefore spans the Tommotian and Atdabanian stages.

My colleagues have dated the base of the Manakayan at 544 million years ago (with potential error of only a few hundred thousand years), and have determined that this initial stage lasted some 14 million years. The Tommotian began about 530 million years ago and—get this, for now the intellectual impact occurs—the subsequent Atdabanian stage ended only 5 to 6 (at the very most, 10) million years later. Thus the entire Cambrian explosion, previously allowed up to 30 or even 40 million years, must now fit into 5 to 10 (and almost surely nearer the lower limit), from the base of the Tommotian to the end of the Atdabanian. In other words, fast is much, much faster than we ever thought.

This story rocked the airwaves (insofar as any scientific tale merits the cliché). *The New York Times* awarded front-page billing in its weekly science section; National Public Radio featured my colleagues on their weekly science talk show. The primary theme was intense surprise. Evolution means slow; how could so much happen so fast? Was the entire conceptual world of evolutionary theory about to be undermined? I was absolutely delighted by my colleagues' result, but I was not surprised. I have believed for many years that fast was at least this fast. (I had regarded the old limits of 30 to 40 million years merely as an upper bound, and had assumed that the Cambrian explosion occupied only a small segment at the beginning of this full interval.) Why such a difference between public perception and my personal reaction?

2. Insects and flowers. Nothing displays human hubris more than the old textbook designation of recent geological times as the "age of man." First of all, if we must use an eponymous designation, we live today, and have always lived, in the "age of bacteria." Second, if we insist on multicellular parochialism, modern times must surely be called the "age of insects." *Homo sapiens* is one species, mammals include about four thousand. By contrast, nearly a million species of insects have been described (and several millions more remain undiscovered or uncatalogued—see essay 29). Insects represent more than 70 percent of all named animal species.

So why are insects so diverse? Many answers have been offered, and the solution will be some complex combination of the good arguments. Small size, great ecological variety, rapid geographic dispersal have all been mentioned, and are probably valid as partial explanations, but one other factor always stands out in the conventional list of reasons: coevolution with flowering plants. The angiosperms, or flowering plants, are by far the most diverse group in their kingdom. Many species are fertilized by insects in a mutually beneficial arrangement that supplies food to the insects while transporting pollen from flower to flower.

So intricate, and so mutually adapted, are the features of both flower and insect in many cases—special colors and odors to attract the insects, exquisitely fashioned mouthparts to extract the nectar, for example—that this pairing has become our classical example of coevolution, or promotion of adaptation and diversity by evolved interaction among organisms. (Darwin wrote an entire book on the subject, using the classic case of intricately coadapted orchids and their insect pollinators.) Thus a received truth of evolutionary biology has proclaimed that insects are so diverse, in no small part, because flowering plants are so varied—and each plant evolves its pollinator (and vice versa).

Sounds good, but is it true? The fossil record suggests an obvious test, but, curiously, no one had ever carried out the protocol until my colleagues Conrad Labandeira and Jack Sepkoski published a paper in July 1993. Insects arose in the Devonian period, but began a major radiation in diversity during subsequent Carboniferous times, some 325 million years ago. Angiosperms, by contrast, arose much later. Their first fossils are found in early Cretaceous strata, some 140 million years ago. (If they arose earlier, as some scientists speculate, they could not have been very abundant.) But angiosperms didn't really flower (pardon the irresistible, if unoriginal, pun) until the Albian and Cenomanian stages of the middle Cretaceous, some 100 million years ago, where their explosive evolutionary radiation stands out as one of the great events of our fossil record.

If insect diversity is tied to the radiation of flowering plants, as traditional views proclaim, then this burst of angiosperms should be matched by a similar explosion of insects in the fossil record. Why has such an obvious test of an important evolutionary hypothesis not been made before? The reason may lie in a common misconception about the fossil record of insects. Many people suppose that this record is exceptionally poor, with so few insects preserved as fossils that we would never be able to get a good enough count to assess the hypothesis of a sharp Cretaceous increase when the angiosperms radiated.

To be sure, insects do not fossilize as readily as clams or trilobites, but their record is by no means so sparse as common impressions hold. Jack Sepkoski has spent most of his twenty-year career (he was my graduate student just before then, so I confess my familial bias toward his work) engaged in an enterprise that some traditional paleontologists dismiss with the epithet of "taxon counting"—that is, he sits in the library (which he describes as his "field area") and tabulates the ranges of all fossil genera and families in all the world's literature in all languages. (This activity is neither so simple nor so automatic as the uninitiated might imagine. First of all, you must know where to find, and how to recognize, obscure sources in publications with non-Roman alphabets. Second, you do not merely list what you find, but must make judgments about the numerous taxonomic and geological errors in such publications. I have never understood why some traditionalists disparage this work. They, after all, published the literature that Sepkoski uses; don't they want their work so honored and well employed? Through Sepkoski's painstaking effort in full and standardized tabulation, we have, for the first time, a usable compendium of changing diversity throughout the history of life, and for all groups.)

Labandeira and Sepkoski found that the insect record is better than anyone thought (once you add all the Russian and Chinese publications). In fact, insects are more diverse than that other famous terrestrial group, for which no one has ever been shy about offering conclusions—the tetrapods, or terrestrial vertebrates (amphibians, reptiles, birds, and mammals combined). The fossil record of insects includes 1,263 families; that of tetrapods, 825 families. Moreover, except for the latest Devonian, when insects were young and hadn't yet taken off on an evolutionary radiation, insect diversity has always exceeded tetrapod diversity in every geological epoch.

Looking at the taxonomic level of insect families, Labandeira and Sepkoski could find no evidence for any positive impact of the angiosperm radiation upon insect diversity. The insect radiation began in the early Carboniferous, some 325 million years ago, got derailed once in the greatest of all mass extinctions at the end of the Permian (when eight of twenty-seven insect orders died), began again in the subsequent Triassic period, and has never stopped since. In fact, and if anything, increase in number of families actually seems to have dropped somewhat during the Cretaceous, as the angiosperms flowered!

Labandeira and Sepkoski then tried a different approach and also found no relationship with angiosperms. Instead of taxonomic diversity, they tabulated ecological variety by dividing insects into thirty-four "mouthpart" categories—that is, different ways of making an ecological living based on modes of feeding. (Many of these categories include insects from several different taxonomic lineages, so my colleagues are measuring ecological disparity, not just numerical abundance.) They found that insects had already filled 65–88 percent of these categories by the middle Jurassic, the period before angiosperms arose. Only between one and seven new categories arose after the angiosperms evolved, but most of these have especially poor fossil records, and may well have originated earlier. Only one category can plausibly be linked to life with flowering plants. Thus, angiosperms are also not responsible for the morphological variety of insect feeding mechanisms.

Again, the newswires buzzed (more punning apologies) with this story, and *The New York Times* again awarded front-page billing. Again, expressions of profound surprise dominated the coverage. How could insects evolve independently of the flowering plants to which many are now so strongly tied? Doesn't Darwinism proclaim that organisms change within webs of competition and interaction toward mutually beneficial states? Again, I was pleased but not surprised at all. For I have long felt that images of balance and optimizing competition have been greatly oversold, that important and effectively random forces buffet the history of life, that most groups of organisms make their own way according to their own attributes, and that interactions among most groups are, on the broad scale of time in millions of years, more like Longfellow's "ships that pass in the night" than the Book of Ruth's "whither thou goest, I will go."

3. WHERE DID *HOMO SAPIENS* ORIGINATE? My last issue is a carry-over from previous years. Nothing decisive happened in 1993 to resolve this hot debate of the last decade or so. Rather, I am amazed that the story has such fantastic "legs," and remains both the hottest item on the paleoanthropological newswire, and a source of dichotomization that has forced a more complex issue into two warring camps (at least in public perception).

One position has been dubbed the "multiregionalist model," or the "candelabra" or "menorah" theory (depending on your ethnic preferences) of recent human evolution. Everyone agrees that our immediately ancestral species, *Homo erectus*, moved out of Africa into Europe and Asia more than a million years ago (where they became "Java Man" and "Peking Man" of the old textbooks). Multiregionalists argue that *Homo sapiens* evolved simultaneously from *Homo erectus* populations on all three continents (with necessary maintenance of some gene flow among the populations, for they could not otherwise have evolved in such a coordinated way).

The other side has been called the "out of Africa" or "Noah's ark" school of human evolution. They argue that *Homo sapiens* arose in one place as a small population, and then spread throughout the world to produce all our modern diversity. If Africa was the single place, then European and Asian *Homo erectus*, and the later European Neanderthals as well, played little or no role in our origin, but were replaced by later invaders in a second and much later wave of human migration.

(The most famous version of the "Noah's ark" theory, the poorly named "mitochondrial Eve" hypothesis of modern human origins in Africa, suffered a blow in 1993, when discovery of an important technical fallacy in the computer program used to generate and assess evolutionary trees debunked the supposed evidence for an African source. But in so disproving the original claim, correction only dictated agnosticism, not a contrary conclusion—that is, the new trees are consistent with origin in a single place, but Africa cannot be affirmed as the clearly preferred spot, though Africa remains as plausible as any other place by this criterion. Other independent sources of evidence, especially the greater genetic diversity measured among African peoples, continue, in my view, to favor an African origin—see Stoneking, in the bibliography, for a thorough and fair review.)

As a student of snails, I have no great personal stake in this argument, though I would be willing to wager that our newfangled Noah's ark will one day find its Ararat (though I won't be devastated if the boat sinks and multiregionalism triumphs). But I am intrigued by journalists' representation of this debate—particularly in their attribution of surprise to one side and expectation to the other (thus linking this tale, through the theme of misplaced surprise, to my previous two stories). Newspapers and science magazines invariably present multiregionalism as the orthodox or expected view, and out-of-Africa (or any other single place) as the surprising new kid on the block.

But this assessment is ass-backwards by any standard rendering of evolutionary theory (divorced from the distortions that intrude upon us whenever we consider something so close to us as human ancestry). Origin in a single place is the expectation of ordinary evolutionary theory, and utterly unsurprising. Species are unitary populations of organisms that split off from their ancestral groups in a limited part of the parental range. Species arise as historical entities in particular places and then spread, if successful, as far as their adaptations and ecological propensities allow. Rats and pigeons live all over the world, just as humans do. Yet we are not tempted to argue that rats evolved in parallel, on all continents simultaneously, toward greater ratitude. We suppose that, like most species, rats arose in a single region and then spread out. Why, then, does origin in a single place surprise us when we, rather than pigeons, represent the subject? Why do we devise an entirely idiosyncratic and unusual multiregional hypothesis, and then proclaim it orthodox and expected?

I can only suppose that we want to segregate humans off as something special. We wish to see our evolution, particularly the late expansion of our brain to current size, as an event of more than merely local significance. We do not wish to view our global triumph as so fortuitously dependent upon the contingent history of a small African population; we would rather conceive our exalted intellect as so generally advantageous that all populations, in all places, must move, in adaptive unison, toward the same desired goal.

I must try to understand the contrast of public surprise with my personal expectation for these three disparate stories by seeking a difference in worldviews, or general models of reality, between me and most of thee. Under what common paradigm, rejected by me, does a shorter Cambrian explosion, a lack of lockstep evolution between flowering plants and insects, and a single place of origin for *Homo sapiens* seem so surprising? I can only observe that all three contraries—a more leisurely origin for anatomical designs, a coordinated evolution of co-adapted groups, and an intercontinental origin of our most valued features—fit well with a more stately, predictable, and comforting view of life's history than I can see in the fossil record. Traditional concepts of evolution, at least in their translation to popular culture, favor a slow and stately process, ruled by sensible adaptation along its pathways, and expanding out toward both greater complexity of the highest forms and more bountiful diversity throughout. Such a view would coordinate all three surprises in my three stories, for the newly shortened Cambrian explosion is decidedly unstately, the independence of insects and flowers seems chaotically uncoordinated, and the emergence of *Homo sapiens*, if viewed as a historical event in a single place, becomes quirky and chancy.

But my worldview accommodates and anticipates all these phenomena of rate, interaction, and place. I have come to see stability as the norm for most times, and evolutionary change as a relatively rapid event punctuating the stillness and bringing systems to new states (see essays 10 and 11). A faster Cambrian explosion feeds this expectation. I view lineages as evolving largely independently of each other. I do not deny, of course, that species interact in adaptively intricate ways. But each lineage is a unique entity with its own idiosyncrasies; and each evolutionary trajectory through a temporal series of environments encounters so many random effects of great magnitude, that I expect historical individuality to overwhelm coordination. Grand-scale independence of insects and flowers (despite the tight linkage of so many species pairs today) conforms to this view. Finally, I regard each species as a contingent item of history with an unpredictable future. I anticipate that a species will arise in a single place and then move along an unexpected pathway. In short, all my non-surprises are coordinated by a worldview that celebrates quick and unpredictable changes in a fossil record featuring lineages construed as largely independent historical entities. I should also add that I find such a world stunning and fascinating in its chaotic complexity and historical genesis—and I happily trade the comforts of the older view for the joys of contemplating and struggling with such multifarious intrigue.

I've put myself in a tough spot. This essay has veered dangerously close to unseemly self-congratulation. But I do not write to claim that I have a "better" worldview, more attuned to solving the outstanding problems of life's history. Nor do I assert the correctness of my position on the three stories, for truth is the daughter of time, and I may be proved wrong about all of them. I developed this topic because I regard the subject of worldviews, or paradigms, as so important for the unification of all creative human thought, and I wrote of my own experience because personal testimony has been an accepted staple of the essay ever since Montaigne invented the genre. (And now I must halt, lest you parry with Shakespeare's observation that the author "doth protest too much, methinks.")

Maybe my worldview, shared by many scholars these days (for I developed it by assimilation, not personal invention), has power as a more fruitful outlook upon reality than previous paradigms provided. Maybe my horse is coming in. But maybe I am only riding a gelding named "fashion," a nag destined to stumble at the gate next season at Hialeah as the Seabiscuit or Secretariat of deterministic gradualism comes thundering down the home stretch.

EPILOGUE

I wrote this essay at the end of 1993. I am now, in late May 1995, revising it for publication. I often ask myself why I remain so pleased that I stuck with my early childhood desire to become a scientist. And I always return to the same primary answer: exciting fields in science grow and change so rapidly; intellectual stimulation is inherent in the dynamics of research; one can never become complacent. A year and a half is a short period, even in that geologically insignificant eyeblink known as a human lifetime. But in this tiny interval between composition and revision of my essay, important new information has been reported for all three subjects highlighted therein. And if I may be permitted one more gloat—Lord only knows that the essay contains enough self-congratulation already—all three items affirm the discoveries here reported, and enhance the fruitfulness of a worldview that emphasizes rapid, unpredictable, and historically contingent events as the key to evolution. My horse is still out in front, and accelerating.

On the Cambrian explosion, even more phyla have now been traced back to this initial geological period of multicellular animal life—and these developments are described in the next essay.

The *New York Times* science section for May 23, 1995, features, as a headline for its lead story, "Which came first: bees or flowers? Find points to bees." The text reports a fascinating discovery by Stephen T. Hasiotis in 220-million-year-old logs (Triassic on our geological time scale) from a celebrated place, the Petrified Forest of eastern Arizona. Hasiotis has found persuasive evidence for complex and distinctive nests of bees and wasps within these logs (although fossils of the insects themselves have not yet been discovered—scarcely surprising, given the difficulties of preservation in this environment). The bees' nests, for example, are excavated in shallow hollows reached through knotholes. Each nest consists of fifteen to thirty chambers less than an inch long and shaped like flasks.

This discovery extends the fossil record of bees back 140 million years, for the earliest known specimen lies in 80-million-year-old amber. But even more surprisingly, according to the author of the article, the *Times*'s leading science reporter, John Noble Wilford, the origin of flowering plants (angiosperms) lay 100 million years in the future when these bees excavated their nests in the logs of these gymnosperms (cone-bearing woody plants without flowers). Perhaps bees first pollinated gymnosperms, and therefore lived for most of their history in a world without flowers, and only much later developed an evolutionary relationship with newfangled angiosperms.

Surprise, and downright astonishment, forms the primary theme of Wilford's article. He writes:

> The problem is that flowers date from only half as long ago. Could bees have lived before flowers? The very idea, once unthinkable, is upsetting traditional theory about the early history of bees and their supposed coevolution with flowering plants . . . The discovery casts serious doubt on the standard theory that flowering plants and social insects like bees more or less evolved together, with the spread of flowers presumably influencing the development and proliferation of the bees.

But why should Wilford have been astonished in the light of Labandeira and Sepkoski's work, published two years earlier (and known and cited by Wilford)? I can conclude only that traditions die hard (and slowly). Wilford interviewed Sepkoski for his article, and my former student responded with a most appropriate one-liner: "It's exactly what we would have expected."

On the third subject of human origins, support for out-of-Africa (and recently) has cascaded during the past two years. First of all, many more genes have been sequenced and studied for their variation across human racial groups—and amounts of change, in each case, indicate a very recent common ancestry inconsistent with the age of our separate populations under the multiregionalist view. (Another discovery of the past year extends the age of *Homo erectus* in Asia to about 1.6 million years, so the multiple regions must be even older than previously imagined—and genetic variation among races should be even greater if our major populations have been distinct for so long.) Study after study (many reported at the annual meeting of the American Association for the Advancement of Science in Atlanta during February 1995) places the exodus from Africa (measured as the common origin of all non-African racial diversity) at 100,000 to 150,000 years, with the latest and most sophisticated analyses pointing toward the younger date. Moreover, several studies continue to affirm the greater genetic diversity of African peoples vs. all other humans combined—a fact hard to understand unless out-of-Africa is correct, and modern humans have inhabited this continent far longer than any other place (thus providing time for the evolution of such genetic diversity).

But the most satisfying, and long-awaited, results have just been announced (last week of May 1995). The original "Eve" hypothesis of recent human origins was based on a study of variation in mitochondrial DNA, a part of the genome with simple inheritance along strictly maternal lines. (Sperm contribute no mitochondria to the fertilized egg. Most genes have more-complex inheritance because maternal and paternal copies are complexly snipped apart and recombined during meiosis and subsequent sexual reproduction. For issues as precise and intricate as the timing of such a recent event as human common ancestry, genes with simple and uninterrupted inheritance along either maternal or paternal lines offer immense advantages.)

But, just as mitochondria have strictly maternal inheritance, a small part of the genome runs with equal exclusivity through paternal lines, from father to son. The sex-determining Y-chromosome is a tiny fragment of DNA with no pairing to any part of the maternal genome. Consequently, genetic diversity of the Y-chromosome should identify the human "Adam," just as mitochondrial DNA can reach "Eve." By the way, I have long hated these cute biblical metaphors, because they provoke the wrong impression that we all descend from a primal pair. Of course we have no such minimal descent; we evolved from a small population that split from an ancestral group, and not from a pair in a cave or garden (though only a limited subset from this initial population probably left descendants—scarcely surprising since all genealogy works this way, with most folks leaving no ultimate descendants, and a few prolific characters responsible for most of the patrimony [and matrimony]).

The Y-chromosome test is obvious and exciting; my colleagues have been after this plum for some time, but results hadn't been reported before, simply because this intricate and complex work requires so much labor and can't be rushed. But now my colleague Wally Gilbert and my former student Rob Dorit (paleontologist turned renegade molecular biologist) have announced their stunning results on variation in a 729-base-pair sequence of the Y-chromosome for thirty-eight males spanning the extremes of human racial diversity. They find almost no genetic variation—a strong affirmation for recency of common ancestry in all humans (their results don't indicate a place for this commonality, but Africa seems strongly favored on other grounds, previously mentioned). They arrive at a date of 270,000 years for the initial split of *Homo sapiens* from an ancestral population—a date beautifully consonant with the usual "Eve" estimate of about 250,000 years. (Thus, under this evidence, humans originated, presumably in Africa, about a quarter of a million years ago, and began to spread into other continents about 100,000 years ago.) Gilbert and colleagues also felt that they could estimate the size of the founding human population at about 7,500 males—a particularly gratifying affirmation of my own prejudices, under punctuated equilibrium, that new species arise quickly by branching from parental groups, and not by gradualistic transformation of entire ancestral stocks.

I will leave the last word (for now) to Gilbert and his powerful methodological point that controversial discoveries are best affirmed when entirely independent sources of data achieve the same result: "The date for our Adam is a good match for that of the African Eve . . . The exciting fact is that our experiment is based on different principles from previous work. This strengthens the finding that there was a common origin for modern humans."

9

Of Tongue Worms, Velvet Worms, and Water Bears

I BELONG TO the last generation of students formally educated, in large measure, by the practice of rote memorization. Hence, I know the Gettysburg Address by heart. (And who can ever expunge the arcana learned at age ten, while who, at age fifty, can retain the important items encountered last week?) At least I know Lincoln's line for paleontologists: "we here highly resolve that these dead shall not have died in vain." And, when Stephen Dedalus, in Joyce's novels, routes "on old Olympus's topmost tops" into his stream of consciousness, I know that he is musing upon the standard mnemonic for remembering the names of cranial nerves in proper sequence front to back—olfactory, optic, oculomotor . . .

Among the classical items of rote memorization in early schooling, two stand out for later utility to paleontologists like me—the geological time scale, and the list of animal phyla, the major taxonomic divisions of life in our kingdom (some twenty to forty, depending on the version you learned). Most of my fellow students didn't complain too much about the dozen or so major groups, for everyone should know a vertebrate from an arthropod from a mollusk from an echinoderm, if only because we do encounter such creatures in our daily lives. But for the larger number of so-called "minor phyla"—the unrememberable whatchamacallits with such funny names as Ctenophora, or comb jellies, and Priapulida, or little penis worms—most of us had only contempt and loathing, for we couldn't recall them on exams, and we never encountered them in Central Park or at Jones Beach ("nature" to New York City kids).

Yet these "minor" phyla embody some of the most fascinating problems of natural history, and should not be ranked with the unknown and the unloved. They are, first of all, "minor" only in the sense of current membership (few species alive today), though some, brachiopods and bryozoans in particular, dominated the early fossil record of multicellular animal life. Moreover, these groups are decidedly not minor in degree of anatomical distinctness, for they are as different, one from the next, as a fish from a fly, or a clam from a sea cucumber.

The minor phyla must play a crucial role in unraveling the greatest of all mysteries surrounding the history and fossil record of animal life. I have often written, in these essays, about the "Cambrian explosion," the extremely restricted time that encompasses the first appearance in the fossil record of nearly all basic anatomical designs for animal life. According to a recent study (see the preceding essay), the first ever based upon rigorously determined radiometric age dates, this episode lasted an astonishingly short 5 million years, from about 535 to 530 million years ago.

Since then, only one new phylum with a prominent fossil record has been added to life's archives—the Bryozoa, a group of small colonial organisms, which, like reef-building corals, secrete calcified skeletons surrounding the individual animals of an aggregate. (The bryozoans arose at the beginning of the very next, or Ordovician, period, and their Cambrian absence may be an artifact of our failure to find earlier representatives.) One might argue, without great exaggeration, that 530 million years of subsequent evolution has produced no more than a set of variations upon themes established during this initial explosion—though some of these little fillips, including human consciousness and insect flight, had quite an impact upon the history of life!

The minor phyla provide a key to the Cambrian explosion because they represent a potential exception and softening. This episode, as stated above, is enormously puzzling and quite contrary to preferred assumptions about the generally slow and steady character of evolutionary change. Therefore, paleontologists have sought (largely unconsciously, for thus do we act upon our deepest biases) mitigating circumstances or arguments that might either diminish or spread out the Cambrian explosion.

Among such sops to our uniformitarian preferences, none has been more common—I can hear the words in my mind as stated by a consensus of professors and read in dozens of books—than the following potential invocation of minor phyla to make the Cambrian explosion merely an intensification of ordinary possibilities, rather than an exclusivity: "But how can you claim that all the phyla originated during this minimal beginning interval? After all, about half the animal phyla contain no soft parts at all and therefore have no fossil record. How do you know that these groups haven't been arising throughout the 530 million years since the Cambrian explosion? Moreover, most of these phyla contain very few species. Doesn't their rarity indicate a potentially recent origin, leaving little time for gradual spread and speciation?"

This argument is not unreasonable, and seems particularly strong under certain circumstances. Consider, for example, the case of a classic minor phylum of small membership—the Pentastomida, or tongue worms. (Their name, literally, means "five-mouthed," in reference to the two pairs of limbs surrounding the true mouth at the anterior end. In some species—see the accompanying illustration—the mouth resides at the end of a stalk comparable in length to the surrounding four legs, thus giving the appearance of a five-pointed star. The usual name of "tongue worm" commemorates the more common species that resemble a vertebrate tongue in miniature—also on the accompanying figure.)

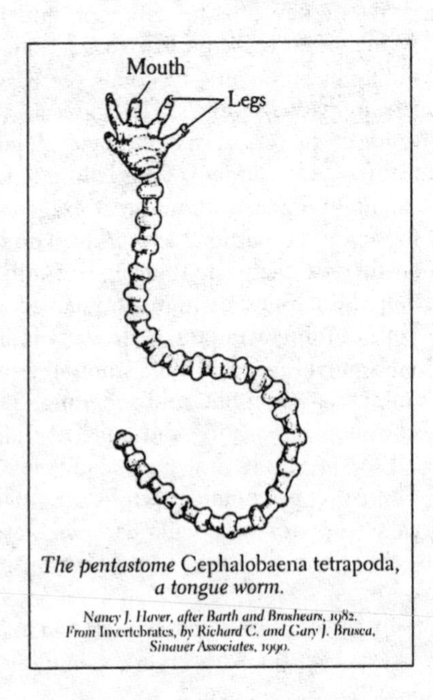

The pentastome Cephalobaena tetrapoda, *a tongue worm.*

Nancy J. Haver, after Barth and Broshears, 1982. From Invertebrates, *by Richard C. and Gary J. Brusca, Sinauer Associates, 1990.*

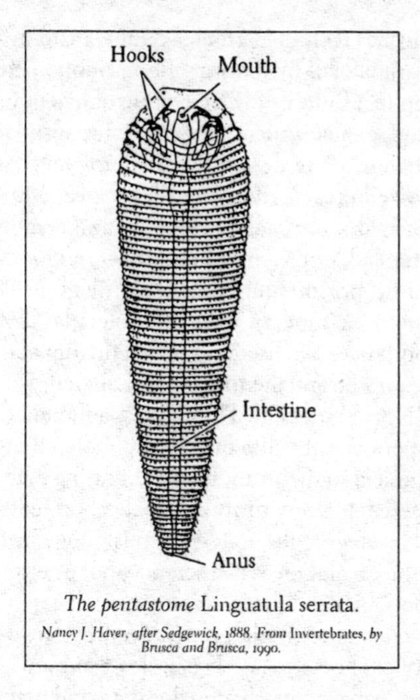

The pentastome Linguatula serrata.

Nancy J. Haver, after Sedgewick, 1888. From Invertebrates, *by Brusca and Brusca, 1990.*

Pentastomes are parasites, and they live almost exclusively upon terrestrial vertebrates—a group that did not evolve until well after the Cambrian explosion. Pentastomes also resemble crustacean arthropods in a few important features. Therefore, following a century of intense debate and a range of hypotheses that either allied pentastomes with other major phyla or gave them separate status on their own, a recent consensus had emerged for viewing these enigmatic creatures as a phylum evolved from a crustacean stock much later than the Cambrian explosion. And if the unique pentastomes could evolve more recently from a well-established group, why not most of the other minor phyla as well? The Cambrian explosion would then lose its exclusivity, and the "phylum making machine" of evolution would continue to operate throughout geological time.

I write this essay to present some freshly published data leading to an opposite conclusion: the Cambrian explosion is even more extensive in scope and exclusive in effect than heretofore recognized even by its partisans. These data have been presented in two papers, published in 1994 by my German paleontological colleagues Dieter Walossek and Klaus J. Müller of the Rheinische Friedrich-Wilhelms Universität in

Bonn. Such long and technical articles on the anatomy of small Cambrian fossils invariably fall by the wayside in public perception (while often creating quite a buzz in the tiny circle of paleontological professionals). Very few "science writers" from the journalistic side have much patience with the arcana of descriptive anatomy (and professional traditions of jargonized presentation contribute greatly to the impasse as well). Moreover, and more sadly, taxonomy and anatomy occupy the lowest rung in the ladder of scientific status—an old-fashioned, albeit harmless and gentle, pursuit more suited to the eighteenth-century days of Linnaeus than to the modern world of molecular biology.

Yet the importance of a discovery lies in the impact and reforming power of ideas expressed and theories thereby altered, not in the "modernity" of methodology employed. Thoughtless adherence to fashion can blind us to the permanent value of things unnoticed or abandoned as outdated. Just think of such luminous and enduring figures as Bach and Brahms, dismissed in their maturity as hopelessly antediluvian by a gaggle of forgotten devotees to the "latest" trends. Judge by quality and engagement with the central ideas of a science, not by employment of the maximally fashionable machinery or jargon.

Arthropods constitute the largest phylum by far, with major subgroups of insects, chelicerates (spiders, mites, scorpions, and horseshoe crabs), and crustaceans (crabs, shrimp, lobsters, and many other marine forms). Three minor phyla are conventionally placed near the arthropods because they possess a few key anatomical features that suggest possible genealogical proximity to this greatest of all groups: the Onychophora, or velvet worms; the Tardigrada, or water bears; and the Pentastomida, or tongue worms. For example, the most popular recent textbook of invertebrate biology (*Invertebrates*, by Richard C. and Gary J. Brusca) grants them an entire chapter, under the heading "Three Enigmatic Groups and a Review of Arthropod Phylogeny."

These three lineages provide an excellent case for testing the hypothesis that phyla continue to originate throughout time, and that the Cambrian explosion is not so exclusive as the fossil record, read literally, might suggest—for they are usually discussed together and share all key features for putative origin in later geological time (small modern membership and absence of hard parts, leaving little opportunity for preservation of fossils).

We have known for the last few years that the Onychophora have a fossil record extending right back to the Cambrian explosion. Modern velvet worms live in moist terrestrial habitats, usually in wet leaves or rotting wood. Most of the eighty or so species are one to three inches in

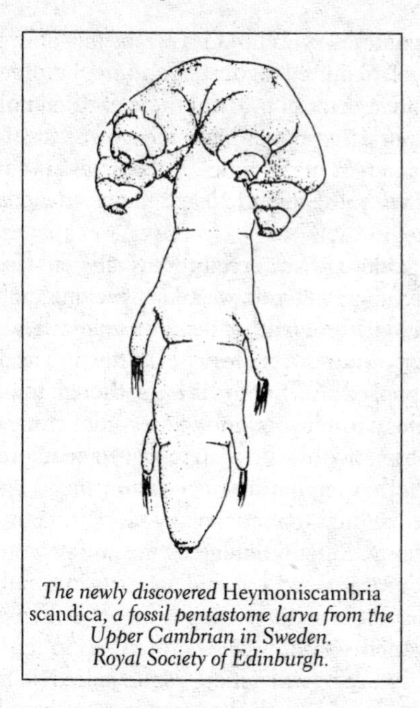

The newly discovered Heymoniscambria scandica, *a fossil pentastome larva from the Upper Cambrian in Sweden. Royal Society of Edinburgh.*

length and have an elongated body with fourteen to forty-three pairs of stumpy legs (called lobopods) behind a head bearing three pairs of appendages (antennae, jaws, and slime papillae, used by these carnivores to shoot a sticky substance at prey). The entire animal looks vaguely like a caterpillar, though onychophorans have no close genealogical relationship with these larvae of moths and butterflies.

Cambrian onychophorans have been proposed for eighty years, since the discovery of *Aysheaia*, a soft-bodied fossil from the famous Burgess Shale in British Columbia. But *Aysheaia* had remained in limbo, with many paleontologists doubting its proper inclusion among the velvet worms, until the discovery of at least four more Cambrian genera during the past ten years, including a reinterpretation of *Hallucigenia*, once the most enigmatic of all Burgess Shale creatures, as an onychophoran originally misconstrued as upside down. These exciting finds (see essay 24 in my previous book *Eight Little Piggies*) show that velvet worms are only minor in current membership and restricted ecological spread. The Onychophora began as a diverse and important group of Cambrian marine invertebrates.

Walossek and Müller have now discovered Cambrian fossils of tardigrades and pentastomes as well—so we can now affirm that all three phyla go right back to the initial diversification of multicellular animal life in the Cambrian explosion. Until these exciting publications, neither tardigrades nor pentastomes had a recognized fossil record at all—not a single specimen from any time. So these new discoveries from the dawn of phyla allow both groups to leapfrog from the present right back to the beginning.

Living tardigrades are tiny creatures, usually only one tenth to one half millimeter in length (hardly visible, since one inch contains 25.4 millimeters), though the "giant" of the phylum reaches 1.7 millimeters. Most of the four hundred species live in films of water on mosses, lichens, flowering plants, soil, or forest litter, though some inhabit freshwater or marine environments on the sediment surfaces of ponds or ocean basins. They look like tiny, eight-legged bears (see illustration), and they move with a lumbering gait—hence their common designation "water bear," though the etymology of their technical name also invokes form and motion, for *tardigrade* means "slow stepper."

Tardigrades have attained a certain notoriety in popular science writing for a suite of unusual features—not to mention their charmingly humorous appearance. Some species indulge in an odd form of reproduction by indirect fertilization. Males penetrate the female's cuticle and deposit sperm beneath. The female then sheds her cuticle (for tardigrades, like insects, grow by molting an external covering and secreting a new and larger version) and lays eggs into this discarded outer shell, already well supplied with previously extruded sperm.

But tardigrades are most famous for their astonishing capacity to shut off metabolism and endure long periods of dormancy—a condition known as cryptobiosis, and defined as a state of dormancy so extreme that no external sign of metabolic activity can be detected at all.

If their habitat dries up (and life in terrestrial water films can be precarious), tardigrades can pull in their legs and secrete a cuticle about their withered body. In this "tun" stage, with undetectable metabolism, tardigrades can survive incredible insults perpetrated by nature or invented by human experimenters—such as immersion in absolute alcohol, ether, or liquid helium, and exposure to temperatures ranging from 149° Centigrade (well above the boiling point of water) to −272° Centigrade (not so far from absolute zero)! When water becomes available again, the animal swells up and returns to activity within a few hours. No one knows how long a tardigrade tun might survive. Brusca's text recounts one tale (likely apocryphal) of living tardigrades emerging from a museum specimen of moss that had been moistened after 120 years of

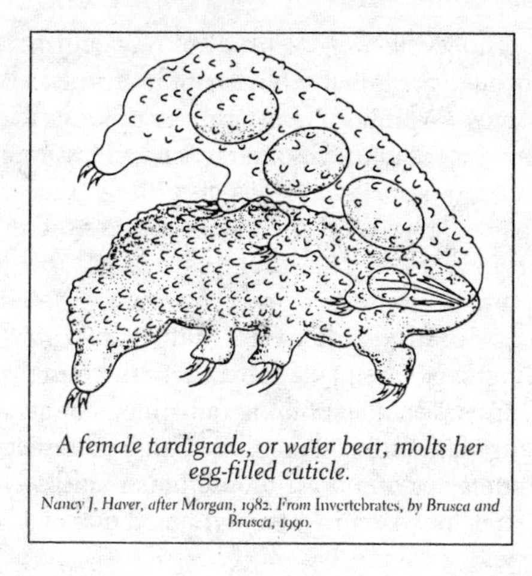

A female tardigrade, or water bear, molts her egg-filled cuticle.

Nancy J. Haver, after Morgan, 1982. From Invertebrates, *by Brusca and Brusca, 1990.*

dry shelf life. Unsurprisingly, tardigrades have been of special experimental interest to students of aging.

The Sixth International Symposium on Tardigrada met in Cambridge, England, on August 22–26, 1994 (my heart warms to the thought that even minor phyla can spawn such multiple and cosmopolitan celebrations by their human devotees). At this meeting, Walossek, Müller, and R. M. Kristensen of the Zoological Museum of the University of Copenhagen presented a showstopping paper titled "A More Than Half a Billion Years Old Stem-Group Tardigrade from Siberia." They had found the first indisputable tardigrade fossil, and this species dated right back to the verge of the Cambrian explosion.

These specimens look just like tardigrades and range from 0.25 to 0.35 millimeter in length (a middling size for modern forms). The key to identification lies not in similar sizes or general appearances, however, for such basic features can be evolved by convergence in independent lineages, but in a large and striking set of unique and complex traits found only in these fossils and in living tardigrades. These marks of genealogical affinity include a distinctive, pitlike mouth, limbs with paired hooks or claws that can be withdrawn at their outer edges, and minute, plate-shaped knobs between the limbs.

The pentastomes provide a much better potential case—the classic putative instance among minor phyla—for continuing origin of major groups after the Cambrian explosion. All of the 100 to 110 species of this phylum are obligatory parasites of vertebrates (almost all terrestrial, though a few species live on fishes). Like many parasites, pentastomes have a complex life cycle, moving from an intermediate to a final host. Larvae bore through the gut wall of the first host, where they mature to their infective stage. When another vertebrate eats this first host, the mature pentastome moves to the respiratory tract, either by crawling from the stomach to the esophagus and boring through, or by tunneling through the intestinal wall and into the bloodstream. The parasite then attaches to the lungs or the nasal or oral cavity (some pentastomes have even been reported from human eyes), by means of hooks at the end of the two pairs of limbs surrounding the mouth. In this (now) permanently attached feeding stage, the pentastome sucks the host's blood through its mouth. (For most people, nothing in biology sounds more, well, to use the contemporary vernacular, yucky than the lifestyles of parasites; but such creatures do form a major component of life's diversity and ecology, and we do need to understand them, though I advance no case for loving them.)

Like many parasites, pentastomes are extremely simplified in anatomy (for the safe and sheltered environment of a host specifies little advantage in retaining the complex features needed for life in the tougher external world). The specific organs of parasitic life—the means of finding, attaching to, and exploiting hosts—are present and complex (in this case, the "five-pointed star" arrangement of the stalked mouth and two pairs of legs at the front end), but the rest of the body has been secondarily simplified. Pentastomes have, for example, no internal organs for respiration, circulation, or excretion. The gut is a simple straight tube, with a muscular pumping apparatus at the front end, obviously useful in extracting the host's blood.

This extreme anatomical simplification of ordinary organs, combined with elaboration of highly specific devices for exploiting hosts, makes the taxonomy of parasites, and their genealogical placement into the evolutionary tree of free-living forms, particularly difficult. Pentastomes have long provided a classic example of this general nightmare in taxonomy. The range of available hypotheses spans nearly all conceivable solutions, with links to annelids (segmented worms), arthropods of one subgroup or another, and separate status (often joined with onychophores and tardigrades) as the favored solutions.

In recent years, however, a consensus arose for allying pentastomes with crustaceans of the arthropod phylum. Several authors presented evidence of similarity between larvae of pentastomids and a group of crustaceans known as branchiurans. Fine structure of the external cuticle and morphology of the sperm cells also seemed to affirm a crustacean link. Then, in 1989, a sealing argument seemed to emerge from the laboratory of my friend and colleague Larry Abele of Florida State University (see bibliography). Abele and associates used the most powerful and appropriately fashionable technique of comparing DNA sequences (in the commonly used and highly informative molecule 18S ribosomal RNA) in pentastomes and representatives of several candidate phyla for relationships—segmented worms and all major groups of arthropods, including insects, horseshoe crabs, millipedes, and crustaceans. The evolutionary tree reconstructed from molecular distances revealed a closest tie of pentastomes with crustaceans. These data led Brusca and Brusca to argue in their textbook for "convincing cases that pentastomids are actually highly modified crustacean parasites."

Moreover, the current life of pentastomes in terrestrial vertebrates led all these exponents of a crustacean link to the reasonable hypothesis that pentastomes originated long after the Cambrian explosion, thus affirming the conventional view that fundamentally new body plans may originate throughout geological time, and that the Cambrian explosion is a great episode of intensification, but not an exclusivity. In fact, Abele and colleagues proposed a divergence date of 350 to 225 million years ago for pentastomes and crustaceans.

But Walossek and Müller (see bibliography) have now described indisputable pentastomes—a whole fauna of several species, not just a single example—from Upper Cambrian strata. These fossils come from the Upper Cambrian "Orsten" beds of Sweden, an extraordinary deposit that has, over the last two decades, yielded a vitally important fauna of tiny, exquisitely beautiful fossils from this early time in the history of multicellular animal life. (The fossils are phosphatized and preserved in tiny calcareous nodules. The nodules can be dissolved away in acid, leaving a perfectly formed and fully three-dimensional fossil, hollow on the inside but recording all elaborate and intimate detail of the surface architecture. Unfortunately, since the nodules are so tiny, no fossils of more conventional, and larger, marine invertebrates can be preserved, and the Orsten fauna consists primarily of arthropod larvae and other tiny adult creatures, including these newly described pentastomes.)

In all such cases of proposed linkage between ancient fossils and living forms (especially in the face of such a temporal gap, for no later pentastome fossils have been found between these ancient Orsten specimens and the living species), one must consider the obvious alternative that the fossils are merely convergent upon modern pentastomes and represent an entirely independent genealogical line. The history of life is replete with examples of astonishing similarity separately evolved—fish and ichthyosaurs, marsupial and placental moles, eyes of squid and vertebrates. But convergence, however stunning in general adaptive features of basic form and function, can never be intricately precise in hundreds of detailed and highly particular parts—because converging lines begin from such different antecedents and must craft similarities from disparate starting points. Thus, ichthyosaur paddles may be dead ringers for fish fins in external form, but they are built of finger bones from a terrestrial past; and the eyes of squid and vertebrates, though so similar in final form, follow markedly different embryological routes in their construction.

We may be confident that Walossek and Müller's specimens are true pentastomes because (as with their tardigrades) the similarities are so numerous, detailed, and pervasive. These features include the basic body plan of a globular head with two pairs of limbs suited for attachment to a host, and a thin, tapering, wormlike body behind. Fossils and moderns also share the basic embryological design of "segment constancy," with growth in size through successive molts, but no addition of further segments in growth. (The Orsten fossils include both larvae and adults, so even these details of growth may be inferred.)

Beyond this identity of basic form and growth, fossils and moderns also match part for apparently trivial part. Both have distinctive pores on the inner edges of the limbs; both can withdraw the limbs partially into their sockets; both grow a pair of papillae or nodes at the rear end surrounding the anus. Such a suite of distinct and apparently minor features would not evolve twice in such detailed similarity of form and position.

Moreover, in one prominent feature, the fossils teach us something important by revealing a structure unknown in modern forms. The body of modern pentastomes (behind the head) seems to consist of four segments. But these divisions are not marked by clearly repeating structures in each zone—the usual sign of true segmentation in several invertebrate phyla. Nerve ganglia are separated and repeated, but since pentastomes are so morphologically degenerate (with no respiratory, circulatory, or excretory organs), few other possibilities exist for crucial evidence of true segmentation. In particular, the most telling of all features—limbs on each segment—do not exist in any modern pentastome. But several of the fossils contain small paired limbs on the second and third body segments! In fact, one might say that the fossils are entirely comparable to the moderns, with this one added (and highly informative) feature.

These fossils clearly disprove the favored hypothesis of later derivation for pentastomes after the evolution of terrestrial vertebrates. Of their profound stability for more than 500 million years, Walossek and Müller conclude, "The long history of the group and its remarkable morphological stasis invalidates any hypothesis of their evolution from terrestrial arthropods" (for arthropods did not invade the land until long after the Cambrian).

The existence of Cambrian pentastomes raises an obvious question about their original hosts, since terrestrial vertebrates had not yet evolved. Switching of hosts, even from one phylum to another, occurs frequently in the evolution of parasites, so the need to postulate such a transition raises no theoretical problems. But we still want to know the potential candidates. The original Cambrian hosts need not have been closely related to vertebrates, but one prominent fossil group, the conodonts, has been enigmatic throughout the history of paleontology (for their soft bodies provide little opportunity for fossilization, and only their microscopic "tooth elements" are generally preserved). But soft-bodied remains of entire animals have been found in the last decade (see essay 16 in my book *The Flamingo's Smile*), and latest evidence indicates that conodonts belonged to the vertebrate line—so pentastomes may always have parasitized vertebrates after all, as Walossek and Müller suggest. Conodont fossils are common in all Cambrian localities that have yielded pentastomes.

But what of the biochemical claims for crustacean affinities, and origin much later than the Cambrian? Molecular data have won such prestige over the last few years that such a contention might seem indisputable—yet the hard evidence of Cambrian pentastomes seems even less subject to refutation. But close reading of Abele's 1989 paper provides a lovely resolution.

I have often pointed out in these essays how theories strongly constrain (often unconsciously) our interpretation of data. (For this reason we must be particularly vigilant and probing as we explicitly consider the consequences of our theoretical preferences.) The solution lies in the last sentence of the paper by Abele and colleagues—but they didn't see it, presumably because later origin from crustaceans fit the usual assumptions of evolutionary theory and its preference for continuous origin of major groups. The last line reads (I will quote verbatim and then translate), "Thus, over a period of time very roughly estimated to be 287 . . . million years, the 18S rRNA of these two groups has diverged about 10.8 percent, or about 1.9 percent per 50 million years, a higher rate than the 1 percent per 50 million years previously reported for eucaryote 18S rRNA. Given the potential errors in making such estimates the significance of this difference remains unknown." In other words, assuming that pentastomes arose only 287 million years ago from crustaceans, the rate of evolution for their RNA is nearly twice as fast as average rates calculated for other multicellular organisms. This disparity doesn't trouble the authors unduly (as their very last sentence states), because rates of change in RNA are so variable, and the techniques for measuring them subject to so much error.

But Abele and colleagues never even mention the obvious alternative hypothesis—which now turns out to be true with high probability. If the pentastomes really diverged from crustaceans (or from some other group) during the Cambrian, some 530 rather than 287 million years ago, then the total measured difference does not translate to an unusually high rate of change, but to an average rate after all—for the 10.8 percent difference, spread over 530 million rather than 287 million years, works out to just about the average of 1 percent per 50 million years! In other words, the molecular data and the fossil evidence coincide and remove an anomaly in the molecular data considered alone under the conventional (and false) assumption of later origin for pentastomes.

The Cambrian explosion is the key event in the history of multicellular animal life. The more we study this episode, the more we are impressed by the evidence of its uniqueness and of its determining effect on the subsequent pattern of life's history. The basic anatomies that arose during this episode have dominated life ever since, with no major additions—and with subtractions imposed for reasons that may more resemble the luck of the draw than the predictable survival of superior lines (see my book *Wonderful Life*). The pattern of life's history has followed from the origins and successes of this great initiating episode. I can, therefore, only end with another well-remembered line from Gettysburg, when Lincoln so misspoke in the first quoted phrase (a cruel irony for schoolchildren forced to memorize against their will), but stated so truly and eloquently in the second—and which applies so well to the extraordinary influence of successful Cambrian groups: "The world will little note nor long remember what we say here, but it can never forget what they did here."

Stability

10

Cordelia's Dilemma

While Goneril and Regan jockey for their father's wealth by proclaiming their love for him in false and fulsome tones, King Lear's third daughter, Cordelia, fears the accounting that her father will soon demand: "What shall Cordelia do? Love, and be silent . . . since, I am sure, my love's more ponderous than my tongue."

Lear then entices Cordelia into this game of escalating professions of love: "What can you say to draw a third more opulent than your sisters?" When the honorable Cordelia, refusing to play falsely for gain, says nothing, Lear cuts her off from all inheritance, proclaiming that "nothing will come of nothing."

Lear's tragic error, which shall lead to blinding, madness, and death, lies in not recognizing that silence—overt nothing—can embody the deepest and most important meaning of all. What, in all our history and literature, has been more eloquent than the silence of Jesus before Pilate, or Saint Thomas More's date with the headsman because he acknowledged that fealty forbade criticism of Henry VIII's marriage to Anne Boleyn, but maintained, literally to the death, his right to remain silent and not to approve.

The importance of negative results—nature's apparent silence or nonacquiescence to our expectations—is also a major concern in science. Of course, scientists acknowledge the vitality of a negative outcome and often try to generate such a result actively—as in trying to disprove a colleague's favored hypothesis. But the prevalence of negative results does pose an enormous, and largely unaddressed, problem in the reporting of scientific information. I do not speak of fraud, cover-up, finagling, or any other manifestation of pathological science (though such phenomena exist at a frequency that, in all honesty, we just do not know). I refer, rather, to the all too wonderfully human love of a good tale—and our simple and utterly reasonable tendency to shun the inconclusive and the boring.

The great bulk of daily scientific work never sees the light of a published day (and who would wish for changes here, as the ever-increasing glut of journals makes keeping up in one's own field impossible and exploration of others inconceivable). Truly false starts are deposited in circular files—fair enough. But experiments fully carried forth and leading to negative results end up, all too often, unpublished in manila folders within steel-drawer files, known only to those who did the work and quickly forgotten even by them. We all know that thousands of novels, considered substandard by their authors, lie in drawers throughout the world. Do we also understand that experiments with negative results fill even more scientific cabinets?

Positive results, on the other hand, tell interesting stories, and are usually written up for publication. Consequently, the available literature may present a strongly biased impression of efficacy and achieved understanding. Such biases, produced by the underreporting of negative results, are not confined to the arcana and abstractions of academic science. Serious, even tragic, practical consequences often ensue. For example, spectacular medical claims for the efficacy of certain treatments (particularly for chronic and fatal illnesses like cancer and AIDS) may be promoted after a positive result (often obtained in a study based upon a very small sample). Later and larger trials may all fail to duplicate the positive results, effectively disproving the value of the treatment. But these subsequent negative results often appear in highly technical journals read by more-restricted audiences and, as non-stories, do not come so readily to the attention of media—and people may continue to squander hope and waste precious time following useless procedures.

Statistics often gets a bum rap in our epithets and editorials. But I am both a champion and a frequent user of statistical procedures—for the science exists largely to identify and root out hopes and misperceptions falsely read into numerical data. Statistics can tell us when published numbers truly point to the probability of a negative result, even though we, in our hopes, have mistakenly conferred a positive interpretation. But statistics cannot rescue us when we hide our non-lights under a bushel (with apologies to Matthew 5:15)—that is, when we publish only positive results and consign our probable negativities to non-scrutiny in our file drawers.

I had thought about this problem a great deal (especially when writing *The Mismeasure of Man*), but I had not realized that this special sort of bias had both a name and a small literature devoted to its weighty problems, until I came upon a paper written in 1988 by Colin B. Begg and Jesse A. Berlin entitled "Publication bias: a problem in interpreting medical data."

Begg and Berlin begin their paper with a wonderful quotation from Sir Francis Bacon (*The Advancement of Learning*, 1605) on the psychological basis for what he calls a "contract of error" between author and reader in the perpetration of publication bias—the tendency to expose only positive results that tell good stories:

> For as knowledges are now delivered, there is a kind of contract of error between the deliverer and the receiver; for he that delivereth knowledge desireth to deliver it in such form as may be best believed, and not as may be best examined; and he that receiveth knowledge desireth rather present satisfaction than expectant inquiry.

Begg and Berlin then cite several documented cases of publication bias. We can hardly doubt, for example, that a correlation exists between socioeconomic status and academic achievement, but the strength and nature of this association can provide important information for both political practice and social theory. A 1982 study by K. R. White revealed a progressively increasing intensity of correlation with prestige and permanence of the published source. Studies published in books reported an average correlation coefficient of 0.51 between academic achievement and socioeconomic status; articles in journals gave an average of 0.34, while unpublished studies yielded a value of 0.24. Similarly, in a 1986 article, A. Coursol and E. E. Wagner found publication bias both in the decision to submit an article at all, and in the probability for acceptance. In a survey of outcomes in psychotherapy, they found that 82 percent of studies with positive outcomes led to submission of papers to a journal, while only 43 percent of negative outcomes provoked an attempt at publication. Of papers submitted, 80 percent reporting positive outcomes were accepted for publication, but only 50 percent of papers claiming negative results.

My favorite study of publication bias may be found in Anne Fausto-Sterling's *Myths of Gender*, a unique and important contribution to feminism for this reason. In tabulating claims in the literature for consistent differences in cognitive and emotional styles between men and women, Fausto-Sterling does not deny that genuine differences often exist, and in the direction conventionally reported. But she then, so to speak, surveys her colleagues' file drawers for studies not published, or for negative results published and then ignored, and often finds that a great majority report either a smaller and insignificant disparity between sexes, or find no differences at all. When all studies, rather than only those published, are collated, the much-vaunted differences often devolve into triviality. Natural history, after all (as I have argued so often in these essays), is preeminently a study of relative frequency, not of absolute yeses or noes. If a claim based on published literature states that "women in all studies strongly . . ."—and the addition of unpublished data changes the claim to "in a minority of studies, a weak effect suggests that women . . ."—then meaning is effectively reversed (even though positive outcomes, when rarely found, show a consistent direction).

For example, a recent favorite in pop psychology (though waning of late, I think) has attributed different cognitive styles in men and women to the less lateralized brains of women (less specialization between right and left hemispheres of the cerebral cortex). Some studies have indeed reported a small effect of greater male lateralization; none has found more lateralized brains in women. But most experiments, as Fausto-Sterling shows, detected no measurable differences in lateralization—and this dominant relative frequency (found even in published literature) should be prominently reported, but tends to be ignored as "no story."

Publication bias is serious enough in its promotion of a false impression based on a small and skewed subset of the total number of studies. But at least the right questions are being asked and negative results can be conceptualized and obtained—even if they then tend to be massively underreported. But consider the far more insidious problem closer to Cordelia's dilemma with her father: What if our conceptual world excludes the possibility of acknowledging a negative result as a phenomenon at all? What if we simply can't see, or even think about, a different and meaningful alternative?

Cordelia's plight is a dilemma in the literal sense—a choice between two equally undesirable alternatives: she either remains honorable, says nothing, and incurs her father's wrath; or she plays an immoral game to dissemble and win his affection. She tumbles into this plight because Lear cannot conceptualize the proposition that Cordelia's silence might signify her greater love—that nothing can be the biggest something.

Cordelia's dilemma is therefore deeper and more interesting than publication bias, as we glimpse the constraining role of neurological, social, and psychological conditioning in our struggle to grasp this complex universe into which we have been so recently thrust. Publication bias only acts as a guard at the party door, giving passage to those with the right stamp on their hands. At least the guard can see all the people and make his unfair decisions. Those rejected can gripe, foment revolution, or start a different and better party. The victims of Cordelia's dilemma are "unpersoned" in the most Orwellian sense. They are residents in the last gulag in inaccessible Siberia, the farthest outpost of Ultima Thule. They are not conceptualized and therefore do not exist as available explanations.

These two forms of nonreporting require different solutions. Publication bias demands, for its correction, an explicit commitment to report negative results that appear less interesting or more inconclusive than the "good story" of positive outcomes. The solution to Cordelia's dilemma—the promotion of her nothing to a meaningful something—requires the more extensive revision of conceptual overhaul. Cordelia's dilemma cannot be resolved from within, for the existing theory has defined her action as a denial or non-phenomenon. A different theory must be imported from another context to change conceptual categories and make her response meaningful. In this sense, Cordelia's dilemma best illustrates the dynamic interaction of theory and fact in science. Correction of error cannot always arise from new discovery within an accepted conceptual system. Sometimes the theory has to crumble first, and a new framework be adopted, before the crucial facts can be seen at all. We needed to suspect that evolution might be true in order to see variation among individuals in a population as the dynamic stuff of historical change, and not as trivial or accidental deviation from a created archetype.

I am especially interested in Cordelia's dilemma, and the role of new theories in promoting previously ignored phenomena to conceivability and interest, because the "main event" of my early career included an example that taught me a great deal about the operations of science. Before Niles Eldredge and I proposed the theory of punctuated equilibrium in 1972, the stasis or nonchange of most fossil species during their lengthy geological life spans had been tacitly acknowledged by all paleontologists, but almost never studied explicitly because prevailing theory treated stasis as uninteresting nonevidence for nonevolution. Evolution had been defined as gradual transformation in extended fossil sequences, and the overwhelming prevalence of stasis became an embarrassing feature of the fossil record, best ignored as a manifestation of nothing (that is, nonevolution). My own thesis adviser had mastered statistics in the hopes of detecting subtle gradualism not visually evident in fossil sequences. He applied his techniques to some fifty brachiopod lineages in Devonian rocks of the Michigan Basin, found no evidence for gradual change (but stasis in all lineages with one ambiguous exception), considered his work as a disappointment not even worth publishing, and left the field soon thereafter (for a brilliant career in another domain of geology, so our loss was their gain).

But Eldredge and I proposed that stasis should be an expected and interesting norm (not an embarrassing failure to detect change), and that evolution should be concentrated in brief episodes of branching speciation. Under our theory, stasis became interesting and worthy of documentation—as a norm disrupted by rare events of change. We took as the motto of punctuated equilibrium: "Stasis is data." (One might quibble about the grammar, but I think we won the conceptual battle.) Punctuated equilibrium is still a subject of lively debate, and some (or most) of its claims may end up on the ash heap of history, but I take pride in one success relevant to Cordelia's dilemma: our theory has brought stasis out of the conceptual closet. Twenty-five years ago, stasis was a nonsubject—a "nothing" under prevailing theory. No one would have published, or even proposed, an active study of lineages known not to change. Now such studies are routinely pursued and published, and a burgeoning literature has documented the character and extent of stasis in quantitative terms.

Punctuated equilibrium is a theory about the origin and history of species—that is, the stability of individual species counts as the "nothing" that our theory promoted to the interest and attention of researchers. A different kind of "nothing" permeates, and also biases, our consideration of the next most inclusive level of evolutionary stories—the history of phyletic bushes, or groups of species sharing a common ancestry: the evolution of horses, of dinosaurs, of humans, for example. This literature is dominated by the study of trends—directional changes through time in average characteristics of species within the bush. Trends surely exist in abundance, and they do form the stuff of conventional good stories. Brain size does increase in the human bush; and toes do get fewer, and bodies bigger, as we move up the bush of horses.

But the vast majority of bushes display no persistent trends through time. All paleontologists know this, but few would ever think of actively studying a bush with no directional growth. We accept that the history of continents and oceans presents no progressive pattern most of the time—"the seas come in and the seas go out," to cite an old cliché of geology teachers from time immemorial. But we expect life's bushes to grow toward the light, to tell some story of directional change. We cannot accept for life the preacher's assessment of earthly time (Ecclesiastes 1:9): "The thing that hath been, it is that which shall be; and that which is done, is that which shall be done; and there is no new thing under the sun."

Yet we must study bushes with no prominent directional change if we are to gain any proper sense of the full range and character of life's history. Even if we believe (and I will confess to holding this conventional bias myself) that trends, however rare, are the most interesting of phyletic phenomena—for they do supply the direction that makes evolution a pageant rather than a tableau—we still need to know the relative frequency of nonprogressive evolution, if only to grasp the prevailing substrate from which rare trendiness builds interesting history. How can we claim to understand evolution if we only study the percent or two of phenomena that construct life's directional history, and leave the vast field of straight-growing bushes—the story of most lineages most of the time—in a limbo of conceptual oblivion?

I see some happy signs of redress, as paleontologists are now beginning to study this higher-order stasis, or nondirectional history of entire bushes. An excellent and pathbreaking case for Cretaceous corals has recently been published by Ann F. Budd and Anthony G. Coates in our leading trade journal, *Paleobiology*. Budd and Coates state their aim in an introduction, and I could not agree more:

> Just as the study of stasis within species has facilitated understanding of morphologic change associated with speciation, we show that study of nonprogressive evolution offers valuable insight into how the causes of trends interact and thereby produce complex evolutionary patterns within clades [evolutionary bushes], regardless of their overall direction.

Montastraea is a genus of massive colonial reef-building corals, still important in our modern faunas (many readers undoubtedly have a piece of *Montastraea* on their mantelpieces). Budd and Coates studied the earlier history of the *Montastraea* bush during the long span of Cretaceous time—some 80 million years' duration, and representing the last period of dinosaurian domination on land. They found little evidence of directional change, but rather a story of oscillation within a range set by minimal and maximal size of corallites (individual coral animals within the colony). At one end, "large-corallite" species (3.5–8.0 millimeters in corallite diameter) are more efficient in removal of sediment and tend to be more common in regions of turbid water; "small-corallite" species (2.0–3.5 millimeters in diameter) tend to dominate in clearer waters near the reef crest. In addition, large-corallite species tend to feed actively on small planktonic animals, while small-corallite species derive more nutrition directly from the zooxanthellae (photosynthetic algae) that live symbiotically within their tissues.

Budd and Coates conjecture that corallite diameters may be held within these limits by some ecological or developmental constraint at the low end (implying that still smaller corallites could neither develop nor function adequately), and by a limit to the number of septa at the high end. (Septa are the radiating series of plates that form the skeletal framework of a corallite. The size of corallites might be limited if new septa could not form beyond a certain number—though this argument is frankly speculative. The "astraea" in *Montastraea*'s name refers to the star-shaped pattern of these radiating septa in cross section.) If such constraints limit the range of corallite size, and if each end enjoys advantages in different environments always available in some parts of the habitat, then evolution might oscillate back and forth, with no persistent directional component, through the bush of time.

Budd and Coates found just such an oscillation, hence their well-chosen title of "nonprogressive evolution." They divided the Cretaceous into four intervals and then traced the pattern of change through these times. (Most of their long paper presents technical details of defining species and inferring genealogical connections among them.) They found that the transition from interval 1 to interval 2 featured a differential production of small-corallite species from large-corallite ancestors, and a southward spread of the bush's geographic range. "Limited speciation and stasis" then predominated within intervals 2 and 3. Later, between intervals 3 and 4, large-corallite species tended to radiate from small-corallite ancestors as the bush became restricted in range to the Caribbean. The end, in other words, did not leave the bush very different from its beginnings—the seas came in and the seas went out, and *Montastraea* oscillated between prevalence of small and large-corallite species within its restricted range. And so it goes for most groups in most long segments of geological time—lots of evolutionary change, but no story of clear and persistent direction.

I do feel the force of Cordelia's dilemma as I write these words. Budd and Coates's article inspired my decision to write this essay. Yet my description of their results occupies only a small portion of this text, because nondirectional evolution doesn't elicit the stories that stir our blood and incite our interest. This is the bias of literary convention that we must struggle to overcome. How can we interest ourselves sufficiently in the ordinary and the quotidian? Nearly all of our life so passes nearly all the time (and thank goodness, lest we all be psychological basket cases). Shall we not find fascination in the earth's daily doings? How can we hope to understand the rarer moments that manufacture history's pageant, if we do not recognize and revel in the pervasive substrate?

No one has illustrated the dilemma better than Cordelia and Lear themselves, in their last appearance as prisoners in act 5, scene 3. They are about to be taken away and Lear, through a veil of madness, speaks of forthcoming time in jail, made almost delightful by the prospect of telling stories in the heroic and directional mode:

> Come, let's away to prison.
> . . . So we'll live,
> And pray, and sing, and tell old tales, and laugh
> At gilded butterflies, and hear poor rogues
> Talk of court news. And we'll talk with them too,
> Who loses and who wins, who's in, who's out,
> And take upon's the mystery of things
> As if we were God's spies. And we'll wear out
> In a walled prison, packs and sects of great ones
> That ebb and flow by the moon.

Sean O'Casey said that "the stage must be larger than life," for how can we make adequate drama from the daily doings of shopping, eating, sleeping, and urinating (in no particular order). If this be so, then our biases in storytelling augur poorly for an adequate account of life's real history, for how shall we ever promote the "nothing" that surrounds us to adequate fascination for notice and documentation? But, then, one of O'Casey's countrymen solved this problem in the greatest novel of the twentieth century. James Joyce's *Ulysses* treats one day in the life of a few ordinary people in 1904, yet no work of literature has ever taught us more about the nature of humanity and the structure of thought. May I then close with a kind of literary sacrilege and borrow the famous last line of *Ulysses* for a totally different purpose? Molly Bloom, in her celebrated soliloquy, is, of course, speaking of something entirely different! But her words provide a good answer to a pledge we should all take: Shall I promise to pay attention to the little, accumulating events of daily life and not treat them as nothing against the rare and grandiose moments of history? "yes I said yes I will Yes."

11

Lucy on the Earth in Stasis

QUEEN VICTORIA, just a bit behind the times as usual, took her first journey by railroad in 1842—from Windsor to London (by 1840 the United States already had 2,816 miles of track in operation, while England boasted 1,331 miles). Beyond this royal symbol, 1842 was a good year for change in general. Darwin composed his first sketch of the theory of natural selection (followed, in 1844, by an expanded draft and finally, in 1859, by a published version, *The Origin of Species*). And Alfred, Lord Tennyson wrote, in *Locksley Hall*, the most famous of all Victorian lines about the inevitability of change: "Let the great world spin for ever down the ringing grooves of change."

I unite Tennyson's line with Victoria and rail transport for several reasons, most literally because Tennyson himself later wrote that his striking, though peculiar, metaphor for change (both visual and aural) arose from a misperception during his own first journey by rail: "When I went by the first train from Liverpool to Manchester (1830), I thought that the wheels ran in a groove. It was a black night and there was such a vast crowd round the train at the station that we could not see the wheels. Then I made this line."

We are beset by dualities, perhaps because nature favors pairings, but more, I suspect, because our mind works as a dichotomizing machine. (See further discussion of this issue in essay 4.) Among the organizing dualities of our consciousness, change and constancy stands out as perhaps the deepest and most pervasive. Heraclitus said that we can't step twice into the same river, while his contemporary Pythagoras tried to extract invariance from the world's overt complexity by discovering simple regularities in number and geometry—a scholar's dream still pursued, as by Bertrand Russell in our day, when he included among the three passions of his life: "I have tried to apprehend the Pythagorean power by which number holds sway above the flux."

These deep dualities cannot be analyzed in terms of truth and falsity, for the two sides are both and neither. In our struggles to comprehend this immensely puzzling and amazingly intricate universe, themes of change and themes of constancy both yield crucial insights for different questions and different scales. Since the two sides of this duality are equally true and useful, favoring one or the other at various, fluctuating times in the history of science becomes our best illustration of social impact upon a process that mythology regards as free of personal preference and driven exclusively by observation—for no organizing construct of the mind can be more socially and politically influenced than our transient preference for either change or stability as the essential nature of the universe.

Many periods of Western history have favored stability, if only as a supposed natural buttress to a ruling political hierarchy of kings and nobles, or popes and bishops. But a fundamental tenet of Western life, at least since the late eighteenth century, has proclaimed change as natural, constant, and inevitable. Social conservatives may rail and moan, visionaries and romantics may dance and sing, but the ringing grooves have dominated our view of the world for the past two centuries at least. Belief in change as nature's essential way blossomed in the eighteenth-century age of revolutions, with America and France leading a sometimes ambiguous way, flourished with the subsequent wave of romanticism in the arts, and reached an apogee (for Tennyson chose his metaphor wisely) with the even more ambiguous Victorian triumph of industrial and colonial expansion.

Evolution is a fact of nature—one that could probably not have been perceived, and certainly not widely promulgated, before preference for change in this cardinal duality swept the Western world. But evolution also enjoyed a much easier path to acceptance in Darwin's century because its central theme of change meshed so well with prevailing social context. Biological evolution, with its unbeatable combination of empirical truth and social fit, therefore became the quintessential theory of change within Western science.

Obviously, I do not write this essay to challenge evolutionary change because one aspect of its popularity has a social root. But I do wish to stress the importance of acknowledging social influence as the best possible antidote to overconfidence about our perception of truth, and the best spur to healthy skepticism and self-examination. Much of what we regard as empirically proven, or logically necessary, may only be a contingent reflection of transient social preferences. And if notions of change as nature's essence rank among the strongest of these social preferences, then we need to be especially skeptical when we weigh our assumptions about the character of change.

Social preference extends beyond a simple belief in change as essential, to a set of assumptions about the nature of transformation. In particular, we usually view change as intrinsic and continuous, not rare and episodic. That is, we wish to conceptualize change as its own form of constancy, to define systems by their changes, and to view constant alteration as a normal state—particularly of systems undergoing biological evolution.

But other theories of change are just as consistent with the general view of a universe driven by alteration. For example, stability might reign most of the time, and change might be a rare event, usually of substantial magnitude and occurring only when stresses impact a system beyond its capacity to absorb without substantial modification. In this alternative view, stability is the norm for most systems most of the time—and change, while driving the universe in the fullness of vast scales and long times, is absent at almost any given moment—see the preceding essay.

This conflict between change as continuous and steady vs. rapid and episodic underlies many debates in the history of science—the great late-eighteenth through early-nineteenth-century struggle (when general theories of change had just become dominant and were therefore flexing muscles and dividing turf) between uniformitarianism and catastrophism for the physical history of the earth and the biological alteration of faunas, or (to cite a contemporary skirmish of much smaller scale) the debate between punctuated equilibrium and gradualism for the process of speciation in biological lineages.

I shall not hide my preferences and biases. I helped to devise the theory of punctuated equilibrium with Niles Eldredge in 1972. I have cheered from the sidelines (and occasionally given a boost in these essays) as catastrophic theories of mass extinction make their comeback in the virtual proof now available for extraterrestrial impact as the trigger of the Cretaceous-Tertiary dying—see next essay.

I am not a foe of gradual change; I believe that this style of alteration often prevails. But I do think that punctuational change writes nature's primary signature—and I am convinced that our difficulty in conceptualizing this style of alteration arises from social and psychological bias rather than from any shyness of nature in printing its John Hancock (so conspicuously that the king might read it without his spectacles—though we poor ordinary mortals often seem blind, however prominent the signature).

I have come to understand, in a different and personal way, that an equation of evolution with a belief in continuous change as nature's norm sets the most pervasive misconception of life's history in the general culture of intelligent and well-educated lay audiences. At this point in my mid-career as a writer and lecturer, I have given so many hundreds of talks, and received so many thousands of letters, that I have a good sense of recurring themes and their relative frequencies. Some questions arise rarely; a few are unique and wonderfully idiosyncratic or challenging. But other questions occur with such predictable regularity that they inspire clichéd comments of the variety, "If only I had a dollar for each time I've heard that one, I could retire to a life of indolent luxury."

I do not regard these inevitable questions as stupid in any way. Quite the contrary: I hear them every time I speak because they are good questions coming from the heart of human concern, interest, and puzzlement. But such questions are often based on deep misconceptions about the nature of evolution. In fact, people get stuck on these issues precisely because they grasp (however dimly) an inconsistency between the empirical world and a formulation that seems exclusive or inevitable according to their understanding of evolutionary theory. The solution does not lie in revising facts, but in forcing a conceptual reformulation that switches the facts from anomaly to expectation.

The two most common questions (really a less and more sophisticated version of the same concern) are rooted in the fallacy of assuming that evolution means continuous change, and that stability must therefore count as the most puzzling of anomalies. The first, less sophisticated version simply asks, "Where is human evolution going in the future?"

Questions are not neutral; they presuppose a list of assumptions that may be long and complex. This query begins with an unstated belief that evolution is always going somewhere and that we would especially like to know where such a universal process will lead parochial little us. I feel that I can respond only with a question of my own: "Why do you think that we are or ought to be going anywhere?" I then try to explain that human bodily form has been stable for tens of thousands of years (and that everything we call civilization has therefore been built without substantial alteration in any physical aspect of brains or bodies that the fossil record might preserve). I then add that stability on scales of hundreds of thousands to millions of years is a norm and expectation for large, successful, geographically widespread populations. Evolution tends to be concentrated in events of branching speciation, and such events usually occur in small and isolated populations. Humans live all over the world, move vigorously from place to place, and maintain an apparently unstoppable habit of interbreeding everywhere they go—therefore no opportunity for isolation and speciation (unless you want to construct a science-fiction scenario about space colonies). Thus, I can answer this most inevitable of questions only by saying that we are unlikely to be going anywhere in the natural course of events (all bets are off with such culturally devised phenomena as genetic engineering), and that evolutionary theory predicts and expects such stability (see essay 25).

The more sophisticated version comes from listeners who already know the facts of long-term recent stability, regard the situation as strongly anomalous, have thought about it, and have devised a potential explanation (which would be quite sensible if the misconceived equation of evolution with continuous change were valid): "Does the recent stability of human bodily form arise because culture has suppressed the ruthless action of natural selection and halted the process of weeding out the unfit, thus blocking adaptive evolutionary change?"

I try to answer this version in two parts. I first identify this question as a vestigial holdover from old-style eugenics and its false assumption—the bastion of the misnamed and discredited doctrine of "social Darwinism"—that human "progress" requires a relentless struggle in the overt gladiatorial mode, with victors rising to positions of power, and inferior folks either put to the wall or precipitated into the lower classes. In this view, culture stymies nature by permitting the unfit to survive (through such derailments of Darwinian order as manufacturing eyeglasses, hearing aids, and wheelchairs). "Bad" genes accumulate and evolutionary toughening grinds to a halt.

I confess that I do get cross in noting the astonishing persistence of such a badly formulated and socially pernicious argument. Genes leading to eyes requiring corrective lenses are not "bad" in any absolute sense; they do increase our dependence upon culture (to supply the needed assistance), but human life is now so inextricably dependent upon culture for a thousand other reasons that I cannot imagine why we would choose to lament this additional link. As the only evolutionary consequence I can imagine, such a cultural "softening" of natural selection may slightly boost our genetic variability as a species, but I cannot regard such an increase as anything but neutral or favorable.

But I then point out that the initial question rests upon what logicians label an "inarticulated major premise" and we ordinary folk call a "hidden assumption"—the same one that motivated this essay. If we suggest that cultural "softening" caused human stability, then we imply a prior (though unstated) belief that evolutionary change is a natural norm—and that any failure to note such change requires a special explanation. But if stability is really the norm for species like *Homo sapiens,* then the anomaly vanishes, and the question resolves itself into a nonissue.

As another cultural test of the prejudicial hold imposed upon our understanding of evolution by the doctrine of continuous change as a defining norm, we might consider press reporting of discoveries that affirm substantial intervals of stasis on the human family tree. Are such findings reported as affirming an expectation or presenting a strong surprise? I have long noted that surprise always dominates, and I decided to write this essay because such a fine example recently appeared in newspapers and magazines throughout the world.

The March 31, 1994, issue of *Nature,* Britain's leading professional journal of science, featured a strikingly apelike human fossil skull on its cover, above the heading "Son of Lucy." The technical article within, by William H. Kimbel, Donald C. Johanson, and Yoel Rak, bore the less titillating title "The first skull and other new discoveries of *Australopithecus afarensis* at Hadar, Ethiopia."

The human lineage branched off from the clade of our closest cousins, chimpanzees and gorillas, about 6–8 million years ago—a date inferred from genetic distances among living species, not from direct evidence of fossils. The first well-dated and clearly accepted human fossils are 3.9 million years old and come from strata in Ethiopia.[1] All fossil humans spanning the first million years of our recorded history (3.9 to 3.0 million years ago) belong to the single species *Australopithecus afarensis,* named by D. C. Johanson, T. D. White, and Y. Coppens in 1978. (The name *Australopithecus* means "southern ape" and honors the first discoveries of later species in this genus from South Africa in the 1920s; *afarensis* refers to the Afar region of Ethiopia, where this earlier species was found.)

During the 1970s, nearly 250 fossils of this species were recovered

1. Since I wrote this essay, an older putative ancestor has been found in Ethiopia in sediments some 4.4 million years old, and named *Australopithecus ramidus.*

from the main site at Hadar by a team under Don Johanson's leadership. This trove included a 40 percent complete female skeleton now famous throughout the world by its field name "Lucy," given to honor the Beatles' famous and somewhat cryptic song about a hallucinogenic substance once popular in certain segments of society. (The coining of informal and irreverent field names is a hallowed pastime among paleontologists, though few find their way into popular speech; I will not bore you with the names of various snails I have collected.)

Nature often plays cruel jokes upon us, if only to keep this little evolutionary twig in its proper place. Johanson's 250 fossils constituted one of the richest finds in the history of human paleontology. Our skeleton includes about two hundred bones, and many fossilize poorly. We are intrigued and informed, above all, by skulls—not only for prejudicial reasons of traditional overemphasis upon brainpower (or lack thereof in our earliest ancestors), but also for the more legitimate reasons that skulls are so complex and therefore so informative and diagnostic. With so much material, we might have expected a good skull or two. But, alas, not a single skull, or even a really good fragment, emerged from this magnificent collection. Lucy remained headless. (Johanson and his team did try to piece a skull together from numerous fragments found in distant places, but that reconstruction was too partial and conjectural to win much approval.)

Moreover, this rich material provoked as many puzzles and controversies as it provided new and clear information. In particular, the large difference in body size between two groups of bones included within these fossils sparked a lively debate between two interpretations: Do these groups represent males and females of a single species, or might *two* species be hiding under the single name *Australopithecus afarensis*? Modern humans average about 11 percent difference in length of arm bones between males and females, while the two groups included in A. *afarensis* average 22 to 24 percent for the same measures. Proponents of the "two species" theory argue that this difference is too great for sexual dimorphism in a single species, but proponents of the "two sexes" theory (I will not hide my allegiance with this school) reply that many primates, including gorillas, equal or exceed this degree of sexual divergence, and that other and later species of the genus *Australopithecus* also exhibit nearly as high a degree of difference between groups accepted as sexes of a single species.

The obvious best way to resolve such controversies demands an exit from armchair and polemic factory of academic publication, and reentry into the field to search for more fossils. Johanson and colleagues have been following this excellent strategy for several years, and have been richly repaid with fifty-three new specimens from the Hadar region, including the best possible reward of an excellent skull—a large male dubbed, unsurprisingly, "son of Lucy."

I was delighted to note the theoretical emphasis that Kimbel, Johanson, and Rak chose to place upon their skull and related finds. Of all the issues raised by these important fossils, the three authors emphasized evidence for prolonged stasis within A. *afarensis* as their primary and most interesting conclusion. This evidence includes two parts: first, the further affirmation that only one species, with strong dimorphism between sexes, lived in this region (and perhaps anywhere on the human family tree) during this formative interval of nearly a million years; and, second, the strong evidence for morphological stability in A. *afarensis* throughout this long time. The three authors roll both conclusions into the final sentence of their abstract: "They [the new fossils] confirm the taxonomic unity of A. *afarensis* and constitute the largest body of evidence for about 0.9 million years of stasis in the earliest known hominid species."

The new finds provide evidence for prolonged stasis in A. *afarensis* by extending the geological range of this species in both directions. Heretofore, firmly identified specimens occupied only the short interval between 3.18 million years old (the best date for Lucy herself) to 3.4 million for the oldest material from Hadar. (At 3.5 million years old, the famous footprints at Laetoli, presumably of a male and female walking in tandem, probably represent A. *afarensis* as well, but footprints are impressions, however stunning, not bones.) The new skull, at 3.0 million years old, represents the youngest known material of A. *afarensis*. Since the bones are indistinguishable from skull pieces found earlier among the older specimens, Lucy's "son" demonstrates nearly half a million years of stasis in the first species of our distinctive evolutionary bush.

Extension to older times rests on more tenuous inference, but still represents our best tentative conclusion based on limited evidence. In 1987, B. J. Asfaw described a large fragment of the diagnostic and taxonomically important frontal region of the skull (including brow ridges) from substantially older rocks at the nearby site of Belohdelie. He tentatively attributed this 3.9-million-year-old skull piece to A. *afarensis*, but could not be sure because Johanson's main trove of younger material from Hadar included no well-preserved frontal bone for comparison. The Belohdelie frontal has therefore rested in limbo for several years. But the new skull of Lucy's son includes a complete frontal region—and it is indistinguishable from the 0.9-million-year-older material at

Belohdelie. Admittedly, identity in the frontal region is not proof of stasis throughout the skeleton, but Belohdelie is all we have of Lucy's earliest years, and stasis does prevail for the information available.

Thus, strong evidence from much of the skeleton indicates stasis in A. *afarensis* for nearly half a million years (quite a good chunk of time already) from the oldest Hadar specimens at 3.4 million years to the skull and associated bones of Lucy's son at 3.0 million. Limited material from part of the skull also shows no change in recorded morphology right back to the earliest specimen of A. *afarensis* at 3.9 million years.

The first specimens on the hominid bush therefore persisted in stasis, as illustrated by all available positive evidence, for its entire recorded range of nearly a million years. (Some people have a false impression that claims for stasis rely on negative evidence, or absence of demonstrated changes. On the contrary, stasis should be a positive conclusion based upon hard anatomical evidence of nonchange through substantial time. We must also remember that the oldest and youngest specimens are only the first and last so far found, not the full range of the species. A. *afarensis* might have lived even longer in stability—but now I am speculating with negative evidence, and I had best shut up!)

Nature had put a press embargo upon the story of Lucy's son until their official publication date of March 31, and journalists do respect these fair conventions. Thus, press reports of the discovery all appeared in a single whoosh (for reporters had enjoyed ample lead times to prepare their stories) in newspapers for March 31, or April 1 (no joke)—thereby enhancing the force of a global "experiment" to test whether reported stability surprises even well-informed people because they equate evolution with continuous change.

I was delighted to note that two articles did describe Lucy's son in the light of punctuated equilibrium, therefore recognizing stasis as a prediction of this theory, rather than as a surprising anomaly, disembodied from any proposed explanation. The *Miami Herald* wrote: "Experts in human origins . . . said the new skull is a compelling argument for the theory that the evolution of human life on Earth proceeded in fits and starts, with long periods of stasis punctuated by sudden periods of change." Giles Whittell, writing in the *Times* (London) under the headline "Skull find backs evolution leap," stated, "The 3 million year old skull . . . lends weight to the view that evolution was not gradual but sporadic, involving long periods of no progress at all . . . A. *afarensis* flourished unchanged for almost a million years."

But the great majority of stories professed pure surprise that our evolutionary adventure should have begun with a million years of stability.

"Remarkable" surely led the pack as an adjective to modify stasis. (J. N. Wilford used it in *The New York Times*, as did R. C. Cowen in the *Christian Science Monitor*, who wrote, "What's remarkable about this 3 million year old fossil is not that it is so old but that it's so young. It is 200,000 years younger than the famous Lucy . . . and a million years younger than the oldest specimen. Yet it looks like those ancestors.") But Tim Friend in *USA Today* favored "unexpected," while Mr. Cowen also proclaimed the new skull "astonishing" in its demonstration of stasis.

Most revealing are the more subtle linguistic clues that betray an expectation (or even a belief in the higher virtue!) of continuous change. Do you not, for example, sense disparagement in Keay Davidson's description of Lucy's stasis (from the *San Francisco Examiner*)—as though our earliest ancestor didn't quite cut the mustard in delaying progress for so long: "The skull strengthens scientists' belief that Lucy was part of a single species that puttered around Ethiopia, evolving very little, over at least 900,000 years." In an even more revealing passage, Boyce Rensberger, writing in *The Washington Post*, expresses surprise that Lucy's brain may be no larger than an ancestral ape's, even though she lived nearly a million years after the split of our lineage from the ape bush. For here we encounter the unstated implication (another "inarticulated major premise") that a million years damned well ought to be enough time for accumulating notable change in a world of continuous alteration—though no such expectation arises in a world of stasis and punctuation, for such an interval may well and fairly lie within a period of stability. Rensberger wrote: "The newly found skull's brain capacity has not yet been measured. But it is not expected to be much more than that of a large ape, even though the creature lived at least 900,000 years after its ancestors diverged from the ape lineage."

Obviously, for a revised view about the general tempo of evolutionary change, stasis can provide only one side of a story, lest we be left with no evolution at all! The opposite and integrated side (the punctuation in punctuated equilibrium) proposes a concentration of change into relatively short episodes—jabs of reorganization in a world of generally stable systems. Enter this world at any random moment and, as an overwhelming probability, nothing much will be happening in a history of change. But survey the totality over millions of years, and these episodes of punctuation, though they may occupy only a percent or two of time, build the signature of historical alteration. Scale is everything in history and geology.

The punctuations in evolutionary change are usually events of branching speciation, generally occurring in small and isolated popu-

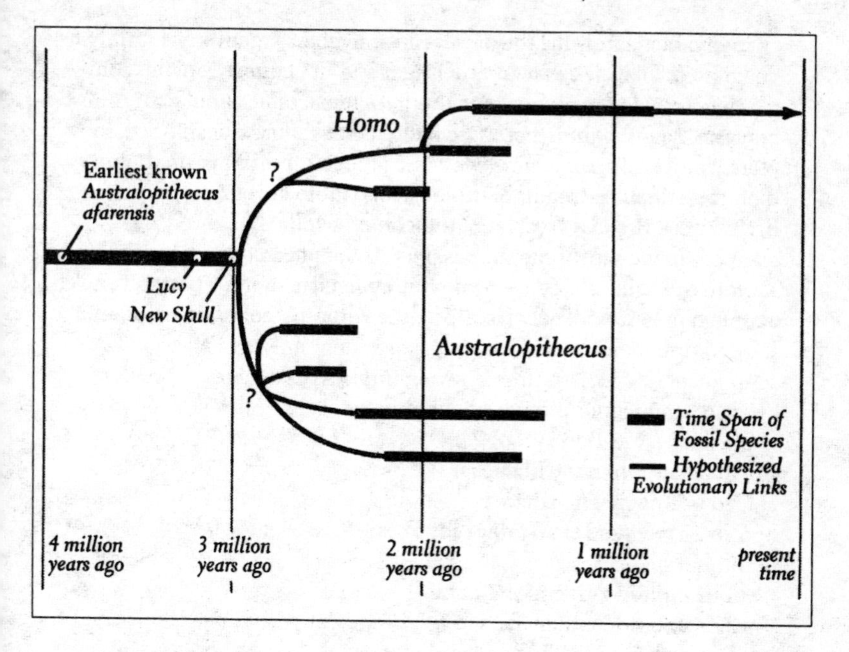

lations within an interval (many thousands of years) that appears glacially slow at the inappropriate scale of a human lifetime, but resolves to a moment at geology's proper scale of millions. (Remember that 10,000 years—a period encompassing all of written human history—equals only 1 percent in Lucy's million-year epoch of stasis.)

In this light, continuing studies on the million-year period following Lucy's tenure point more and more, as data on bones and dates accumulate, to a veritable forest of rapid speciation events, leading to several additional members of the genus *Australopithecus*, and also to the first representatives of our own genus *Homo*. A chart (reproduced here) distributed by Johanson's Institute of Human Origins to accompany their find of Lucy's son proposes as many as seven branching events within a restricted interval following Lucy's demise—a period shorter than the interval of Lucy's own stasis.

This flowering may correspond to a time of rapid and strongly fluctuating environmental change coincident with the beginnings of glaciation at higher latitudes. My colleague Elisabeth Vrba of Yale University has used such evidence to unite the pattern of punctuated equilibrium with the idea that events of speciation are not evenly distributed through time, but concentrated into episodes accompanied by substantial environmental change—the "turnover-pulse hypothesis," in her formulation. Nearly all the news reports of Lucy's son also emphasized this complementary side of copious branching following the stable reign of A. *afarensis*. Many of these stories quoted W. H. Kimbel, first author of the *Nature* article: "There is no obvious sign of evolution in this pre-human species for about a million years. Yet later, within only a fraction of that time, it gave rise to a great branching of the family tree."

I began with the most famous poetic metaphor about change from Victorian England. Let me end with an even more celebrated verse about leaping from Tennyson's predecessor as poet laureate, William Wordsworth:

> My heart leaps up when I behold
> A rainbow in the sky:
> So was it when my life began;
> So is it now I am a man;
> So be it when I shall grow old,
> Or let me die!
> The child is father of the man;
> And I could wish my days to be
> Bound each to each by natural piety.

If you are puzzled, I did intend this implied contradiction. Wordsworth's leap is only metaphorical; the poem expresses a hope for lifetime stability in aesthetic perception and moral value. Try his contemporary William Blake for the other side of life as rupture:

> My mother groaned! my father wept.
> Into the dangerous world I leapt.

Duality may be a conceptual prison, but if we must live with such a mental strategy, we might maximize our opportunity to grasp some of nature's complexity by hitching our star to the dyad of change and constancy. Slow and steady does not always win the race.

Extinction

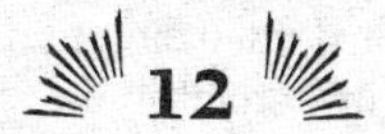

12

Dinosaur in a Haystack

THE FASHION INDUSTRY thrives on our need to proclaim an identity from our most personal space. For academics, who by stereotype (though not always in actuality) scorn the sartorial mode, office doors serve the same function. Professorial entranceways are festooned with testimonies of deepest beliefs and strongest commitments. We may, as a profession, have a deserved reputation for lengthy and tendentious proclamation, but our office doors feature the gentler approach of humor or epigram. The staples of this genre are cartoons (with Gary Larson as the unchallenged *numero uno* for scientific doors), and quotations from gurus of the profession.

Somehow, I have never been able to put someone else's cleverness so close to my heart or soul. I wear white T-shirts and, though I wrote the preface to one of Gary Larson's *Far Side* collections, I would never identify my portal with his brilliance. But I do have a favorite quotation—one worth dying for, one fit for shouting from the housetops (if not for inscription on the doorway).

My favorite line, from Darwin of course, requires a little explication. Geology, in the late eighteenth century, had been deluged with a rash of comprehensive, but mostly fatuous, "theories of the earth"—extended speculations about everything, generated largely from armchairs. When the Geological Society of London was inaugurated in the early nineteenth century, the founding members overreacted to this admitted blight by banning theoretical discussion from their proceedings. Geologists, they ruled, should first establish the facts of our planet's history by direct observation—and then, at some future time when the bulk of accumulated information becomes sufficiently dense, move to theories and explanations.

Darwin, who had such a keen understanding of fruitful procedure in science, knew in his guts that theory and observation are Siamese twins, inextricably intertwined and continually interacting. One cannot perform first, while the other waits in the wings. In mid-career, in 1861, in a letter to Henry Fawcett, Darwin reflected on the false view of earlier geologists. In so doing, he outlined his own conception of proper scientific procedure in the best one-liner ever penned. The last sentence is indelibly impressed on the portal to my psyche.

> About thirty years ago there was much talk that geologists ought only to observe and not theorize; and I well remember someone saying that at this rate a man might as well go into a gravel-pit and count the pebbles and describe the colors. How odd it is that anyone should not see that all observation must be for or against some view if it is to be of any service!

The point should be obvious. Immanuel Kant, in a famous quip, said that concepts without percepts are empty, whereas percepts without concepts are blind. The world is so complex; why should we strive to comprehend with only half our tools. Let our minds play with ideas; let our senses gather information; and let the rich interaction proceed as it must (for the mind processes what the senses gather, while a disembodied brain, devoid of all external input, would be a sorry instrument indeed).

Yet scientists have a peculiar stake in emphasizing fact over theory, percept over concept—and Darwin wrote to Fawcett to counteract this odd but effective mythology. Scientists often strive for special status by claiming a unique form of "objectivity" inherent in a supposedly universal procedure called *the* scientific method. We attain this objectivity by clearing the mind of all preconception and then simply seeing, in a pure and unfettered way, what nature presents. This image may be beguiling, but the claim is chimerical, and ultimately haughty and divisive. For the myth of pure perception raises scientists to a pinnacle above all other struggling intellectuals, who must remain mired in constraints of culture and psyche.

But followers of the myth are ultimately hurt and limited, for the immense complexity of the world cannot be grasped or ordered without concepts. "All observation must be for or against some view if it is to be of any service." Objectivity is not an unobtainable emptying of mind, but a willingness to abandon a set of preferences—for *or against* some view, as Darwin said—when the world seems to work in a contrary way.

This Darwinian theme of necessary interaction between theory and observation gains strong support from a scientist's standard "take" on the value of original theories. Sure, we love them for the usual "big" reasons—because they change our interpretation of the world, and lead us to order things differently. But ask any practicing scientist, and you will probably get a different primary answer—for we are too busy with the details and rhythms of our daily work to think about ultimates very often. We love original theories because they suggest new, different, and tractable ways to make observations. By posing different questions, novel theories expand our range of daily activity. Theories impel us to seek new information that becomes relevant only as data either "for or against" a hot idea. Data adjudicates theory, but theory also drives and inspires data. Both Kant and Darwin were right.

I bring up this personal favorite among quotations because my profession of paleontology has recently witnessed such a fine example of theory confirmed by data that no one ever thought of collecting before the theory itself demanded such a test. (Please note the fundamental difference between demanding a test and guaranteeing the result. The test might just as well have failed, thus dooming the theory. Good theories invite a challenge but do not bias the outcome. In this case, the test succeeded twice, and the theory has gained strength.) Ironically, this particular new theory would have been anathema to Darwin himself, but such a genial and generous man would gladly, I am sure, have taken his immediate lumps in exchange for such a fine example of his generality about theory and observation, and for the excitement of any idea so full of juicy implications.

We have known since the dawn of modern paleontology that short stretches of geological time feature extinctions in substantial percentages of life—up to 96 percent of marine invertebrate species in the granddaddy of all such events, the late Permian debacle, some 225 million years ago. These "mass extinctions" were originally explained, in a literal and commonsense sort of way, as products of catastrophic events, and therefore truly sudden. As Darwin's idea of gradualistic evolution replaced this earlier catastrophism, paleontologists sought to mitigate the evidence of mass dying with a reading more congenial to Darwin's preference for the slow and steady. The periods of enhanced extinction were not denied—impossible in the face of such evidence—but they were reinterpreted as more spread out in time and less intense in effect, in short as intensifications of ordinary processes, rather than impositions of true and rare catastrophes.

In his *Origin of Species* (1859), Darwin rejected "the old notion of all the inhabitants of the earth having been swept away at successive periods by catastrophes"—as well he might, given the extreme view of total annihilation, with its antievolutionary implication of a new creation to start life again. But Darwin's own preferences for gradualism were equally extreme and false: "We have every reason to believe . . . that species and groups of species gradually disappear, one after another, first from one spot, then from another, and finally from the world." Yet Darwin had to admit the apparent exceptions: "In some cases, however, the extermination of whole groups of beings, as of the ammonites towards the close of the secondary period, has been wonderfully sudden."

We now come to the central irony that inspired this essay. So long as Darwin's gradualistic view of mass extinction prevailed, paleontological data, read literally, could not refute the basic premise of gradualism—the "spreading out" of extinctions over a good stretch of time before the boundary, rather than a sharp concentration of disappearances right at the boundary itself. For the geological record is highly imperfect, and only a tiny fraction of living creatures ever become fossils. As a consequence of this imperfection, even a truly sudden and simultaneous extinction of numerous species will be recorded as a more gradual decline in the fossil record. This claim may sound paradoxical, but consider the following argument and circumstance:

Some species are very common and easily preserved as fossils; we may, on average, find specimens in every inch of strata. But other species will be rare and poorly preserved, and we might encounter their fossils only once every hundred feet or so. Now suppose that all these species died suddenly at the same time, after four hundred feet of sediment had been deposited in an ocean basin. Would we expect to find the most direct evidence for mass extinction—that is, fossils of all species through all four hundred feet of strata right up to the very top of the sequence? Of course not.

Common species would pervade the strata, for we expect to find their fossils in every inch of sediment. But even if rare species live right to the end, they contribute a fossil only every hundred feet or so. In other words, a rare species may have lived through four hundred feet, but its last fossil may be entombed one hundred feet below the upper boundary. We might then falsely assume that this rare species died out after three fourths of the total time had elapsed.

Generalizing this argument, we may assert that the rarer the species, the more likely that its last fossil appears in older sediments even if the species actually lived to the upper boundary. If all species died at once, we will still find a graded and apparently gradualistic sequence of disappearances, the rare species going first and the common forms persisting as fossils right to the upper boundary. This phenomenon—a classic example of the old principle that things are seldom what they seem and that literal appearances often obscure reality—even has a name: the Signor-Lipps effect, to honor two of my paleontological buddies, Phil Signor and Jere Lipps, who first worked out the mathematical details of how a literal petering out might represent a truly sudden and simultaneous disappearance.

We can now sense the power of Darwin's argument about needing theories to guide observations. We say, in our mythology, that old theories die when new observations derail them. But too often—I would say usually—theories act as straitjackets to channel observations toward their support and to forestall potentially refuting data. Such theories cannot be rejected from within, for we will not conceptualize the disproving observations. If we accept Darwinian gradualism in mass extinction, and therefore never realize that a graded series of fossil disappearances might, by the Signor-Lipps effect, actually represent a sudden wipeout, how will we ever come to consider the catastrophic alternative? For we will be smugly satisfied that we have "hard" data to prove gradualistic decline in species numbers.

New theories work upon this conceptual lock as Harry Houdini reacted to literal straitjackets. We escape by importing a new theory and by making the different kinds of observations that any novel outlook must suggest. I am not making an abstract point or waving arms for my favorite Darwinian motto. Two lovely examples with the same message have recently been published by a pair of my closest colleagues: studies of ammonites and dinosaurs through the last great extinction.

Anyone who keeps up with press reports on hot items in science knows that a new catastrophic theory of mass extinction has illuminated the paleontological world (and graced the cover of *Time* magazine) during the past decade. In 1979 the father-son (and physicist-geologist) team of Luis and Walter Alvarez published, with colleagues Frank Asaro and Helen Michel, their argument and supporting data for extraterrestrial impact of an asteroid or a comet as the cause of the Cretaceous-

Tertiary extinction, the most recent of the great mass dyings, and the time of departure for dinosaurs along with some 50 percent of marine invertebrate species.

This proposal unleashed a furious debate that cannot be summarized in a page, much less an entire essay, or even a book. Yet I think it fair to say that the idea of extraterrestrial impact has weathered this storm splendidly and continually increased in strength and supporting evidence. At this point, very few scientists deny that an impact occurred, and debate has largely shifted to whether the impact caused the extinction *in toto* (or only acted as a coup de grace for a process already in the works), and whether other mass extinctions may have a similar cause.

But paleontologists, with very few exceptions, reacted negatively at first (to say the least)—and Luis Alvarez, a virtual model for the stereotype of the self-assured physicist, was fit to be tied. (Luis, in retrospect, was also mostly right, so I forgive his fulminations against my profession. I, if I may toot my horn, was among his few initial supporters, but not for the right reason of better insight into the evidence. Catastrophic extinction simply matched my idiosyncratic preference for rapidity, born of the debate over punctuated equilibrium—see essay 8.) After all, my colleagues had been supporting Darwinian gradualism for a century, and the fossil record, read literally, seemed to indicate a petering out of most groups before the boundary. How could an impact cause the extinction if most species were dead already?

But the extraterrestrial impact theory soon proved its mettle in the most sublime way of all—by Darwin's criterion of provoking new observations that no one had thought of making under old views. The theory, in short, engendered its own test and broke the straitjacket of previous certainty.

My colleagues may have disliked the Alvarez hypothesis with unconcealed vigor, but we are an honorable lot and, as debate intensified and favorable evidence accumulated, paleontologists had to take another look at their previous convictions. Many new kinds of observations can be made, but let us focus on the simplest, most obvious, and most literal example. In the light of new prestige for impact and sudden termination, the Signor-Lipps argument began to sink in, and paleontologists realized that catastrophic wipeouts might be recorded as gradual declines in the fossil record.

How, then, to break the impasse produced by this indecisive literal appearance of petering out? Many procedures, some rather subtle and mathematical, have been proposed and pursued, but why not start with the most direct approach? If rare species actually lived right to the impact boundary, but have not yet been recorded from the uppermost strata, why not look a whole lot harder? The obvious analogy to the usual cliché suggests itself. If I search for a single needle in a haystack by sampling ten handfuls of hay, I have very little chance of locating the object. But if I take apart the stack, straw by straw, I will recover the needle. Similarly, if I really search every inch of sediment in every known locality, I might eventually find even the rarest species right near the boundary—if it truly survived.

This all seems rather obvious. I cannot possibly argue that such an approach could not have been conceptualized before the Alvarez hypothesis. I cannot claim that ideological blinders of gradualism made it impossible even to imagine pulling apart the haystack rather than sampling a few handfuls. But this example becomes so appealing precisely through its entirely pedestrian character. I could cite many fancy cases of original theories that open entirely new worlds of observation; think of Galileo's telescope and all the impossible phenomena thus revealed. In this case, the Alvarez theory suggested little more than hard work.

So why wasn't the effort expended before? Paleontologists are an industrious lot; we have faults aplenty, but laziness in the field does not lie among them. We do love to find fossils—the reason why most of us entered the profession in the first place. We didn't scrutinize every inch of sediment for the most basic of all scientific reasons. Life is short and the world is immense; you can't spend your career on a single cliff-face. The essence of science is intelligent sampling, not sitting in a single place and trying to get every last one. Under Darwinian gradualism, intelligent sampling followed the usual method of handful-from-the-haystack.

The results obtained matched the expectations of theory, and conceptual satisfaction (one might say "sloth" in retrospect) set in. No impetus existed for the much more laborious technique of dismember-the-entire-haystack, a quite unusual approach in science. We could have worked by dismemberment, but we didn't, and had no reason to do so. The Alvarez theory made this unusual approach necessary. The new idea forced us to observe in a different way. "All observation must be for or against some view if it is to be of any service."

Consider two premier examples—the best known marine and the best known terrestrial groups to disappear in the Cretaceous-Tertiary extinction: ammonites and dinosaurs. Both had been prominently cited as support for gradual extinction toward the boundary. In each case, the Alvarez hypothesis inspired a closer look using the inch-by-inch method;

and in each case this greater scrutiny yielded evidence of persistence to the boundary, and potentially catastrophic death.

Ammonites are cephalopods (mollusks classified in the same group as squids and octopuses) with coiled external shells closely resembling those of their nearest living relative, the chambered nautilus. Ammonites were a prominent, often dominant, group of marine predators, and their beautiful fossil shells have always been prized by collectors. They arose in mid-Paleozoic times and had nearly become extinct twice before—in two other mass dyings at the end of the Permian and the close of the Triassic periods. But a lineage or two had scraped by each time. At the Cretaceous-Tertiary boundary, however, all lineages succumbed and, to cite Wordsworth from another context, there "passed away a glory from the earth."

My friend and colleague Peter Ward, paleontologist from the University of Washington, is one of the world's experts on ammonite extinction, and a vigorous, committed man who adores fieldwork and could never be accused of laziness on the outcrop. Peter didn't care much for Alvarez at first, largely because his ammonites seemed to fade out and disappear entirely some thirty feet below the boundary at his favorite site, the cliffs of Zumaya on the Bay of Biscay in Spain. In 1983, Peter wrote an article for *Scientific American* titled "The extinction of the ammonites." He stated his opposition to the Alvarez theory, then so new and controversial, at least as an explanation for the death of ammonites:

> The fossil record suggests, however, that the extinction of the ammonites was a consequence not of this catastrophe but of sweeping changes in the late Cretaceous marine ecosystem . . . Studies of the fossils from the stratigraphic sections at Zumaya in Spain suggest they became extinct long before the proposed impact of the meteoritic body.

But Peter, as one of the smartest and most honorable men I know, also acknowledged the limits of such "negative evidence." A conclusion based on *not* finding something provides the great virtue of unambiguous potential refutation. Peter wrote: "This evidence is negative and could be overturned by the finding of a single new ammonite specimen."

Without the impact hypothesis, Peter would have had no reason to search those upper thirty feet of section with any more care. Extinctions were supposed to be gradual, and thirty feet of missing ammonites made perfect sense, so why look any further? But the impact hypothesis, with its clear prediction of ammonite survival right up to the boundary itself, demanded more intense scrutiny of the thirty-foot haystack. In 1986, Peter was still touting sequential disappearance: "Ammonites . . . appear to have become extinct in this basin well before the K/T [Cretaceous-Tertiary] boundary, supporting a more gradualistic view of the K/T extinctions."

But Peter and his field partners, inspired by Alvarez (if only by a hope of disproving the impact hypothesis), worked on through the haystack: "The remaining part of the Cretaceous section was well exposed and vigorously searched and quarried." Finally, later in 1986, they found a single specimen just three feet below the boundary. The fossil was crushed, and they couldn't tell for certain whether it was an ammonite or a nautiloid, but this specimen did proclaim a need for even more careful search. (Since nautiloids obviously survived the extinction—the chambered nautilus still lives today—a fossil nautiloid right at the boundary would occasion no surprise.)

Peter started a much more intense search in 1987, and the ammonites began to turn up—mostly lousy specimens, and very rare, but clearly present right up to the boundary. Peter wrote in a book published in early 1992: "Finally, on a rainy day, I found a fragment of an ammonite within inches of the clay layer marking the boundary. Slowly, over the years, several more were found in the highest levels of Cretaceous strata at Zumaya. Ammonites appeared to have been present for Armageddon after all."

Peter then took the obvious next step: look elsewhere. Zumaya contained ammonites right to the end, but not copiously, perhaps for reasons of local habitat rather than global abundance. Peter had looked in sections west of Zumaya and found no latest Cretaceous ammonites (another reason for his earlier acceptance of gradual extinction). But now he extended his fieldwork to the east, toward the border of Spain and France. (Again, these eastern sections were known and had always been available for study, but Peter needed the impetus of Alvarez to ask the right questions and to develop a need for making these further observations.) Peter studied two new sections, at Hendaye on the Spanish-French border, and right on the yuppie beaches of Biarritz in France. Here he found numerous and abundant ammonites just below the boundary line of the great extinction. He writes in his 1992 book:

> After my experience at Zumaya, where years of searching yielded only the slightest evidence . . . near the Cretaceous-

Tertiary boundary, I was overjoyed to find a score of ammonites within the last meter of Cretaceous rock during the first hour at Hendaye.

We professionals may care as much about ammonites, but dinosaurs fire the popular imagination. No argument against Alvarez has therefore been more prominent, or more persuasive, than the persistent claim by most (but not all) dinosaur specialists that the great beasts, with the possible exception of a straggler or two, had died long before the supposed impact.

I well remember the dinosaur men advancing their supposed smoking gun of a "three-meter gap"—the barren strata between the last known dinosaur bone and the impact boundary. And I recall Luis Alvarez exploding in rage, and with ample justice (for I felt a bit ashamed of my paleontological colleagues and their very bad argument). The last bone, after all, is not the last animal, but rather a sample from which we might be able to estimate the probable later survival of creatures not yet found as fossils. If my buddy throws a thousand bottles overboard and I later pick up one on an island fifty miles away, I do not assume that he only tossed a single bottle. But if I know the time of his throw and the pattern of currents, I might be able to make a rough estimate of how many he originally dropped overboard. The chance of any single animal becoming a fossil is surely much smaller than the probability of my finding even one bottle. All science is intelligent inference; excessive literalism is a delusion, not a humble bowing to evidence.

Again, as with Peter Ward and the ammonites, the best empirical approach would order a stop to the shouting and organize a massive effort to dismember the haystack by looking for dinosaur bones in every inch of latest Cretaceous rocks. *Peter* means "rock" in Latin, so maybe men of this name are predisposed to a paleontological career. Another Peter, my friend and colleague Peter Sheehan of the Milwaukee Public Museum, has been guiding such a project for years. In late 1991 he published his much-awaited results.

Dinosaurs are almost always rarer than marine creatures, and this haystack really has to be pulled apart fragment by fragment, and over a broad area. The National Science Foundation and other funding agencies simply do not supply grant money at such a scale for projects that lack experimental glamour, whatever their importance. So Peter (Sheehan this time) cleverly availed himself of a wonderful resource that mere ammonites could never command. I will tell the story in his words:

> We co-opted the longstanding volunteer-based "Dig-a-Dinosaur" program at the Milwaukee Public Museum. Sixteen to twenty-five carefully trained and closely supervised volunteers and ten to twelve staff members were present during each of seven two-week field sessions during three summers. The primary objective of each volunteer was to search a predetermined area for all bone visible on the surface. The volunteers were arrayed in "search party" fashion across exposures so that all outcrops were surveyed systematically. Associated with the field parties were geologists whose function was to measure stratigraphic sections and identify facies.

I cannot think of a more efficient and effective way to tackle a geological haystack. Peter's personnel logged fifteen thousand hours of fieldwork and have provided our first adequate sampling of dinosaur fossils in uppermost Cretaceous rocks. Working in the Hell Creek Formation in Montana and North Dakota, they studied each environment separately, with best evidence available from stream channels and floodplains. They divided the entire section into thirds, with the upper third extending right to the impact boundary, and asked whether a steady decline occurred through the three units, leaving an impoverished fauna when the asteroid struck. Again, I will let their terse conclusion, summarizing so much intense effort, speak for itself:

> Because there is no significant change between the lower, middle, and upper thirds of the formation, we reject the hypothesis that the dinosaurian part of the ecosystem was deteriorating during the latest Cretaceous. These findings are consistent with an abrupt extinction scenario.

You can always say, "So what? T. S. Eliot was wrong; some worlds at least end with a bang, not a whimper." But such a distinction makes all the difference, for bangs and whimpers have such disparate consequences. Peter Ward sets the right theme in his final statement on the non-necessary demise of ammonites:

> Their history was one of such uncommon and clever adaptation that they should have survived, somewhere, at some great depth. The nautiloids did. It is my prejudice that the ammonites would have, save for a catastrophe that changed the rules 66 mil-

> lion years ago. In their long history they survived everything else the earth threw at them. Perhaps it was something from outer space, not the earth, that finally brought them down.

The true philistine may still say, "So what? If no impact had occurred, both ammonites and nautiloids would still be alive. What do I care? I had never even heard of nautiloids before reading this essay." Think about dinosaurs and start caring. No impact to terminate their still-vigorous diversity, and perhaps dinosaurs survive to the present. (Why not? They had done well for more than 100 million years, and the earth has only added another 65 million since then.) If dinosaurs survive, mammals almost surely remain small and insignificant (as they were during the entire 100 million years of dinosaurian domination). And if mammals stay so small, restricted, and unendowed with consciousness, then surely no humans emerge to proclaim their indifference. Or to name their boys Peter. Or to wonder about the heavens and the earth. Or to ponder the nature of science and the proper interaction between fact and theory. Too dumb to try; too busy scrounging for the next meal and hiding from that nasty *Velociraptor*.

13

Jove's Thunderbolts

ONE NIGHT IN 1847, partly to escape her parents' noisy dinner party, Maria Mitchell of Nantucket lugged her telescope to the roof of the Pacific National Bank (where her father worked as chief cashier) and discovered a comet five degrees from Polaris, the North Star. For this discovery, the first comet found by an American woman, Mitchell received many honors, including a gold medal from the King of Denmark and election, as the first woman ever so recognized, to the American Academy of Arts and Sciences in Boston.

(Mitchell's certificate of election still hangs on the wall of her birth house in Nantucket, and it is a painful and ambiguous thing to see. Two statements are crossed out: the printed salutation "Sir" has been altered to "Madam" by hand, and the designation of "fellow" has been replaced by "honorary member," meaning that Mitchell had not been granted voting privileges. The document is signed by Harvard's great professor of botany, Asa Gray, later one of Darwin's stoutest supporters. More than ninety years would pass before another woman won election. Today, I am happy to report, for I am a member and the Academy's house lies just around the corner from my own, people of all shapes, sexes, colors, and backgrounds cavort in one of the oldest intellectual societies of our land with liberty and justice for all.)

A few other women excelled in astronomy during this age of nearly total exclusion for one sex from science, but most, like Caroline Herschel, gained access as sisters, wives, or daughters of male astronomers. Maria Mitchell, however, succeeded on her own. Her father, William, was an amateur astronomer, and he did rate ship's chronometers for part of his living (a vital activity in the great whaling port of Nantucket at a time when ships reckoned longitude by maintaining extremely accurate clocks on board). But William Mitchell was not a professional scientist, and he was working primarily as head cashier of a bank when Maria discovered her comet.

Maria Mitchell, who earned no college degree for want of opportunity, became the first female professor of astronomy in America, serving at Vassar College from 1865 to 1888, where she vigorously promoted scientific education at America's premier college for women. She received honorary doctorates from Columbia, Hanover, and Rutgers. Maria Mitchell died in 1889. In 1902, family members and former students established the Maria Mitchell Association in Nantucket, an organization dedicated to astronomical observation and science education.

I gave the annual address at the Maria Mitchell Association on July 21, 1994. In lieu of an honorarium, I made one strong request: after my evening talk, I wanted to see Jupiter through Maria Mitchell's own telescope (not the original instrument of her 1847 discovery, for this machine resides in the Association's museum, but through the still-functioning five-inch refractor made by Alvin Clark of Boston and given to Mitchell in 1857 by an organization called the Women of America).

Call me foolish if you wish, but a professional lifetime as a paleontologist and evolutionist has led me to view connectivity and return to original sources as imbued with the highest value, both intellectual *and* ethical (or aesthetic). I wanted to see Jupiter that night through Maria Mitchell's telescope because she had been the great American pioneer in cometary discovery—and Jupiter, at that very moment, was being bombarded by a succession of more than twenty fragments from comet Shoemaker-Levy 9. I was unable, I must confess, to make out the scars on Jupiter's surface through Maria Mitchell's telescope, for Jupiter had already set behind a clump of trees. But I waited patiently and finally saw the flicker of Jupiter among the leaves. Mission adequately accomplished. Meanwhile, all the astronomical enthusiasts of Nantucket crowded around the adjacent computer, watching the latest information about Shoemaker-Levy coming in from all over the world by Internet.

In March 1993, my friend and colleague Gene Shoemaker (world's expert on the geology of impact structures), along with Caroline Shoemaker and David Levy, discovered a linear array of about twenty cometary fragments, stretched out in a trail nearly 200,000 kilometers long, near Jupiter. They determined that Jupiter's gravity had previously pulled the entire comet apart when it had passed within 100,000 kilometers of the giant planet's surface. They also recognized—and now the excitement started—that all these fragments would inevitably crash into Jupiter in July 1994.

What a show, and what timing—precisely at the twenty-fifth anniversary of Neil Armstrong's first step upon the lunar surface. But an epidemic of cold feet then gripped the astronomical community, and the spin doctors of diminished expectation raised their voices high. The impact would yield results of great scientific value, to be sure, they said, but viewers shouldn't expect to see much of any visceral interest. The fragments would probably be pulled farther apart, and burn into nothingness in the high reaches of Jupiter's atmosphere. Or, given our ignorance of cometary composition, perhaps the fragments contained little more than dust and ice—not the right stuff to punch holes in a planet's surface. One commentator predicted impact flashes so small and fast that each would appear in no more than a single camera pixel from the *Galileo* spacecraft, poised near Jupiter. The semiofficial report in the July 14 issue of *Nature*, just a week before impact, went forth under the headline, "Comet Shoemaker-Levy 9: the big fizzle is coming."

I certainly understand the reasons for toning down expectations. The general ethos of science tends toward caution, especially before the fact—and astronomers had been badly burned twice in recent years, both times by comets, as Halley and Kohoutek truly fizzled relative to most people's expectations. In addition, scientists are still reluctant to grant much of a role to catastrophic agents as major players in the recent history of the solar system, and particularly the earth and its life. A whimper for Shoemaker-Levy 9, rather than a bang, would surely reinforce this old prejudice.

All the world now knows that Shoemaker-Levy 9 put on the show of shows, vastly exceeding all publicly expressed expectations, both in stunning visual terms and in a plethora of scientific data that will keep us busy for years to come. I am writing this essay on the weekend after the bombardment, and cannot begin to make any sensible assessment of causes and reasons, but we need no more than a quick review of phenomena to illustrate the difference between a rave and a fizzle.

All twenty-one fragments of Shoemaker-Levy 9 hit Jupiter along a single latitude of the southern hemisphere. Impacts occurred during the Jovian night, and were not directly visible from earth. But Jupiter rotates in about nine hours, rather than our twenty-four, and the sites of impact moved quickly into view, forming a linear chain of circular scars, some larger than the entire earth! We do not know how long the scars will last, or even whether their darkness records shadow or substance (blackened carbon, or some forms of sulfur, have been suggested). We do not know how far the fragments penetrated into Jupiter, or even what penetration means on a gaseous planet that has no discrete rocky surface—but white-hot gases, presumably from the planet's interior, erupted through holes of impact "in great fireballs as if from cosmic cannons" (to quote Malcolm W. Browne in *The New York Times*, July 26).

The largest chunk of comet, called Fragment G, hit Jupiter at 3:30 A.M. on July 18, producing a flash and fireball that briefly exceeded the entire planet in brightness. (So much for single pixels!) Fragment G may have plunged nearly forty miles into Jupiter's atmosphere. A plume of superheated gas then erupted above the impact site to a height of 1,300 miles above the Jovian surface, yielding some of the most spectacular pictures of the entire week.

I have read various estimates of the size and power of this largest strike, but taking the most conservative figure reported in major newspapers and scientific journals (*USA Today*, July 22), Fragment G measured some 2.0 to 2.5 miles in diameter and hit with an explosive energy of about 6 million megatons of TNT. Taken all together, the twenty-one fragments, by this same estimate, released an equivalent of about 40 million megatons of TNT—or some five hundred times the power of all the earth's nuclear weapons combined!

This comparison to the megatonnage of our entire nuclear arsenal sent shivers up my spine and brought forth from my memory a similar figure that recalled the key debate within paleontology during the past decade, and also crystallized the various themes for this essay. In 1979, Luis Alvarez and his collaborators first published their theory that a large extraterrestrial object, a comet or an asteroid some six miles in diameter, struck the earth at the end of the Cretaceous period, 65 million years ago, triggering one of the five great mass extinctions of life's history—the latest and, parochially for us at least, the most important, as this event wiped out dinosaurs and gave mammals their opportunity—see the preceding essay.

I knew nothing about the physics of impact at the time, and I remember harboring some strong initial doubts about the efficacy of such an event. I did understand how a crashing comet or asteroid might be decidedly unpleasant for a *Tyrannosaurus* caught directly in the path of descent, but why should an object only six miles across wreak such havoc upon a planet with a diameter of eight thousand miles? I expressed my naive skepticism to Luis Alvarez, a Nobel Prize physicist who had worked on the Hiroshima bomb and certainly understood such matters, and he shut me up with the following estimate: a bolide six miles in diameter would strike the earth with ten thousand times the megatonnage of all the earth's nuclear weapons combined! Fragment G, in comparison, was a pipsqueak—yet this bolide tore an earth-sized hole in Jupiter, and produced a fireball equal in brightness to the entire planet. If Fragment G could so impact giant Jupiter, shall we doubt the catastrophic effect of a much larger strike upon our much smaller planet?

When Alvarez and company first proposed their radical hypothesis of catastrophic extinction, most paleontologists rejected the idea with ridicule and vehemence. Since then, however, evidence for impact has accumulated to virtual proof—first the initial discovery of iridium at high concentration in strata marking the extinction (for iridium is almost absent in indigenous sediments of the earth, but present at normal cosmic abundances in comets and asteroids); then the finding of shocked quartz in the same sediments (for this unusual form of a common mineral can be generated, so far as we know, only by high pressures associated with impacts, and not by any internal process acting at or near the earth's surface); and finally the apparent "smoking gun" itself, a massive crater of the right age, up to two hundred miles in diameter, off the Yucatán Peninsula in Mexico.

When strong hostility greets an interesting theory, and continues with substantial vigor even after the theory has been effectively validated, we must seek deeper causes rooted in general philosophies and methodologies. No theoretical preference has been so strong and dominating in the earth sciences as Charles Lyell's doctrine of uniformitarianism—a complex set of ideas centered upon the notion that current and observable causes, acting at characteristically minute and gradual rates, can produce all the grand effects of earth's history by accumulating their tiny increments through the immensity of geological time. To produce the Grand Canyon over time's vastness, erode the Colorado River valley grain by grain. To populate the earth with a novel fauna, extirpate old species and evolve new forms one by one, so that any observer at a moment in time would notice nothing. At most, under the uniformitarian perspective, climates and topographies might occasionally alter at especially high rates, promoting more species deaths than usual over a relatively short interval—but never a truly catastrophic mass extinction. Lyell and Darwin both strenuously argued that so-called mass extinctions must really extend over several million years (with the appearance of suddenness arising as an artifact of an imperfect geological record), and that the causes of such episodes could reside only in an intensification of ordinary processes.

This issue of uniformitarian vs. catastrophic change stands as one of the grand questions of science, for the debate pervades so many disciplines and bears so strongly upon some of the most profound puzzles of our lives. Consider just two of the deep issues: *The nature of change itself*. Is human culture, life, the physical universe indefinitely mutable and subject to continuous and usually insensible change (the uniformitarian view), or does stable structure characterize most forms and institutions, with change therefore concentrated in rare and rapid episodes of transition between stable states, often initiated by catastrophic disturbances to which existing systems cannot adjust? *The nature of causality*. Do large effects arise as simple extensions of small changes produced by the ordinary, deterministic causes that we can study every day, or do occasional catastrophes introduce strong elements of capriciousness and unpredictability to the pathways of planetary history?

William Glen, a distinguished geologist and historian of science at the United States Geological Survey in Menlo Park, California, has spent the last fifteen years working in the interesting area of scientific revolutions in the making. He first wrote a fine book *(The Road to Jaramillo)* on what most people have judged as the greatest revolution in geology since the discovery of time's vastness in the late eighteenth and early nineteenth centuries: the development of the theory of plate tectonics and the consequent validation of continental drift. During the past decade, Glen has been tracing the movement of Alvarez's asteroid from heresy to orthodoxy, and chronicling the regrowth of catastrophist thinking in general, particularly in the context of discussion about mass extinction (see his recent book *The Mass Extinction Debates: How Science Works in a Crisis*).

Glen surprised the hell out of me one day by saying that he regarded the debate over catastrophic mass extinction as potentially more important in the history of science than plate tectonics. Much as I might like to accept such an argument, I initially cringed and strongly objected, for I had so long accepted the virtual mantra among geologists that no reformulation could be more profound than plate tectonics—the earth's surface broken into thin plates in motion, with new crust welling up at oceanic ridges, spreading out to form and push the plates, and eventually descending back into the earth at subduction zones.

But Glen reminded me of the famous statement by Freud that I have often quoted in these essays: The most important scientific revolutions all include, as their only common feature, the dethronement of human arrogance from one pedestal after another of previous convictions about our centrality in the cosmos (see essay 25). At the very least, great revolutions must alter some central concept about our lives or the workings of the universe. Plate tectonics radically changed the physics of the earth, but few fundamental tenets of human life or physical causality were altered thereby. We believed, before, that the earth's crust could move up and down to form mountains and ocean basins; we know, after, that planetary real estate flows laterally as well.

But catastrophic impact theory, as Bill Glen argued with great force, has much broader implications if collision can be established as a generality in the mechanisms of planetary history, and not just as an explanation for a few peculiar events like the Cretaceous-Tertiary extinction. For if impacts shape much that matters in geological time, then catastrophism becomes at least coequal with a previously dominant or nearly exclusive uniformitarianism—and our fundamental views on the nature of change and causality must be revised. Moreover, these potential revisions speak to the tensions that Freud identified as both scary and liberating—for catastrophism supports themes that many of us would rather not acknowledge about chance and unpredictability in the evolution of all lineages, including our own. No random bolt from the blue 65.3 million years ago, no extinction of dinosaurs, no mammalian dominance, no human life today—whereas the uniformitarian perspective harmonizes much better with traditional ideas of a gradual rise to inevitable success by mammals.

As a believer in connectivity with initial sources, illustrated earlier by my tale of Maria Mitchell's telescope, I then decided to excavate the original definition of geology's great dichotomy in an attempt to understand the importance of catastrophism's resurgence and the role of Shoemaker-Levy 9 as a contribution to the argument. The terms "uniformitarian" and "catastrophist" were coined by William Whewell, England's leading philosopher of science, in a review (published in the *Quarterly Review*, 1832) of the second volume of Lyell's *Principles of Geology*. Whewell wrote:

> Have the changes which lead us from one geological state to another been, on a long average, uniform in their intensity, or have they consisted of epochs of paroxysmal and catastrophic action, interposed between periods of comparative tranquility. These two opinions will probably for some time divide the geological world into two sects, which may perhaps be designated as the *Uniformitarians* and the *Catastrophists*.

G. P. Scrope, one of Lyell's leading geological colleagues, read Whewell's review and wrote to Lyell suggesting that the dichotomy might easily be resolved or compromised:

> As to the dispute he speaks of, which I know has now and then raged pretty warmly between you as a Uniformitarian and the Catastrophists I do not see any but an imaginary line of separation between you. It is only a dispute about degree, a plus or minus affair; a little concession on either side will unite you in perfect cordiality.

But Lyell firmly rejected this mediation and stuck by Whewell's division. He wrote to Whewell, quoting Scrope's opinion and then adding his own support for a dichotomy of principle:

> On this point I cannot budge an inch for reasons to be developed in v. 3 [that is, the third volume of his *Principles of Geology*, published the next year, 1833]. It is of course a question of probability to some extent in the present state of our knowledge, but I consider it a most important question of principle, whether we incline to the probabilities as seen in the last 3,000 years or possibilities which I hold to be uncalled for by any overwhelming evidence.

In other words, Lyell writes, why invent unseen (and catastrophic) forces when slowly acting present causes, observed for more than three thousand years, suffice to render all geological events in the fullness of time.

The intensity of Lyell's rhetorical support for uniformitarianism may best be judged in the famous first chapter of volume 3, composed in the light of Whewell's review and its codified dichotomy. (Later editions of the *Principles of Geology* dispersed the material of this short seven-page chapter into earlier parts of the book.) Catastrophism, Lyell argues, must be rejected as a speculative system that will degrade geology to a playground of untestable conjecture, whereas uniformitarianism, firmly grounded in observation of present causes, will establish geology as a rigorous, empirically based, mature science. Lyell was a lawyer by profession and a brilliant writer by avocation. No one has ever matched him for persuasive prose:

> Never was there a dogma more calculated to foster indolence, and to blunt the keen edge of curiosity, than this assumption of the discordance between the former and the existing causes of change. It produced a state of mind unfavorable in the highest conceivable degree to the candid reception of the evidence of those minute, but incessant mutations, which every part of the earth's surface is undergoing . . . We hear of sudden and violent revolutions of the globe, of the instantaneous elevation of mountain chains, of paroxysms of volcanic energy . . . We are also told of general catastrophes and a succession of deluges, of the alternation of periods of repose and disorder, of the refrigeration of the globe, of the sudden annihilation of whole races of animals and plants, and other hypotheses, in which we see the ancient spirit of speculation revived, and a desire manifested to cut, rather than patiently to untie, the Gordian knot.

I have long regarded Lyell's argument as deeply unfair in principle, however brilliant in rhetoric—yet he prevailed for nearly 150 years, thus restricting the acceptable range of geological hypotheses. He argues that we should prefer direct observation to any form of inference (labeled as "speculation" in his rhetoric). Catastrophism then loses by unjustly blinkered definition, rather than by fair test. If the catastrophic model has any validity, then natural forces of paroxysmal intensity occasionally impact the globe to great and sudden geological effect, but at very infrequent intervals. If the waiting time between such events usually amounts to millions of years—as must be the case for large impacts producing global mass extinctions—and if human observation of present processes has been limited to only a few thousand years, what chance do we have of ever observing such catastrophes over the course of human history?

We cannot reject plausible forces because we do not see them directly. Most of science relies upon ingenious and rigorous inference, not passive observation alone. We were quite confident about many phenomena—atoms and black holes among them—before we developed technologies for their visualization. How can we ridicule a force that we have not seen for three thousand years, when good theory predicts that its momentary episodes of operation should be separated by many millions of years? Why are we so loath to accept global catastrophes as major actors in the history of earth and life when we have seen the commoner events of smaller scale that occur so much more frequently? The Tunguska object, a cometary fragment that exploded 28,000 feet above Siberia in 1908, flattened a thousand square miles of forest and would have produced the greatest disaster in human history had it struck anywhere near a population center. This object, by best estimate, measured about three hundred feet in diameter. Such small particles hit the earth with waiting times of hundreds to thousands of years and can therefore be studied as uncommon historical events. May we not extrapolate from these to the much rarer giant bolides, with diameters measured in miles, that must strike the earth with waiting times of tens to hundreds of millions of years, but with massive global effects and with sufficient frequency to shape much of planetary history in a half-dozen or so good strikes during the tenure of multicellular animal life?

Lyell had lobbied hard to secure Whewell's review in the most prestigious of British journals. But Whewell, while praising the book and fairly describing Lyell's system, then argued that Lyell had been unduly restrictive in his uniformitarian rigidity, and that the case for catastrophism remained open in principle. In other words, the issue must be settled by scientific study (observation *and* inference), not by *a priori* definition. Whewell wrote in defense of catastrophism, and in the next paragraph following his definitions:

> It seems to us somewhat rash to suppose, as the uniformitarian does, that the information which we at present possess concerning the course of physical occurrences, affecting the earth and its inhabitants, is sufficient to enable us to construct classification, which shall include all that is past under the categories of the present. Limited as our knowledge is in time, in space, in kind, it would be very wonderful if it should have suggested to us all the laws and causes by which the natural history of the globe, viewed on the largest scale, is influenced—it would be strange, if it should not even have left us ignorant of some of the most important of the agents which, since the beginning of time, have been in action; of something, in short, which may manifest itself in great and distant catastrophes.

We didn't need Shoemaker-Levy 9 to validate catastrophism by the most direct route of overt observation. We were fortunate indeed to witness any event at such a scale during the few hundred years that technology and enlightened understanding have combined to give us tools of comprehension. Through inferences based upon fossilized results, the

Cretaceous bolide had pretty much proven itself a few years before Jupiter fractured Shoemaker-Levy into its murderous bits.

Shoemaker-Levy must therefore represent an indulgence of nature—a gift to us, a reward for our proper use of scientific inference to validate at least one global catastrophe (the terminal Cretaceous bolide) against all the prejudices of our uniformitarian training. Nature almost seemed to be saying to her wayward and insignificant child on earth: "You did it the right and hard way, by difficult inference; well done, thou good and faithful servant; now I will show you a cosmic catastrophe directly, and for free, even though such an event might plausibly not occur for many thousands of years."

All this, of course, is metaphor. I have not become an addlepated New Ager, and I do not believe that nature operates by anything akin to our notion of intent. We just caught a break with Shoemaker-Levy. But do consider two more truly wondrous events in closing. Shoemaker-Levy hit Jupiter precisely to the day of the twenty-fifth anniversary of our first human landing on the moon. Is nature returning the compliment of our respectful foray into her realm by teaching us something fundamental about her ways at such an auspicious moment? Let us also, and finally, thank nature for setting this display on a distant and giant planet that can easily bear the shock, and not closer to home, where a similar event might relegate us to the fate of the dinosaurs.

PART FOUR

Writing about Snails

14

Poe's Greatest Hit

POSTHUMOUS RECOGNITION MAY BE GOOD for the soul, but most artists would prefer a bit of bodily succor during their hours of earthly need. Edgar Allan Poe (1809–1849) published his first work, *Tamarlane and Other Poems*, in 1827 at age eighteen. He paid the costs of publication himself, and produced but fifty copies. Only twelve have survived, making this work the rarest and most expensive of all American first editions. The latest auction records may be stupendous, but Poe's pamphlet received neither literary attention nor sales at its debut.

Here is a trivia question on the same subject: Can you name the only book by Poe that was successful enough to be republished during his lifetime? The answer? *The Conchologist's First Book: or, A System of Testaceous Malacology, arranged Expressly for the Use of Schools*, published in 1839. The first edition, at a price of $1.75 per copy, sold out in two months, leading to a second, enlarged edition in 1840, and a third in 1845. For all of Poe's other work, from the House of Usher to the Rue Morgue, we can only intone the entire vocabulary of his most famous character, the Raven: "Nevermore." No second edition for any other book during Poe's lifetime.

We need not always marvel when celebrated people boast accomplishments far from the source of their fame. Henry VIII wrote quite acceptable pieces of music—but why not, for he was intellectually inclined

Cover of the low-priced textbook written by Edgar Allan Poe.

and well educated in the various arts of Renaissance humanism. Leonardo da Vinci designed formidable machines of war—but why not, for princely patrons needed his practical skills even more than his painter's art. But nothing about Edgar Allan Poe's life suggests any abiding concern with natural history. We can picture him drunk and penniless in Philadelphia or New York, but who can conjure up an image of Poe wandering through the woods, or strolling along the seashore "dreaming dreams no mortal ever dared to dream before"? His various literary creatures—the Raven, the ape of the Rue Morgue, or the Black Cat—evince no particular knowledge of zoology. His biographers have had a most difficult time trying to suggest any source for Poe's potential interest in seashells. (Some have suggested that Poe may have known the naturalist Edmund Ravenel while he was stationed on Sullivan's Island in Charleston Harbor during an army hitch in 1827 and 1828. Others point to a description of some shells in one chapter of his novella *The Narrative of Arthur Gordon Pym.*) We may well conclude that *The Conchologist's First Book* is a most "quaint and curious volume," a true anomaly in the life of a literary man.

We may better understand Poe's methods and motivations in writing the utterly forgotten best-seller of his lifetime when we acknowledge the ambiguous status of *The Conchologist's First Book*—basically a scam, but not so badly done. Poe didn't need much scientific knowledge to fulfill his appointed role as part name-lender, part literary man, and part plagiarist. Poe's friend Thomas Wyatt had, in 1838, published a lavish and expensive book on mollusk shells, retailing for eight dollars a copy. Sales were predictably slow, and Wyatt wished to produce a shorter edition with uncolored plates at a much lower price. But his publisher, Harper, objected, citing a reasonable concern that their fancy edition would then become entirely unsellable. Wyatt, still wishing to proceed but fearing legal action should he publish the shorter version under his own name, sought a surrogate to help with his new volume and to serve as name-bearer for a flat fee.

Poor Poe, perennially broke and often drunk, seemed ideally suited for the role. Others have shaken the world for thirty pieces of silver, but Poe apparently received fifty dollars for both fronting and helping in preparation of *The Conchologist's First Book*. (Details of the financial arrangements have been lost, but I rather suspect that the anonymous Wyatt received royalties on the successful product.)

I have tried to survey all the standard biographies of Poe, and they are unanimous in their severe discomfort with *The Conchologist's First Book*. None gives this episode more than a page or two—a quick ac-

CONCHOLOGIST'S FIRST BOOK:

OR,

A SYSTEM

OF

TESTACEOUS MALACOLOGY,

Arranged expressly for the use of Schools,

IN WHICH

THE ANIMALS, ACCORDING TO CUVIER, ARE GIVEN WITH THE SHELLS,

A GREAT NUMBER OF NEW SPECIES ADDED,

AND THE WHOLE BROUGHT UP, AS ACCURATELY AS POSSIBLE, TO THE PRESENT CONDITION OF THE SCIENCE.

BY EDGAR A. POE.

WITH ILLUSTRATIONS OF TWO HUNDRED AND FIFTEEN SHELLS, PRESENTING A CORRECT TYPE OF EACH GENUS.

PHILADELPHIA:

PUBLISHED FOR THE AUTHOR, BY

HASWELL, BARRINGTON, AND HASWELL,

AND FOR SALE BY THE PRINCIPAL BOOKSELLERS IN THE UNITED STATES.

1839.

The title page lists Poe as the only author, although the book was a joint effort.

knowledgment followed by an apparent search for "respite and nepenthe" (the drug of forgetfulness). F. T. Zumbach states that "it didn't bear even the slightest relevance to Poe's literary career." Julian Symons, an excellent writer of detective fiction as well as a literary biographer, writes that Poe "put his name to a piece of hackwork." David Sinclair describes *The Conchologist's First Book* as "a piece of shameful hackwork to which only desperation could have driven him." Jeffry Meyers labels the book as Poe's "grossest piece of hackwork."

I do not write this essay either to exonerate Poe or to revise the usual story about the genesis of *The Conchologist's First Book* in any substantial way. By modern standards and copyright laws, Poe would either be in jail (along with Wyatt and other collaborators), or in the dock awaiting conviction and a fat fine. But I do want to suggest that Poe's biographers need not be so apologetic, and should not treat *The Conchologist's First Book* as a curious and shameful anomaly best passed over in near silence. I want to suggest, rather, that Poe's book succeeded because it filled a need in a competent and at least mildly innovative way (whatever the dubious circumstances of the volume's origin); moreover, when we grasp the reasons for its success, we learn something interesting about popular education in nineteenth-century America.

To analyze *The Conchologist's First Book* properly, we must first understand the genesis of the volume's various parts, and Poe's role in cobbling them together. (Remember that Poe's name fronted for the efforts of a committee. Wyatt certainly worked on the book, and others may have helped as well.) We get our best evidence from Poe's own testimony—a typically self-serving letter with an undeniably positive spin, but true in its particular claims, so far as I can tell.

Poe led a short and tortured life (following an old quip usually made about Mozart, I am tempted to say that Poe was eleven years dead when he reached my age—and I'm still a kid by my own reckoning). Poe created all his own devils—the fights, the near duels, the bouts of drunkenness, the endless and debasing pleadings for money, the vituperative literary disputes. He haunted himself with a line from his most celebrated poem:

> Is there—is there balm in Gilead?—tell me—tell
> me—I implore:
> Quoth the Raven "Nevermore."

Poe counted several charges of plagiarism (some true) among his many troubles, and he spent much of his last decade threatening and bringing legal suits against such accusations. *The Conchologist's First Book* did not escape this pattern. In February 1847, Poe learned that a Philadelphia journal had questioned his foray into natural history—and he responded with a letter to George W. Eveleth:

> What you tell me about the accusation of plagiarism made by the "Phil. Sat. Ev. Post" surprises me. It is the first I heard of it . . . Please let me know as many particulars as you can remember—for I must see into the charge—Who edits the paper?

> Who publishes it? etc. etc.—about what time was the accusation made? I assure you that it is *totally* false. In 1840 [Poe is a year off here] I published a book with this title—The Conchologist's First Book . . . This, I presume, is the work referred to. I wrote it, in conjunction with Professor Thomas Wyatt, and Professor McMurtrie of Ph[iladelphi]a—my name being put to the work, as best known and most likely to aid its circulation. I wrote the Preface and Introduction, and translated from Cuvier the accounts of the animals etc. *All* school-books are necessarily made in a similar way. The very title page acknowledges that the animals are given "according to Cuvier." This charge is infamous and I shall prosecute for it, as soon as I settle my accounts with the Mirror.

The Conchologist's First Book begins with a two-page "Preface," and I have no reason to doubt Poe's claim that he wrote this part all by himself. A four-page "Introduction" then follows—and now the trouble begins. Poe expropriated much of this text from the fourth edition (1836) of a British work by Captain Thomas Brown, titled *The Conchologist's Text-Book*. Some biographers have claimed that Poe's entire text is a paraphrase, if not a direct copy, of Brown's. (F. T. Zumbach, for example, writes that Poe "copied from Brown almost word for word.") In fact, by my own comparison between the two books, only three paragraphs (about one fourth of the text) show extensive "borrowings." (Poe wins no exoneration thereby, for plagiarism, like pregnancy, does not increase in severity by degrees: beyond a point of definition, you either did it or you didn't—and Poe surely did.)

The plot thickens with the next section of twelve plates. The first four, illustrating the parts of shells, are lifted *in toto,* jot for jot and tittle for tittle, from Brown. No fuss, no pretenses, no excuses—just plain stolen. The subsequent eight plates, illustrating the genera of shells in taxonomic order, follow Brown in the more interesting pattern of back to front—that is, Brown's last plate is Poe's first (with considerable rearrangement, reorientation, and switching around of individual figures), and we then move up through Poe, and down in Brown, until Poe's last plate largely reproduces Brown's first.

Others have caught the pattern and even suggested that Poe and Wyatt were now consciously trying to hide their plagiarism. The actual reason is different and more interesting. (What could Poe and Wyatt be trying to hide anyway, after copying the first four plates exactly?) Brown's book follows the pedagogical scheme of the great French naturalist

Lamarck, who always presented his discussions in chain-of-being order, but top down, rather than the more conventional direction of bottom up. That is, Lamarck began with people and ended with amoebae, rather than the usual vice versa. Brown followed Lamarck and therefore started with the most "advanced" mollusks, but Poe and Wyatt obeyed the usual convention and began with the most "primitive"—hence the reversed order.

The choice of Brown as victim best illustrates the nature of the scam, or at least the easy-street methods followed by Wyatt, Poe, and company in producing their cheap edition of a molluscan textbook. The key lies in Glasgow, home of Brown's publisher. No effective international copyright existed in Poe's day, and material published in foreign nations could be plagiarized and expropriated without fear of legal reprisal (moral sanctions existed, just as today, but principles without practical power have never been very effective as instruments of persuasion). Brown was British and therefore eminently exploitable. (For the same reason, exactly forty years later, Gilbert and Sullivan brought their entire troupe to America to present the premiere of *The Pirates of Penzance* in New York, thus to upstage the theatrical pirates who had made so much money, without paying a cent in royalties, from Gilbert and Sullivan's previous smash hit, *H.M.S. Pinafore*.) Poe must have felt at least minimally guilty, for the penciled corrections in his own copy of *The Conchologist's First Book* include the following addition to the acknowledgments: "[Thanks] also to Mr. T. Brown upon whose excelent [*sic*] book he has very largely drawn."

The next section of *The Conchologist's First Book*, a ten-page "explanation of the parts of shells," also derives verbatim from Brown, with a few minimal changes to simplify terminology and eliminate British parochialisms. The body of the book then follows—some 120 pages describing each genus, the animal first and then the shell, and concluding with a list of all species within the genus. This section follows the order of Wyatt's more expensive book. Poe apparently excerpted material from Wyatt on the shell of each genus and then translated passages from Cuvier's French to supply descriptions for the soft anatomy of the animals within. (Poe's redemption, I will argue, comes from this contribution and its rationale.) The book then ends with a glossary and an index, apparently supplied by Wyatt.

Plagiarism? Yes, in several parts. Shameless "hackwork"? I'm not so sure—for Poe did an interesting thing or two that gave the book some value. Evolutionary biologists know, better perhaps than members of any other profession, that historical origin and current utility represent different facets of a biological object, and that one need not be correlated with the other (a frequent theme of these essays). We should apply the same distinction to Poe's book. I will not defend its origin, but the final product had genuine utility (and at least a dollop of innovation)—and its commercial success was no fluke. To appreciate these legitimate reasons for success, we must ask two questions: Why did Wyatt call upon Poe and not someone else, and for whom was the book intended?

1. POE'S GENUINE EXPERTISE. Poe gives one reason for his selection as "front man" in his letter to Eveleth, previously quoted: ". . . my name being put to the work, as best known and most likely to aid its circulation." If this claim can be accepted at face value, then we have some interesting (if also interested) testimony that Poe had achieved some measure of literary recognition, and was not the abject failure or utterly misunderstood genius of so many later portrayals.

But I suggest that Poe's selection also recorded Wyatt's good judgment about particular skills for the job required. *The Conchologist's First Book* presents, as an explicit and fundamental feature of its organization, a progressive, even innovative, arrangement of material. None of Poe's biographers has ever picked up on this theme, probably because the arrangement seems so "obviously" right that we assume its employment in books about mollusks from all times. Not true. The arrangement was unusual in Poe's day—and I suspect that Wyatt needed Poe's skills to implement the reform.

The name of the phylum Mollusca comes from a Greek word for "soft" (as in *mollify* or *mollycoddle*)—a reference to the animal that, in the absence of hard parts within its body, secretes a calcareous shell for an enclosure. Most collectors and natural history buffs, for obvious reasons, focus upon the preservable shell, not the quickly decomposed body. The study of shells and their classification goes by the separate name of conchology—as in Poe's book.

In Poe's day, most popular books on mollusks treated the shells alone. Wyatt's original and expensive version described only the shells, as did Brown's volume, the source of Poe's plagiarism. Linnaeus himself had based his classification of mollusks only upon shells. As an example, in the popular work on mollusks featured in the next essay—*The Conchologist's Companion*, published in 1834 and therefore a contemporary of Poe's volume—author Mary Roberts begins her text by separating the study of animal and shell, and by defending a treatment based only upon shells and ignoring animals entirely: "The elegant science of

Conchology, my friend, comprises the knowledge, arrangement, and description of testaceous animals; a science, according to Linnaeus, which has for its basis the internal form and character of the shell, and is totally independent of the animal enclosed within the calcareous covering." And Thomas Brown added, in 1836, "It is upon the exclusive shape of the shell, not the animal inhabitant, that the Linnaean arrangement of Conchology is formed."

Wyatt and Poe may have plagiarized Brown, but they did improve upon his organization, and this entailed some effort deserving more than an epithet of "hackwork." An arrangement based only upon shells, and ignoring the biological manufacturers of these coverings, must represent an exercise in artificial pigeonholing, and cannot provide an optimal account of molluscan life. A more integrated and fully biological discussion must present information about animals and shells together. Wyatt and Poe decided to provide this dual treatment—even though Wyatt had failed to do so in his more elaborate volume—and not on a whim, or as a nice little sidelight, but as the fundamental feature of their new volume.

Poe's two-page preface is little more than a rationale for this defining feature of the new volume. He begins with definitions, contrasting malacology, or study of all biological aspects of mollusks, with conchology, or consideration of the shells alone. Poe then states that he will keep the more familiar name of conchology, but introduce the important biological innovation of describing both the animals and shells together:

> The common works upon this subject, however, will appear to every person of science very essentially defective, inasmuch as the relations of the animal and shell, with their dependence upon each other, is a radically important consideration in the examination of either . . . There is no good reason why a book upon Conchology (using the common term) may not be malacological as far as it proceeds.

Poe then reinforces his intent by describing the new book's "ruling feature"—"that of giving an anatomical account of each animal, together with a description of the shell which it inhabits." (Incidentally, the most recent biography of Poe, published in 1992, misses this point by failing to recognize the conceptual reform behind Poe's focus upon disciplinary names [malacology vs. conchology]. The author writes that "Poe's boring, pedantic and hair-splitting Preface was absolutely guaranteed to torment and discourage even the most passionately interested schoolboy.")

But where could Poe and Wyatt find the descriptions of animals, since neither Wyatt's nor Brown's volume contained such information? They turned to France, the leading nation by far in natural history during the early nineteenth century, and to the work of Europe's greatest anatomist, Georges Cuvier. And now Poe's particular skills came to the fore.

Poe had no known expertise in natural history, but he was certainly fluent in French. Poe's actress mother had died when Poe was only two, and he had been raised in the home of an intermittently wealthy Richmond businessman, John Allan (from whom Poe took his middle name, though he was never formally adopted). He lived in England and Scotland for five crucial years (1815–20), where he received a classical education in rigorous schools, including a thorough grounding in French. His intense training in ancient and modern languages then continued in Richmond. He attended preparatory schools and then spent a year at the University of Virginia in 1826, where he probably met Thomas Jefferson just before the great statesman's death; and where two ex-presidents, James Madison (who had succeeded Jefferson as rector) and James Monroe, examined him for several hours and gave him highest honors in both classical and modern languages. In short, Poe surely knew his French, probably far better than anyone else engaged in the project of downsizing and upgrading Wyatt's book.

I do not know how much time Poe spent translating Cuvier's accounts of the animals and integrating them with Wyatt's more conventional material on shells alone. But this effort defined the central, and surely admirable, feature of *The Conchologist's First Book*. Poe understood the importance of this task, the key to the book's distinction and success, for he stated in the letter to Eveleth: "I wrote the Preface and Introduction, and translated from Cuvier, the accounts of the animals." I do not claim that Poe spent many a "midnight dreary" pondering himself "weak and weary" over Cuvier, but the effort required some time, demanded a linguistic skill perhaps not widely spread in Wyatt's circle, and provided the central ingredient to the book's legitimate success.

2. THE BOOK'S INTENDED AUDIENCE. But why was Wyatt so eager to produce a cheaper edition of his poorly selling longer book? Why did he think that the new volume would succeed? Why was he so keen that he was even willing to abandon any advantage of explicit authorship

to maneuver around the concerns, and the legal arm, of his original publisher?

To answer these questions, we need to make the right analogy to the most comparable modern phenomenon—musical performers who tote tapes and CDs of their work to flog at intermissions during their concerts. *The Conchologist's First Book* did not sell primarily in bookstores (if many existed in early-nineteenth-century America). Wyatt had a specific, dedicated market—and he really needed an inexpensive product to sell. Presenters of popular science now teach in schools or frequent the true "midnight dreary" of our modern age—late-night radio talk shows. But a nineteenth-century counterpart to Jacques Cousteau or David Attenborough hit the traveling circuit as an itinerant lecturer at the various lyceums, athenaeums, book circles, and ladies' and gentlemen's clubs (usually separate) that kept the fire of learning alive in every substantial town of nineteenth-century America. Thomas Wyatt was such a traveling salesman of science—and he needed a book to accompany his frequent and popular lecture series on mollusks. I don't claim that his motives were entirely, or even to any large extent, idealistic. His need probably translated to dollars for his pocket. Lecture fees were small, and the supplement added by heavily promoted book sales to a willing audience probably meant the difference between penury and solvency. (The glitzy expansion of museum shops in our own times has occurred for the same reason.) Remember that Poe got a flat fee for a little writing, some translation, and the loan of his name; Wyatt, no doubt, was recompensed on a per-copy basis.

Nonetheless, I must again invoke the distinction between reasons for origin and sources of utility. Wyatt may have joined with Poe almost entirely for the money, but they promoted a good and innovative idea—elevating conchology from artificial description to integrative biology—by insisting that animals and shells be described together. Their product was surely worthy, if not brilliant or pathbreaking (for Poe lacked the biological knowledge to integrate his translated material on animals with Wyatt's old descriptions of shells, and therefore did little more than list one source of information above the other).

I have a copy of *The Conchologist's First Book*. (I could never dream of owning something like *Tamarlane and Other Poems*, but Poe's shell book was a success, and therefore survives in many copies at manageable prices.) I never knew the reasons behind the copious pencil scribblings in my copy until I did the research for this essay. One of the blank pages at the back bears the inscription, "Lectures delivered before the

146 CONCHOLOGY.

nated, longitudinally subinvolute in the same plane; aperture very wide, symmetrical, complete, square anteriorly, slightly modified by the turn of the summit, and provided on each side with an earlike appendage having thick and smooth edges. Inhabits the Mediterranean. Three species.

Argonauta argo. Argonauta tuberculosa.
A. nitida.

2. Genus *Carinaria*. Pl. XII.

Animal. Body elongated, prolonged behind the nucleus into a veritable tail edged at its extremity by a vertical fin; head sufficiently distinct, two long conical tentacula; two sessile eyes; the organs of respiration and the nucleus entirely enveloped in a mantle with lobed edges.

Shell. Very thin, symmetrical, a little compressed, without spire, but with the summit a little reflexed posteriorly; aperture oval and entire. Inhabits the African, Mediterranean, and Australian seas. Three species.

Carinaria vitrea. Carinaria fragilis.
C. cymbia.

After a lecture, "Proff Wyatt" (Thomas Wyatt, the expert behind the shell book) probably sold this copy to its first owner for $1.75.

young ladies of Charlestown Female Seminary" (probably in the settlement just east of central Boston and spelled with a "w," and not the city of South Carolina, West Virginia, Illinois, or Mississippi, rendered with a letter less as Charleston). After the last page of Poe's joint anatomical and conchological descriptions, the owner has written: "The end of Lectures on Conchology by Proff. Wyatt"—the clincher to my search for a source (I had never heard of Wyatt before writing this essay, and therefore had no clue to the meaning of the inscriptions). Other jottings largely note the etymology of Linnaean names, or record popular designations under Poe's Latin. The owner, for example, writes "clam" under the genus *Mya*, and "muscle" (for "mussel") under the genus *Mytilus*.

I don't know how else to make the point, except by a simple confession of feelings. I am enormously thrilled to infer that Wyatt carried my copy of *The Conchologist's First Book* to one of his lecture series on mollusks, where he sold the volume to a woman who then attended the lectures and took copious notes. I love to think of the book passing from Wyatt's own hand into the ownership of Ms. Charlestown, perhaps for two dollars forked over, and a quarter given in change. Chuckle if you will at my imagination, but the lyceum movement in America was one of the most worthy institutions ever spawned in an otherwise nonintellectual nation (the precursor to later Chautauquas and to all the popular education of our time). This movement also represented one of the few pathways then available to women for any collective education. *The Conchologist's First Book* was a reasonable contribution to this effort, and my copy comes from the heart of a worthy cause.

I am therefore led to view Poe's most famous image in a different light from the author's intent. Poe opens his window, thus admitting the Raven who perches on his white bust of Pallas and never leaves:

> And the Raven, never flitting, still is sitting, still
> is sitting
> On the pallid bust of Pallas just above my chamber door.

Poe's Raven is an ominous and tragic weight upon hope for psychological peace and dreams of accomplishment. But consider the union in another way. Pallas Athena (equivalent to the Roman Minerva), chaste goddess of practical reason and protector of cities (also namegiver to Athens)—symbol of the urban and civilized, as opposed to Artemis, goddess of the outdoors. Does she not represent Poe, or at least his unrealized hopes? (Athena's two designations, *pallas* and *parthenos*—the Parthenon of Athens is her temple—refer to her purity and virginity. Poe calls her Pallas, and undoubtedly longed for the unsullied life that she symbolized.) The urbane Poe, dweller in cities, scholar of languages, no fan of natural history. And the Raven, symbol of raw nature, imposed from without, but now firmly joined in a union of opposites: black and white, science and literature, nature and culture. Shall we not cherish the juxtaposition?

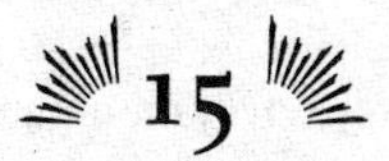

15

The Invisible Woman

FOLLOWING KHRUSHCHEV'S REVELATIONS of Stalin's less than saintly persona and procedures, the Soviet Union revised its official version of Communist Party history during the twentieth century. I bought a copy of this new edition and immediately turned to the index to learn the latest word on Uncle Joe. I found that he had suffered the worst of all fates: he simply wasn't there. And I thought to myself, Love him or hate him, but how in hell can you tell the story of twentieth-century Russia without him? The keepers of official records had used the primary device of excommunicators, anathematizers, and ostracizers throughout history: there is a fate far worse than death or the rack, and its name is oblivion—not the acceptable fading of an honored life that passes from general memory as historical records degrade (for nearly all of us must endure this erasure), but the terror of unpersoning, of being present (either in life or immediate memory) but bypassed as though nonexistent.

Whole groups have suffered this fate as a consequence of general prejudice rather than special excoriation. As a primary example from my own field of evolutionary reconstruction, nearly all older theories for the "ascent of man" limited their concepts by the same prejudice that set their choice of words. Until the feminist movement provoked a salutary expansion to the entire human race, nearly all theories attributed our shared capacities for language, intelligence, and other valued properties of the mind entirely to the activities of prehistoric males. Thus we learned that language arose from the coordination needed to hunt large animals (an all-male activity in conventional reconstructions), or that consciousness itself emerged from the more complex mental functioning required to stalk game (another male preserve). Women, under these theories, remained invisible—sitting in the cave with the kids, I suppose (and so depicted in paintings and museum dioramas), but unmentioned in explicit text.

This ideological invisibility of prehistoric females was bolstered by a sexist sociology of disciplines that prevented living women from practicing the most prestigious parts of science—research and publishing. Only in this generation have women been entering science in substantial numbers. (I am proud that my own lab has included 50 percent or more of women graduate students during the past decade, but I must admit that the first woman teaching fellow in our largest general course for nonscientists did not obtain her position until the early 1970s; she is now a distinguished researcher at the Smithsonian Institution.)

If intellectual women have been so restricted in our own day, consider the even greater limits imposed during the nineteenth century, the subject of this essay. In England, women were excluded from most major scientific organizations. The Geological Society of London did not admit women until 1904 (T. H. Huxley, to his shame, had supported the ban), the Linnean Society until 1905. Women fared better in botany, a subject considered suitable to the tastes and sensibilities of a "weaker sex." But even here, the reasons for limited acceptability took root in discrimination and fostered no egalitarian flowering. In an admirable study of "The Woman Members of the Botanical Society of London, 1836–1856," D. E. Allen wrote:

> Botany could break the rules because it had the great good luck to be in keeping with both of the contemporary alternative ideals of femininity. On the one hand it was able to masquerade as an elegant accomplishment and so found favor with the inheritors of the essentially aristocratic "blue-stocking" creed, with its studied cultivation of an unintense intellectualism. On the other, it passed as acceptable in those far more numerous middle-class circles which subscribed to the new cant of sentimentalized womanhood: the "perfect lady" of a repressive Evangelicism.

Even so, women played only a subsidiary role when admitted. The Botanical Society of London began in 1836 with some 10 percent female membership, but the proportion dropped and remained at about 5 percent during the society's twenty-year life. Only one woman ever contributed a paper to the society's meetings, though she did not read the work herself, but enlisted a male member as a surrogate for the occasion. No woman was ever elected to the council, or served as an officer of the society. Women members could vote at meetings, but only (as the rules stated) after "having previously informed the secretary in writing of their appointment of some gentleman, being a Member of the Society, as their proxy for the occasion." Finally, the Botanical Society was, itself, an iconoclastic organization, and the more established scientific institutions continued their total ban. Allen writes:

> The Botanical Society . . . was one of that tangle of minor bodies which had gradually been springing up to cater for the large under-class of the scientifically inclined who, even if they had the intellectual attainments, could on social grounds scarcely hope for election to the major societies. In short, it was an organization of outsiders. And like its fellows, it reflected this in a self-consciously liberal stance that verged even on radicalism.

Women with scientific interests were therefore confined to a narrow range of marginal activities, away from (or, at best, auxiliary to) the centers of prestige and innovation in research and publishing. Women could illustrate works written by men. The plates for John Gould's *Birds of Europe*, second only to Audubon in desirability and cost for modern book collectors, were drawn largely by his wife, who therefore deserves most credit for the work's reputation—a consequence of the figures, not the text. Incidentally, many of the other plates were done by Edward Lear, one of Europe's best scientific illustrators by profession, but better known to us today for his nonsense verse.

Women could be collectors of specimens then turned over to men for formal description and publication. The beginnings of British vertebrate paleontology in the early nineteenth century owe more to the premier collector of this or any other age—Mary Anning of Lyme Regis—than to Buckland, or Conybeare, or Hawkins, or Owen, or any of the men who then wrote about her ichthyosaurs and plesiosaurs. The greatest collector of marine algae, Mrs. A. W. Griffiths of Torquay, was warmly praised by the man who then wrote the most popular work on seaside botany: "She is worth ten thousand other collectors; she is a trump." Charles Kingsley stated that British marine botany would scarcely exist without her, and, in a revealing choice of words, lauded her "masculine powers of research." Yet, as Lynn Barber writes in her excellent account of British popular science, *The Heyday of Natural History, 1820–1870*: "One genus and several species of seaweed were named after her, and she is mentioned with respect—almost with awe—by every Victorian writer on seaweeds, but she published nothing in her own name, and now survives only as an acknowledgment in other people's prefaces."

The most public pathway for women lay in the writing of popular works in natural history—but only of a definite and characteristic genre: the saccharine and sentimental exaltation of nature's objects as illustrations of divine goodness, and as guides to human reverence and proper behavior. Scores of women wrote an astonishing variety of such books, now almost totally forgotten, but then a conspicuous and profitable staple of publishing. These works are often and all too easily dismissed. Even Lynn Barber follows this tradition of modern denigration in a sharp passage within a chapter devoted to praising those of her gender who persevered in science despite the obstacles:

> Unnumbered tribes of Victorian ladies seem to have written without ever doing an iota of research. Ladies were next in line to clergymen as relentless producers of popular natural history books, able at the rustle of a publisher's contract to launch into endless stories about . . . faithful dogs who rescued their masters from everything under the sun, and elephants Who Never Forgot. Most of their effusions were directed to other ladies, or to children (it is often hard to tell which) and characterized by glutinous sentimentality . . . and an ability to drop into verse at the least provocation.

I do not dispute Barber's assessment, but I believe that we must take this genre more seriously for many reasons, ranging from the scholarly (for insights provided into the history of women's social and intellectual struggles) to the ethical (respect due to marginalized people who bridle and suffer under imposed limitations, but who, even on pain of illustrating a stereotype, must find some expression for creative impulses—blacks who preferred opera but entered the only available world of popular music; Jews like my grandfather who wished to be artists but ended up as skilled garment makers). I am happy to note that several scholars, particularly in feminist circles and programs for women's stud-

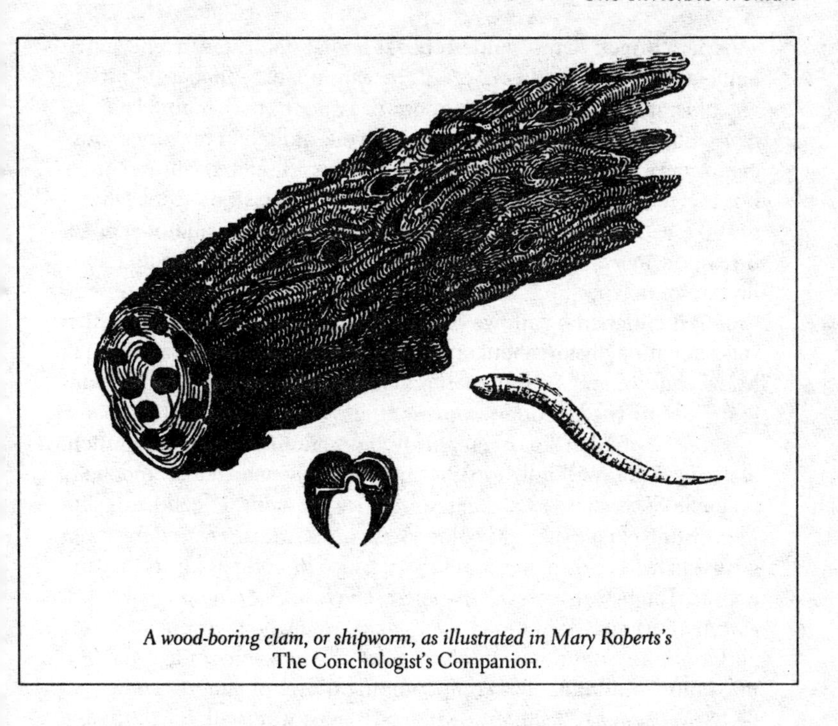

A wood-boring clam, or shipworm, as illustrated in Mary Roberts's The Conchologist's Companion.

ies, are now rediscovering the invisible women who wrote popular books in natural history during the nineteenth century. I do not have the knowledge or experience to make a professional contribution in this specialized area, but I wish to record a personal encounter.

I recently purchased, at the decidedly low-priced end of a catalogue from Britain's major purveyor of antiquarian books in natural history, a copy of a quintessential representative of this genre—*The Conchologist's Companion*, 1834 edition, by one Mary Roberts. I had not heard either of it or her, but I was intrigued because I am a conchologist by specialty (a student of mollusk shells), and because I wanted to learn about this once important but now invisible genre.

As I tried to discover Ms. Roberts, I quickly encountered the bugbear of all scholarly research into the activities of people declared peripheral by society's gatekeepers to intellectual prominence. Such people are nearly invisible today. Nobody wrote about them during their lives, and they never became subjects for later historians. We are often left with the meager evidence of parish birth records, publishers' accounts, and epitaphs.

Mary Roberts wrote a dozen books on natural history, some apparently quite popular in her day, but she wins no more than a column in any biographic source. She was born in London in 1788, the daughter of a Quaker merchant (a common religious affiliation for nineteenth-century women writers). Her family moved to Gloucestershire in 1790, but she returned to London (and left the Quakers) after her father's death. She lived in Brompton Square, London, for the remainder of her life, never married, and died on January 13, 1864. I have been able to find absolutely nothing else about her life. The most common observation, mentioned by all five of our sources, is her frequent confusion "with another Mary Roberts, 1763–1828 . . . who dedicated to Hannah More an ambitious collection of poems." Such are the inevitable fruits of invisibility (and a common name).

The biographical sources also say almost nothing about the content of her work, beyond listing titles—although this evidence is potentially available (if also hard to get, for few libraries maintain collections of popular writing of past ages). My oldest source, *A Critical Dictionary of English Literature*, from 1870, refers to Mary Roberts as "a useful and popular English authoress." My latest, *The Feminist Companion to Literature in English*, from 1990, notes that "even when writing for young children, she carefully names her sources." She wrote some nonscientific books, most notably a compendium on lives of famous women, titled *Select Female Biography* (1821), and an 1823 work with the intriguing title *Sequel to an Unfinished Manuscript of H. Kirke White's to Illustrate the Contrast between the Christian's and the Infidel's Close of Life*. But most prominent are her dozen or so works in popular natural history, including (in chronological order): *The Wonders of the Vegetable Kingdom; Annals of My Village, being a Calendar of Nature for Every Month in the Year; Domesticated Animals considered with reference to Civilization and the Arts; Sister Mary's Tales in Natural History; The Seaside Companion; Wild Animals; Sketches of the Animal and Vegetable Productions of America; Ruins and Old Trees associated with Memorable Events in English History; Flowers of the Matin and Evensong, or Thoughts for those who rise early in prose and poetry; A Popular History of the Mollusca*; and (my favorite title) *Voices from the Woodlands, descriptive of Forest-Trees, Ferns, Mosses and Lichens*.

I cannot deny that *The Conchologist's Companion* is as conventional as conventional could possibly be—as conservative, as mainstream, as establishment, as conformable to all expectations both of general argument and of particular style expected from women authors. (I do sense the contradiction implied by stating that work in a marginalized genre can be mainstream in content, but ideology and respect are different phenomena, and we all know only too well the phenomenon of the slave who lustily mimics his master's voice.)

The conventionality of argument helps us to grasp a superseded conceptual world in which God made all nature in beauty, in harmony, as an illustration of his power and goodness, and to provide the crown of his creation (us, of course) with a bounty of food, fuel, clothing, gems, and building material. A documentation of this system also helps us to understand, in an immediate and visceral way, the depth of the intellectual revolution promulgated by Darwin and his theory of evolution by natural selection.

Nature, to Mary Roberts, is pervaded by divine purpose; every organism works in harmony with all others toward the general good:

> All the various parts of nature are beautifully designed to act in concert. We see the hand of God employed in forming the lowest, and frequently, in our opinion, the most despicable creatures; assigning to each its station, and so admirably adjusting the mighty whole, that every particle of matter, and every living thing that creeps, or moves, upon the surface of the earth, is formed in subserviency to the general good.

God is so attentive that he even created some animals as food for others, being careful to locate them in accessible places. Mollusks exist, in part, to feed higher creatures:

> Some inhabit ditches and stagnant waters, where they afford a constant supply of food to such birds as frequent their banks; others, no doubt with the same benevolent design, incrust marine plants in sandy barren places, near the sea; a large proportion remain concealed in the deep recesses of the ocean, where they furnish food to the finny tribes; others adhere to floating sea-weeds, and abundantly supply the wants of marine birds; and, lastly, exotic snails abound in many uncultivated regions of the globe, where they frequently afford a welcome repast to the fainting traveler.

Roberts illustrates the principle of universal goodness and teleology by the standard device of taking an apparently harmful creature and demonstrating its actual contribution to the general good (and to specific human benefit). Teredos, or wood-boring clams (commonly called shipworms), seem to be noxious in their destruction of ships, piers, and pilings. But look again to a wider beneficence. First of all, they bore very carefully, in a mode divinely calculated to cause minimal harm. I confess that for all my vaunted desire to approach the past on its own terms, without ridicule born of current knowledge, the following passage did make me howl: "But mark the protecting care of Providence. The destructive operations of these insidious animals are in a great degree obviated, by the singular fact of their generally perforating the wood in the direction of the grain."

Moreover, by reducing logs and clumps of vegetation to sawdust, teredos prevent the clogging of rivers and flooding of land. Finally (and here I howled even more), teredos "open a source of considerable riches to the inhabitants of Sweden" by "employing the vigilance of the Dutch." For the teredos, you see, force the Dutch to keep their dikes and ships in continual repair, thus requiring "a perpetual demand for oak, pitch, and fir," largely supplied by Sweden—and so "these apparently pernicious insects are continually at work at Amsterdam, for the advantage of Stockholm." Roberts therefore concludes:

> Cease then, my friend, to regard this creature as decidedly obnoxious . . . The Creator has assigned it an important station among his works. The evil which it produces, is readily obviated by a little care and contrivance; but the good which it is appointed to effect, is incalculably great in the mighty scale of universal nature.

All aspects of nature illustrate and glorify God, even those features that seem to contradict each other. We may think that God made the pretty colors of shells for His (or our) delight, but they actually produce cryptic patterns that camouflage mollusks from their enemies:

> But why, illustrious naturalist, did your observations extend no further? Saw you nothing in these . . . shells, but an arrangement of colors to please the eye . . . Saw you not, that the Almighty Creator of the universe, without whose permission a single hair does not fall from our heads, nor a sparrow from heaven to the ground, nor a shell, nor a pebble, is tossed with the billows on the shore, by investing them in these simple colors . . . provides against their utter extinction, by the depredations of sea-birds and rapacious fishes.

Seashells illustrated in The Conchologist's Companion.

But in other passages we learn that God did make colors only for beauty, and that, as such, they stand in contrast to the utility of shell shapes and sizes. Roberts argues for adaptation in the shape and ornament of the pholad (rock-boring clam) shell, but for decoration alone in the color: "An ovate or oblong form is consequently the very best that could be adopted; and, moreover, the points with which it is covered and adorned, are evidently designed to protect the shell from external injury . . . At the same time a beautiful variety of tints evince that minute attention to the finishing and decorating of his works which the Deity so continually displays."

Finally, nature is not only well designed and full of beauty, but also replete with moral messages for human betterment. The metamorphosis of butterflies symbolizes the liberation of the soul from an earthly body, while the unfolding of a flower represents our hope for mental enlightenment: "The botanist confesses, in the unfolding of the calyx . . . an attractive emblem of the expanding of the human mind, as it emerges from a state of ignorance . . . or in the gradual development of a plant, the progressive advancement of every moral excellence."

Roberts's conventionality of argument fully matches her faithful adherence to the style of presentation expected from women authors in this genre. When, for example, she discusses the utility of mollusks in human life, she emphasizes supposedly "feminine" themes of adornment, rather than "masculine" motifs of immediate sustenance. Her longest chapters treat pearls and purple dyes (classically extracted from snails), but she barely mentions the fact that many people eat clams and snails.

More important, and now getting closer to the bone of our urge to render judgment, even across centuries, Roberts invokes the putative hierarchy and stability of nature to argue that each of us must accept our appointed role in human society, even if we be placed at severe disadvantage by accident of birth as a female or as a member of the working classes. Our ultimate reward, after all, will come in another and better world. Just as God has designed each species of mollusk for its appropriate environment, he has "assigned to every individual being, his respective sphere of action; and happy will it be for us, if we as steadily perform our portion of allotted duty, as these feeble creatures fulfil the purposes for which they are designed, in accordance with their respective instincts." Later, in the book's most uncomfortable passage (for me at least), Mary Roberts echoes Alexander Pope in the *Essay on Man*, by arguing that any change in appointed ranks will cause the exquisitely balanced apparatus of nature to tumble:

> To this splendid superstructure, nothing can be added; neither can any thing be taken from it, without producing a chasm in creation, which, however imperceptible to us, would materially affect the general harmony of nature. All things were made by Him, and without him cannot any thing subsist; besides, it seems as if he designed to teach us by the admirable arrangement of his creatures, that the different gradations in society are designed by his providence, and appointed for our good.

These overtly sexist and politically conservative themes lie exposed at the surface of *The Conchologist's Companion*. But I was struck even more by the pervasive "deep sexism" of Roberts's obedience to what her society viewed (and we, in large and unfortunate measure, continue to regard) as contrasting ideals of the abstract and eternal masculine and feminine—the key, I believe, to a true understanding of *The Conchologist's Companion* and other works of this once influential genre. I suspect that we can best grasp this vital, but largely covert, theme by returning to one of the most important essays in English letters—a document that many of us read (at least in excerpt) in Philosophy I, but have probably never thought about since: Edmund Burke's *Philosophical Inquiry into the Origin of our Ideas of the Sublime and Beautiful*,[1] first published in 1756.

Burke argues that our aesthetic senses are moved by two separate configurations, which he calls the sublime and the beautiful. These are truly distinct, for one is not the negation or the reciprocal of the other. The sublime (which Burke also calls the "great") is based on our instinct for self-preservation and founded in terror. Its themes include vastness, darkness, verticality, massiveness, roughness, infinity, solidity, and mystery. The beautiful, on the other hand, is rooted in pleasure and linked to our instinct for generation (necessary for the preservation of our race, but not so elemental as self-preservation and the sublime). The themes of beauty encompass smallness, smoothness, variety in shape (but only by rounded rather than angular transitions), delicacy, transparency, lack of ambiguity, weakness, and bright colors.

Burke does not dwell on the correlations of these themes with conventionally sexist views of the masculine (sublime) and feminine (beautiful), but this contrast sets the basis of "deep sexism." Burke argues, for example, that women instinctively recognize the necessary link between beauty and weakness: "Women are very sensible of this; for which reason, they learn to lisp, to totter in their walk, to counterfeit weakness, and even sickness. In all they are guided by nature." He also attributes the supposed grandness of male thought, and the timidity of female men-

1. A personal footnote: People often ask how or why I keep writing these monthly essays after more than twenty years without a rest. The answer is so simple. Burke's *Essay* is one of the greatest documents ever written in our language—for all that I find repulsive therein (amid other insights that are truly wonderful). My volume of Burke has always stood accessible on my bookshelf—as part of volume 24 in my set of Harvard Classics. My grandfather bought this set in the 1920s, when he decided that he had to provide, for himself and his children, an opportunity for the education that he had never received. I doubt that he ever consulted more than a volume or two. My mother read and reread the volume on Aesop, Grimm, and Andersen during her childhood. I shot arrows at the set during my youth—and my volume 24 is transfixed with nine holes marking my successes. I always liked this particular volume because the spine says: "On the sublime French revolution." I never would have so characterized this episode in history, so the title always puzzled me. I don't know when I ever thought of actually opening the book, but I do remember my delight in discovering that the title mixed two separate works—Burke's essay on the sublime and his reflections on the revolution!

These generational and ontogenetic continuities give meaning and structure to our lives. Then I shot arrows; now I want to read. I could have pulled this book off the shelf and read Burke's essay anytime—but I never did and, to be honest, I doubt that I ever would have done so amid all the crazy and maniacal crowded business of our lives. My decision to write this essay on Victorian natural history brought me to Burke, and I had the wonderful privilege of reading this great and influential document for pleasure, and with some maturity of judgment, rather than in the rush of a course and for a grade at too young an age. For this privilege I am most profoundly grateful. If I didn't write these monthly essays, Burke would probably have stayed on my shelf until the day I died.

tality, to this distinction: "The sublime . . . always dwells on great objects, and terrible; [beauty] on small ones, and pleasing . . . The beauty of women is considerably owing to their weakness or delicacy, and is even enhanced by their timidity, a quality of mind analogous to it."

I submit that we cannot grasp the essence of Roberts's book, and the genre she represents, until we assimilate this classic distinction of the sublime and the beautiful. We must, above all, recognize that Roberts and her colleagues of like gender accepted this distinction and sought to be completely beautiful and not at all sublime—that is, essentially feminine by their lights. (I also urge readers to acknowledge this distinction as both productive and supportive of the worst aspects of sexism—and to remember that liberation presupposes knowledge of the causes of oppression.)

Burke's criteria of beauty supply a key that opens *The Conchologist's Companion* to our understanding (rather than our ridicule based largely on puzzlement). The conceptual themes are all present: conventionality, timidity, boundedness, lack of surprise, rounded transitions. Even the physical appearances proclaim beauty rather than sublimity. These books by women tended to be small in size—printed at dimensions that publishers call "duodecimo" or "small octavo," rather than the "large octavo" or "quarto" favored for books by male authors. Type sizes are generally small, and the engraved illustrations particularly delicate (see examples). The prose itself proclaims saccharine sentimentality, rather than raw power—particularly in the dum-de-dum of doggerel verses:

> Oh! who that has an eye to see,—
> A heart to feel,—a tongue to bless,
> Can ever undelighted be
> By Nature's magic loveliness!

The choice of subject matter—small and humble mollusks—matches the attributes of a female writer, as Roberts mentions again and again:

> In the prouder forms of animated being, in the towering cedar, or the turret-bearing elephant, nature appears to act in a manner analogous to the grandeur of her designs; whereas these feeble creatures are often passed by, as undeserving the attention of the naturalist; and yet what tokens of beneficence and power, what exquisite perfection is discoverable!

Even the words of praise used by men in their favorable reviews of women's work invoked the standards of circumscribed beauty, rather than awesome sublimity. *The Athenaeum* (journal of the intellectual men's club that included Darwin and Huxley as members) praised Mary Roberts's second book on conchology as "a useful and entertaining volume" (issue of November 22, 1851, page 1224).

I was going to end this essay here, with some words of exculpation for Ms. Roberts, acknowledging her utter submission to conventional expectations, but refusing to judge her too harshly—for the urge to create can be so overpowering, and the pain of self-imposed silence so overwhelming, that we sometimes kowtow to the most iniquitous of limitations. (I dare not, as a white man, criticize Stepin Fetchit or Mantan Moreland for playing the only—albeit degrading—roles that Hollywood then allowed black actors. And I will not castigate any woman who needed to write, but could be published only by adhering to standards of Burkean beauty.)

But nothing in our complex world ever ends so cleanly. As I reread and considered, I began to see more in Ms. Roberts. I began to pick up an undertone of rebellion—muted to be sure, but unmistakably present. I began to realize that Mary Roberts had not totally internalized the norms imposed upon her, and that some flicker of feminine anger smoldered on the pages of her small and beautiful text.

One passage particularly caught my attention. Ms. Roberts often cites the conventional theme that nature will always hide secrets from human probing, and that we should not be too arrogant in our claims for understanding. She usually follows the standard line of attributing this prevailing mystery to masculine power—that is, to God's omniscience as creator of all nature, versus our mental midgetry by comparison. (She writes, for example, "In many instances we are unable to comprehend the intentions of the Deity with regard to the construction of his creatures.") But in one striking passage she identifies the cause of obstruction as *feminine* nature—and she explicitly contrasts the necessary victory of this female power with the conventional image (used explicitly by Bacon and many later writers) of science as male and active, seizing (almost raping) knowledge from passive and feminine nature:

> It seems as if maternal nature delighted to baffle the wisdom of her sons; and to say to the proud assertors of the sufficiency of human reason for comprehending the mysteries of creation and of Providence, "Thus far can you go, and no further"; even

> in the formation of a shell, or its insignificant inhabitant, your arrogant pretensions are completely humbled.

I knew that I had to probe further into Mary Roberts's hidden motives and feelings. So I located her paleontological book, *The Progress of Creation* (1846 edition), in the stacks of Widener Library (actually not in the stacks, but further into purgatory, at the book depository for rarely used volumes, and requiring a two-day wait for delivery—another sign of invisibility for this genre).

I was struck by an immense difference in style within a basic similarity of drearily conservative content. *The Conchologist's Companion* may have made me sad for bowing to limitations of a genre, but *The Progress of Creation* made me mad for its ignorant pugnaciousness in clinging to clearly disproven religious formulations. In her geological book, Mary Roberts takes a hard creationist line in insisting upon biblical literalism—with creation in six days of twenty-four hours, and a total age of only a few thousand years for the earth. She states in no uncertain terms, "Throughout this volume, I have ever kept in view, that the heavens, and earth were finished, and all the host of them in six days; and that no theory, however plausible, can be admitted in opposition to the Divine Record."

She uses the writings of catastrophist geologists to assert the reality of Noah's Flood and to argue that this deluge created the entire geological record in one grand gulp. But her argument is either ignorant or disingenuous. By 1846, all serious catastrophists, including all the men falsely quoted by Mary Roberts, knew that any flood within human history could only represent the latest catastrophe in a long series of earlier crises stretching throughout an immensity of geological time.

Yet Mary Roberts wades in, dukes held high and swinging at all the greatest male scientists of Europe. She thinks that mastodons are carnivorous and lambasts Georges Cuvier, the Newton of natural history, for his (evidently correct) belief that these elephants ate plants: "The carnivorous elephant, or Mastodon of the Ohio, is one of the most remarkable. Cuvier describes this animal as herbivorous, but surely without reason." This is absolute nonsense from Ms. Roberts's pen—but it is sublime nonsense!

What was Mary Roberts really thinking when she wrote *The Conchologist's Companion* by the rules of sweet, delicate, inoffensive prose—when we know that she could also write in pugnacious sublimity? Were her personal views on obedience and natural harmony as conventional as those she presented? Did she accept the limitations placed upon women, or did she seethe inside, yet keep her own counsel? I doubt that the records exist to answer such questions—for a conventional history told largely by males did succeed in rendering her nearly invisible.

I wonder. What really went on behind the mask of acceptance and convention respected by most women writers on natural history? Perhaps we should invoke one of the great women of strength from our literature—Little Buttercup of Gilbert and Sullivan's *H.M.S. Pinafore*. She tries to tell the captain that "things are seldom what they seem"—perhaps also that women often hide the pain and anger of ages under a soft surface of acceptance:

> Though to catch my drift he's striving,
> I'll dissemble—I'll dissemble;
> When he sees at what I'm driving,
> Let him tremble—let him tremble!

16

Left Snails and Right Minds

WHAT IMMORTAL HAND or eye
Could frame thy fearful symmetry?

William Blake's familiar inquiry about the creation of tigers raises a vital question that we may pose literally, though the poet's intention may have been more metaphorical: why does symmetry, particularly our own bilateral style of mirror images around a central axis, predominate among animals of complex anatomical design? Why do we come in equivalent right and left halves? And why do we get so fascinated by the minor departures, usually more of function than overt form, that loom so large in our culture: the predominance of righthandedness and the difference between "right" and "left" brains?

A few major groups of organisms do not present a basically bilateral symmetry, including my own favorite subject for research, the gastropods, or snails. The soft body of a snail is tolerably bilateral when pulled from the shell and stretched out, but the animal houses this body in a shell built by winding a tube in one direction around an axis of coiling. The snail shell may therefore be the most familiar nonbilateral form among so-called higher animals.

A tube can be wound around a vertical axis in either of two directions, designated as right and left handed. If we hold a snail in our conventional position, with the apex at the top and the aperture (or opening for the body) at the bottom, then we call the direction of coiling "right-handed" if the aperture lies to the right of the axis of coiling when we view the specimen face to face, and "left-handed" if the aperture lies to the left of the axis of coiling. (All this should be much clearer in the accompanying illustration than in any words I can supply. Incidentally, we use the same convention for the threads of screws.)

This designation is truly arbitrary, for snails know nothing about apex up and aperture down (in life, most snails carry their shells more or less horizontal to the ground). If we draw the specimen apex down (as French traditions of scientific illustration have always done), then the apertures of "right-handed" specimens open to the left of the axis of coiling.

In India, for example, the conch shell *Turbinella pyrum* is venerated as a symbol of Vishnu. (In the *Bhagavad Gita,* Vishnu, in the form of his most celebrated avatar Krishna, blows his sacred conch shell to call the army of Arjuna into battle.) The exceedingly rare left-handed specimens of this shell are particularly treasured, and used to sell for their weight in gold. But Hindus interpret the apex as the bottom of the shell and therefore call this rare form "right-handed." Perhaps they treasure these rare shells because only these specimens, in the Indian version of an arbitrary decision, match the style of dominant handedness in human beings (and, I suppose, in anthropomorphic deities).

A purist might forgive snails for departure from the bilateral paradigm if only they honored an even more inclusive symmetry by growing right and left-handed spirals in equal numbers. But snails remain twisted and awry on this criterion as well—for right-handed shells vastly outnumber lefties, not only in the sacred conch of India, but in virtually all species and groups. Right-handed shells are called "dextral," from the Latin *dexter,* meaning "right," and memorialized in our language by a host of prejudicial terms invented by the right-handed majority to honor their predominance. Right is dextrous, not to mention "correct" in many languages—awright buddy. The law, by the way, is *droit* in French and *Recht* in German, both meaning "right." (The language police will never regulate these essays, but we may still note, in fairness and for historical interest, that the "rights of man," noble as the sentiments may be, embody two linguistic prejudices of unfairly dominant groups.) Left-handed shells are called "sinistral," from the Latin *sinister,* meaning "left"—also denigrated in our languages as "sinister" or *gauche,* a French lefty. I shall, for the rest of the essay, use this terminology by calling right-handed shells dextral, and lefties sinistral. I also can't help wondering if we didn't make our initial arbitrary decision to

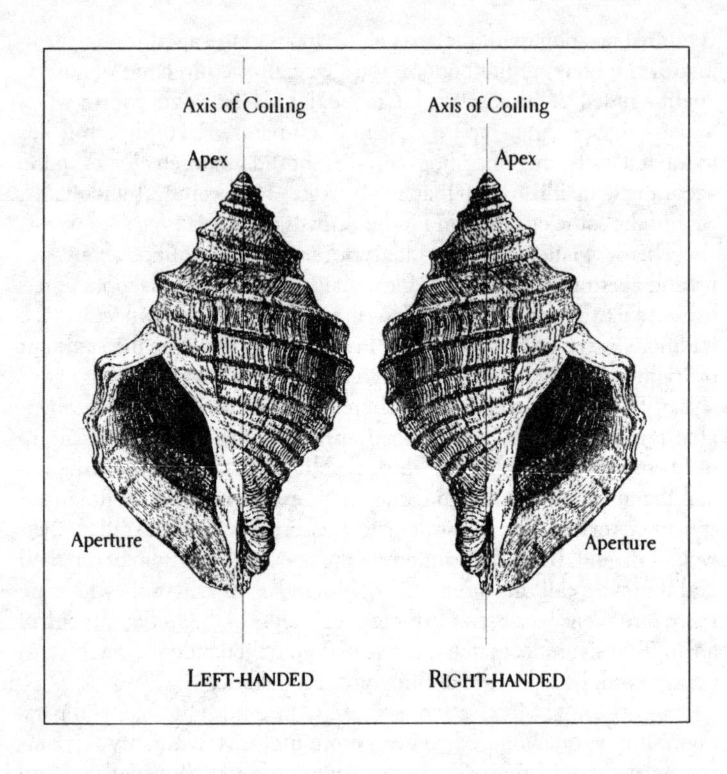

call a snail's apex "up" because this orientation would then allow us to designate the vastly more common direction of coiling as "right."

The vast majority of forms grow a dextral shell, although a few sinistral specimens have been found in most species. For example, in *Cerion*, the West Indian land snail that forms the subject of my own technical research, only six sinistral specimens have ever been found, out of millions examined (while, as stated above, lefty *Turbinellas* in India were literally worth their weight in gold). A few species grow exclusively or predominantly sinistral shells, but related species of the same group are usually dextral. We often exact a price from these rare sinistrals by giving them names to match their apostasy—as in *Busycon contrarium* or *Busycon perversum*, the technical monikers variously awarded to the most common sinistral species of northern Atlantic waters. A few groups of species (notably the family Clausiliidae) are predominantly sinistral, but, again, all closely related lineages are dextral. In short, dextral snails greatly predominate (at a far higher frequency than human righties vs. lefties), and at all levels: individuals within a species, species within a lineage, and lineages within larger groups.

At this point, any astute and inquisitive reader will be asking the obvious question, "Why? What conceivable advantage does dextrality hold over coiling in the other direction?" I can only report that this inquiry is both appropriate and fascinating—and that we don't have a clue about the answer. (I would not even assume that the question should be posed in terms of putative advantages. The two modes might be entirely equivalent in functional terms, with dominant dextrality only a historical legacy of what happened to arise first.) I'm sorry to wimp out on such an interesting question, but I can at least quote, on the same subject and to the same point, that greatest of all prose stylists in natural history, D'Arcy Wentworth Thompson (from his book *Growth and Form*, first published in 1917, and still vigorously in print): "But why, in the general run of shells, all the world over, in the past and in the present, one direction of twist is so overwhelmingly commoner than the other, no man knows."

This essay, instead, shall take another turning on the subject of directionality in coiling—namely, the history of illustrations for snail shells in zoological treatises. Let me begin with a figure that I first considered both anomalous and amusingly in error. The figure following reproduces a plate from a famous work in natural history, published in 1681 by one of Britain's finest physicians and zoologists, Nehemiah Grew: *Musaeum Regalis Societatis, or a description of the natural and artificial rarities belonging to the Royal Society, whereunto is subjoyned the comparative anatomy of stomachs and guts.* (They did love long titles back then, and we will ignore the appendix, with its remarkable illustrations of vertebrate intestines, all stretched out and circling the pages.)

Note that all but one of these shells are sinistral in Grew's engraving. The exception, shown at bottom left, is conventionally dextral. Has the world turned? Those shells are labeled "wilk," or "whelk" in our modern spelling (a common name for conchlike shells)—and nearly all whelks are dextral, including the species shown here. The exception, drawn dextrally, gives the story away by the name imprinted above: "inverted wilk snail." In other words, the shell labeled "inverted" is, in life, a rare sinistral named according to an old tradition for derogatory designation of the unusual.

Obviously, Dr. Grew has printed his snails in mirror image from their actual constitution. I initially assumed that Grew had committed a simple error and laughed at his fellowship with snail men throughout the history of illustration, for we are still making the same mistake today.

In the current version, an offspring of modern technology, a snail may appear with reversed coiling because the photograph has been made from a negative inadvertently turned over before printing. Any expert paying explicit attention will notice the error, but we fallible mortals often let something this global slip by—for a reversed snail doesn't look grievously wrong if you don't have your eye and mind directly attuned to the issue of symmetry.

Any professional snail man can give you his list of embarrassments in this category. A dear late colleague, one of the world's leading experts on snails, published a beautiful wraparound dust-jacket photo of reversed shells for the cover of his popular book. I must also admit—and how wonderfully unburdening after all these years of hiding such a shameful secret—that my own first publication on snails included several photographs of a newly discovered protoconch (embryonic shell) of an important genus—all published from reversed negatives. (I received the sweetest and most diplomatic letter from a colleague asking me if these dextral shells really had sinistral protoconchs, and urging me to publish separately on such an important finding, or suggesting that maybe, just maybe, I had made the old error of reversed printing.) Baseball players make a proper distinction between physical errors, which can happen to anyone at any time and should engender no shame, and mental

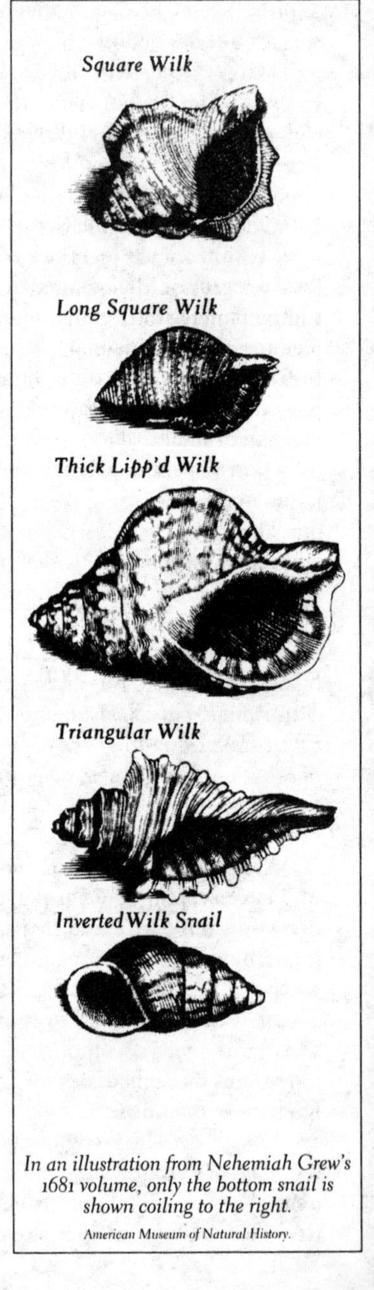

In an illustration from Nehemiah Grew's 1681 volume, only the bottom snail is shown coiling to the right.

American Museum of Natural History.

errors—bonehead judgments, forgetting the rules—which should never occur. Ordinary and honorable errors of fact are unavoidable in science, a field that thrives on self-correction, and properly defines its own progress by such improvement. I have never written an essay, and never will, without this analogue of a physical error. But printing a snail backwards is a mental error. No excuses possible.

So much for my first thoughts about Dr. Grew's mistake. But as soon as I remembered a scholar's first obligation—to drag oneself from judgment within a smug present, considered better, and to place oneself, so far as possible, into the life and times of a person under consideration—I immediately realized that the resolution could not be so simple. All media for printed illustration in pre-nineteenth-century treatises of natural history—woodblock printing, engraving on metal plates, lithography—require the initial production of an inverted image. That is, the engraver must carve a mirror-image figure into his metal plate so that the paper, placed atop the inked plate before pressing down to print, will receive the figure in proper orientation. Needless to say, all printers know this rule to the core of their being; nothing could be more fundamental to their work.

Therefore, printers who want to engrave an ordinary dextral shell must carve a sinistral image upon their metal plate. Clearly, Dr. Grew's printer drew the snails onto his plate as he saw them, rather than reversed—and the result is an inverted image in the resulting book: ordinary dextral snails look sinistral, and the lone sinistral appears as a conventional dextral.

But how did this happen, and why? This oddity cannot be the result of a simple fool's error of the baldest kind, for the engraver surely knew his rules, and must have etched his letters and numbers in proper reversed order onto his plate, for all writing and numeration is correct in the printed version. Many scenarios suggest themselves, and we do not have enough evidence to decide. Perhaps Grew supplied his printer with sketches already reversed, but forgot to pass this information along; perhaps Grew provided all the sketches on a single sheet (without the words), and the printer then erred in pasting the sketch upon his plate *recto* rather than the proper *verso* (I am conjecturing that engravers sometimes worked by affixing a sketch, drawn on transparent paper, directly onto their plate and then carving through).

But we should also consider a hypothesis of a fundamentally different kind. Perhaps we should not be so quick to assume, from our arrogant present, that these "primitives" of the seventeenth century must have been making an error at par with boo-boos still occasionally committed by flopping a photograph. Perhaps the reversed shells of Dr. Grew's illustrations are not errors at all, but representations of a convention then followed and now abandoned.

I shall defend this more generous alternative in concluding my essay, but I had not considered this solution when I first saw Grew's plate, about ten years ago. I simply stored this little "fact" away in my mental file of oddities in natural history. I must have "labeled" the item "Grew's funny mistake," for I never considered the possibility that reversed snails could be anything but an error, however committed.

As their primary virtue and utility, such mental files can lurk in the brain (wherever and however this remarkable organ stores such information) without disturbing one's thinking and planning in any manner. The files just hang around, bothering nothing while they wait for some trigger to transport them into consciousness. (I would, for this reason, defend such ancient practices as rote learning for the basic chronology of human history, and reading the classics, particularly Shakespeare and the Bible, with a view to memorizing key passages.) I love antiquarian books in natural history, and my eyes do inevitably wander, for professional reasons, to pictures of snails. Thus, my "Grew mistake" file has been accessed quite a few times during the past decade. But I never had any project in mind, and I had devised the wrong preliminary conclusion about Grew's reversals. In fact, I needed three or four random repetitions to make the subject explicit as a worthy topic, to force a revision of my own initial error, and to perceive the larger theme about science and human perception that could convert such a trivium (the depiction of snail coiling) into a decent subject for an essay.

A few years later, I purchased a copy of my personal favorite among beautiful and important works in natural history, Michele Mercati's *Metallotheca*. Mercati (1541–1593), director of the Vatican botanical garden, also became curator of the papal collection of minerals and fossils organized under the aegis of the imperial Pope Sixtus V, whose taxes impoverished the papal lands while building Rome in splendor. (I also love the man's name—the fifth instar of a guy named "six"; Sixtus I, a second-century figure, was the sixth bishop of Rome after Peter, and took his name accordingly.) Mercati prepared a series of gorgeous engravings for a catalogue of the Vatican collection, but this work never appeared in his lifetime (perhaps because Sixtus V died unexpectedly in 1590). But the plates hung around in the Vatican's vast storehouses for nearly a century and a half, until J. M. Lancisi finally published them, along with Mercati's text and many new engravings, in 1719 as the *Metallotheca*.

(If a *bibliothèque* is a library, then a *metallothèque* is a collection of metals and other objects of the mineral kingdom.)

The *Metallotheca* contains numerous plates of fossil snails in a chapter called *Lapides Idiomorphoi* (or stones that look like living things—Mercati, along with many sixteenth-century scholars, did not interpret fossils as remains of organisms, but as manifestations of "plastic forces" inherent in rocks). In all plates—so we are in the presence of a conscious generality, not an individual error—dextral snails appear as sinistral engravings (see the accompanying illustration).

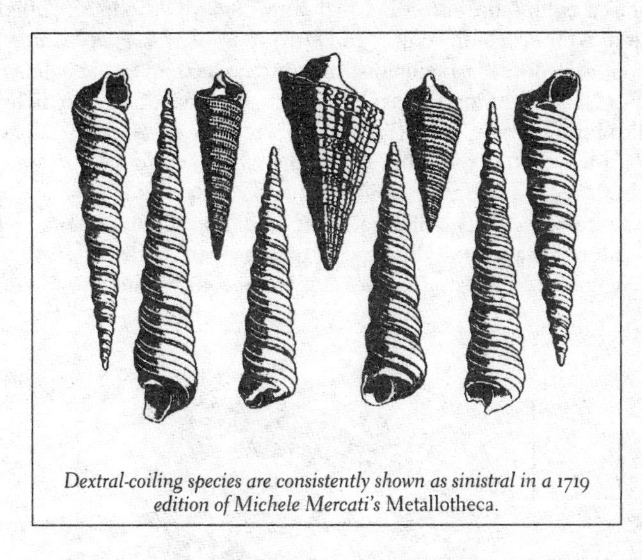

Dextral-coiling species are consistently shown as sinistral in a 1719 edition of Michele Mercati's Metallotheca.

But assumptions die hard, even if never founded on anything sensible. I couldn't call reversed printing a simple error anymore, so I opted for the next line of defense within the bias of progress: I assumed that such indifference to nature's factuality must represent a curious archaicism of the bad old days (for Mercati goes way back to the sixteenth century)—and thus not worthy of much intellectual attention. Again, I stored the observation on the back shelves, in the stacks of my *mentatheca*.

More random encounters since then have finally destroyed my false assumption, for I have noted sinistral illustrations of dextral shells again and again in works published before 1700. In fact, almost all snail illustrations from this period are reversed, so we must be noting a general convention, not an occasional error. By contrast, I have almost never seen a reversed illustration in works, say, from Linnaeus's time (early to mid-eighteenth century) onward, except as real and infrequent errors. Therefore, and interestingly, the obvious hypothesis that photography ushered in the change must be false. I simply do not know (but would dearly love to have the answer) why a convention of drawing snails in reversed coiling yielded to the conviction that we should depict them as we see them.

To shorten my chronicles of personal discovery, two further examples finally convinced me that older illustrations had drawn snails with reversed coiling on purpose. I first consulted as close to an "official" source as the sixteenth century can provide—the *Musaeum Metallicum* (another account of a major fossil and rock collection) by the Italian naturalist Ulisse Aldrovandi (1522–1605), who, in competition with his Swiss colleague Conrad Gesner (1516–1565), wrote the great compendia that pulled together all available knowledge about animals—ancient and modern, story and observation, myth and reality, human use and natural occurrences. My edition of Aldrovandi's posthumous work on fossils dates from 1648 and illustrates all snails as sinistrally coiled, though the figures depict dextral species (see the accompanying illustration).

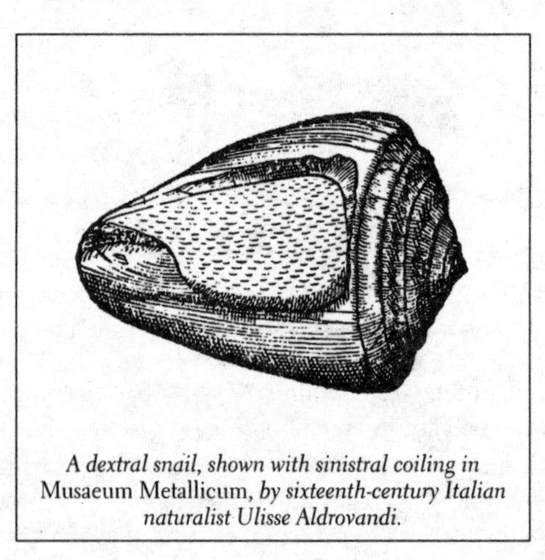

A dextral snail, shown with sinistral coiling in Musaeum Metallicum, *by sixteenth-century Italian naturalist Ulisse Aldrovandi.*

If the standard source still doesn't completely convince, then seek an author with special expertise. I therefore consulted my copy of one of the great works in late-seventeenth-century paleontology, *De corporibus marinis lapidescentibus* (*On Petrified Marine Bodies*) by Augostino Scilla (my Latin edition dates from 1747, but Scilla first published his work in Italian in the 1670s). I decided on Scilla as a final test case because he was a painter by trade, a leading figure of the *seicento* in Sicily, and he engraved his own plates. All his snails are dextral species, and all are engraved with sinistral coiling. Clearly, if standard sources and noted artists all drew snails in mirror image from their natural occurrence, authors and illustrators must have been following a well-accepted convention of the time, not making an error.

But why would earlier centuries have adopted a convention so foreign to our own practices? Why would these older illustrators have chosen to depict specimens in mirror image, when they surely knew the natural appearance of shells? Did they devise this convention in order to make life easier for a profession founded on the principle that one carves in reverse in order to print in the desired orientation? But if so, what aid could be provided by the practice of printing snails in reverse? I suppose that an engraver could then paste a picture directly on his plate, and cut through with maximal visibility, whereas the usual technique forced him to invert the drawing before affixing it to the plate, thus making him view the sketch through the backside of the paper (but papers of adequate transparency must have been available, and I wonder if the usual technique really imposed any great hardship). Or did engravers mechanically copy an original figure in reverse orientation and then paste this copy onto the plate? If so, a convention permitting reversed printing would allow engravers to omit a time-consuming step. Or did illustrators project the image directly onto the plate and then draw? But then one more mirror would project a reversed image—and printers usually carved reverse images in any case.

Whatever the reason, the very existence of the convention does, I think, teach us something important: that the conceptual world of pre-eighteenth-century zoology must have accorded little importance to the orientation of a shell. These men were not stupid, and they were not primitive. If they were willing to sacrifice what we would call "accuracy" for some gain in ease of production (or for some other reason not now apparent to us), then they must have held a notion of "accuracy" quite different from ours. The recovery of "fossil" thought patterns from such intriguing hints as this small, but previously unnoted, change in a practice of illustration provides the kind of intellectual lift that keeps scholars going.

The greatest impediment to such recovery—one that infested my own first thoughts on this issue, and precluded any movement toward a proper solution after I had made my initial and accurate observations—lies in lamentable habits imposed by the twinned biases of

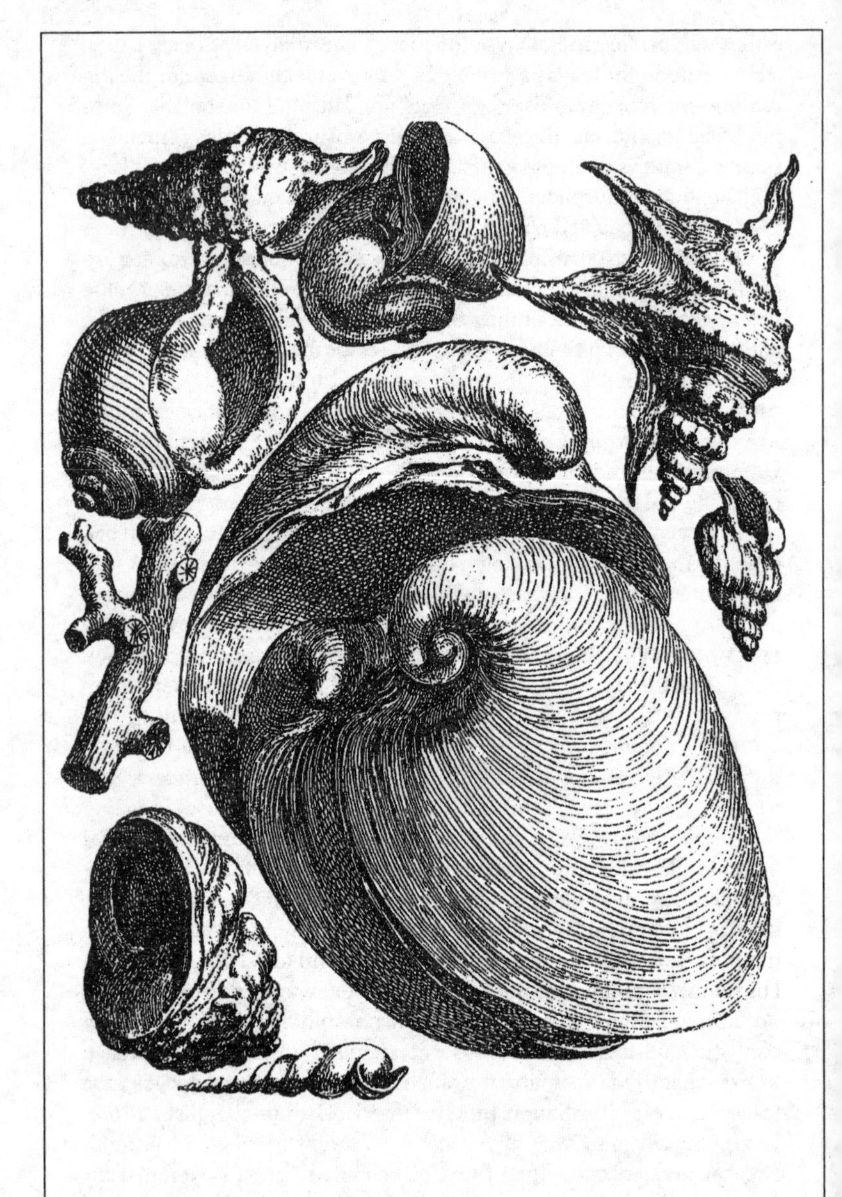

Dextral snail species are depicted with left-handed coiling in a volume on fossil snails by seventeenth-century paleontologist Augostino Scilla.

progress and objectivity. We assume that we now do things better than at any time in the past, and that our improvements record increasing objectivity in shedding old prejudices and learning to view the world more accurately. We therefore interpret our predecessors, especially when their views differ from ours, as weighted down by biases and lacking in data—in short, as pretty darned incompetent compared with us. We therefore do not take them seriously, and we view their differences from us as crudity and error. Thus we cannot understand the interesting reasons for historical changes in practice, and we cannot recover the older systems, coherent in their own terms (and often based on a fascinatingly different philosophy of nature), that made the earlier procedures so reasonable.

The key, in this case, lies in realizing that an apparent error in past practice represents a convention, now foreign to our concepts but evidently pursued for conscious reasons by our predecessors. We must still overcome one obstacle in striving to view the past more sympathetically (thereby gaining insight into present styles of thinking). We might understand that printing snails in reverse represented a convention, not an error, but still hold (via the bias of progress) that the history of changing conventions must record a pathway to greater accuracy in representation. We might, for example, hold that our predecessors once drew what they wanted to see, whereas we now photograph what actually is.

Two arguments should convince us that history marks no path from stilted convention to raw accuracy. First, I have talked with many professional photographers, and all recognize as a canard the old claim that their technology gave us objective precision, where only subjective drawing reigned before. Technological improvements in photography do make older styles of prevarication less possible. (In my book *The Mismeasure of Man*, I showed how one pioneer eugenicist doctored his pictures of supposedly retarded people to make them look more benighted. His retouchings are so crude that no one today, with a lifetime of experience in looking at good photographs, would be fooled. But he got away with his ruse in 1912, for few people then had enough experience to recognize a doctored photo, and retouching represented an accepted art for repairing crude shots in any case.) But other technological improvements make all manner of fooling around with photos ever more possible and elaborate (just think of Woody Allen as Zelig, or Tom Hanks as Forrest Gump, artificially incorporated into the great events of twentieth-century history by trick photography). Who can balance the gains and losses? Why speak of these changes as gains and losses at all?

We have not dispensed with conventions for accuracy; we have only adopted different conventions.

Second—and the clinching argument that made me decide to write this essay—we have not, even today, abandoned all conventions for reversed illustration. In fact, one highly prestigious and technologically "cutting edge" field continues to present upside-down photographs, just as our forebears drew their snails right to left. How many readers realize that conventional photos of moons and planets are upside down? (If you doubt my claim, compare the full moon on a clear night with the photograph in your old astronomy text.) Modern astronomers, of course, are no more fools than the old snail illustrators. They present photographs upside down to match what one sees in a conventional refracting telescope. (Or, rather, they print the photos as they are taken through such telescopes. Is this convention any different from carving the snail as one sees it onto an engraving plate and then producing the paper image in reverse?)

Clearly, astronomers feel that the trouble taken to print photographs from refracting telescopes upside down (thus rendering the object as it appears in the sky) would not be worth the putative gain. In fact, one might argue that reversing the photo would sow confusion rather than provide benefit—for (with the exception of our moon) we cannot see features of other satellites and planets with our naked eye, and we therefore know these bodies primarily as viewed through refracting telescopes—that is, upside down. I must suppose that the old snail illustrators also regarded the direction of coil as unimportant for illustration—and I would like to know why. I would also like to know what triggered the change from an accepted convention to a no-no.

I shall not, either in this forum or anywhere, resolve the age-old riddle of epistemology: How can we "know" the "realities" of nature? I will, rather, simply end by restating a point well recognized by philosophers and self-critical scientists, but all too often disregarded at our peril. Science does progress toward more adequate understanding of the empirical world, but no pristine, objective reality lies "out there" for us to capture as our technologies improve and our concepts mature. The human mind is both an amazing instrument and a fierce impediment—and the mind must be interposed between observation and understanding. Thus we will always "see" with the aid (or detriment) of conventions. All observation is a partnership between mind and nature, and all good partnerships require compromise. The mind, we trust, will be constrained by a genuine external reality; this reality, in turn, must be conveyed to the brain by our equally imperfect senses, all jury-rigged and cobbled together by that maddeningly complex process known as evolution.

EPILOGUE

With 230 essays as a large sample, I know which pieces bring the most frequent and most impassioned responses from readers—and the answer fills me with admiration for human foibles and human good sense. I receive very little correspondence about essays treating the broadest and most troubling issues—the meaning of evolution for human life, the relationship (if any) of science and moral values. But when I write about tiny little puzzles with intriguing subjects and potentially ascertainable resolutions, I am flooded with commentary and passionate claims for closure. Give us a grain of concrete any time in this vale of tears, this realm of such trying uncertainty; let the mountain of dreamy clouds and the mythical pot of gold remain on the unreachable horizon.

Three essays have topped the hit parade of responses: 1. My challenge to readers (with a free copy of all my future books for any genuine solution) to construct a minimal pangram of only twenty-six characters—a sentence without abbreviations or proper nouns, using every letter in our alphabet only once—see essay 4 in *Bully for Brontosaurus*. (No one has yet won the prize, but some readers have spent scores of hours at scratch pads and their own computer programs!) 2. My puzzlement over an old legend about Columbus balancing an egg on its end, with my claim that this generation has forgotten the story—a challenge that did not go unanswered among at least a hundred rememberers. (I later judged this essay as subpar and did not include it in a published volume, for I do try to follow the "Prairie Home Companion" principle that all essays, like the schoolkids of Lake Wobegon, must be above average.) 3. This essay's befuddlement about the whys and wherefores of an old convention for printing inverted snails.

I have been inundated with interesting proposals for why engravers, before the mid-eighteenth century, usually carved pictures of snails as they appear in life on their plates (rather than inverted), thus leading to the printing of reversed images. Most correspondents offered an obvious (but clearly false) suggestion that I thought I had resolved in the essay itself—so I was not clear enough, and should now reinforce the point. In short, many readers suggested that engravers permitted themselves this convention because it made their lives easier by providing at least one small category of images that didn't have to be drawn in reverse on their plates.

This explanation sounds intriguing, but really cannot be right. Drawing in reverse, however difficult the enterprise may appear to us, is the fundamental daily activity of an engraver; how can daily work be deemed troubling after so many years of practice? Let me reinforce the point by admitting to an embarrassing incident of my own youth. My Uncle Mordie played first viola in the Rochester Symphony Orchestra nearly forever (he's still going strong in retirement in his mid-nineties). When I was a teenager, I learned that violists play on the alto clef, a notation unfamiliar to most casual musicians, who never see anything but the treble and bass clefs of piano scores. I was proud of my arcane discovery and wanted to show off. So I asked my uncle, "How does it feel to play on the alto clef?" And he just roared in laughter. After all, he had been doing nothing else for some forty years—so it was supposed to feel strange? Same point for professional engravers.

Other readers reacted to my claim for arbitrariness in the designation of one direction of coiling as dextral (or right-handed) and the reverse as sinistral (or left-handed). They pointed out that we use the same convention for screws and other implements of hardware, so the transfer to snails cannot be deemed arbitrary. But this statement just establishes a recursion—for the designation of right and left for screws is just as arbitrary as for snails. I thank my readers and have added a sentence about hardware.

I have at least convinced myself that the problem is interesting and recalcitrant. Two months after writing the essay, I presented the results as a talk to an annual meeting of antiquarian book dealers. The meeting featured a session on prints and other illustrations—and some fifty experts on printmaking in antiquarian books attended. They were powerfully intrigued, and I was enormously gratified, but no one came up with a good answer.

I have since made one further tiny step in affirming the purposefulness of inverted images for snails. I have noted two books—Mercati's *Metallotheca* discussed in the essay, and Ceruti's *Musaeum Calceolarianum* of 1622, purchased last month—which include snails in a large frontispiece engraving featuring all objects in a museum. These snails are correctly dextral in the illustration because the shells are minor items on a large plate that had to be carved properly in reverse (if only because several items in the museums feature letters and other writing). But, in both books, individual pictures of snails in the main text are sinistral.

My most interesting comment, perhaps pointing the way to a resolution, came from one of my favorite scientists, America's leading malacologist, R. Tucker Abbott (now head of the shell museum on Sanibel Island). Tucker made the following observation about my search for the rationale behind a convention for printing snails as inverted images: "Simply because the guy who carved or etched the shell on wood or copper plate thought a reversal (when printed in ink) did not matter with a gastropod shell."

A convention based on *active* not caring makes a good deal of sense, for the eighteenth-century shift to our modern practice of printing shells as we see them in life would then record a conscious commitment to a new way of interpreting nature. But why should the old engravers have been indifferent only for snails? This got me thinking, and made me realize that indifference could have extended far beyond snails—without any clear evidence in the printed product. After all, most organisms are bilaterally symmetrical, with right and left halves as mirror images (as in our own bodies). Perhaps engravers carved most organisms *recto*, but we notice only the result for snails and for a few other asymmetrical creatures. Such near universality might be regarded as a worthwhile convenience (whereas an indulgence only for a rare snail or two makes no sense).

Since I love my correspondence, I will therefore end by setting a task for readers. Tucker's hypothesis is testable. Look at old engravings of other asymmetrical organisms (crabs with a large claw on one side only, for example)—and see if these creatures are also carved *recto* and printed *verso* before the eighteenth century.

PART FIVE

The Glory of Museums

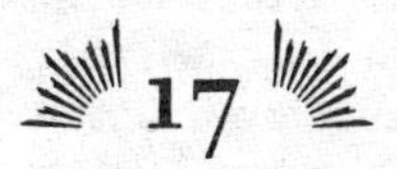

17

Dinomania

I

MACBETH'S SOLILOQUY on his intended murder of King Duncan provides our canonical quotation for the vital theme that deeds spawn unintended consequences in distant futures. "If it were done," Macbeth muses, " 'twere well it were done quickly." The act must be swift, but, even more important, the sequelae must be contained, as Macbeth hopes to "trammel up the consequence, and catch, with his surcease, success; that but this blow might be the be-all and the end-all here." Yet Macbeth fears that big events must unleash all the genies of unknowable futures—for "bloody instructions, which being taught, return to plague the inventor."

I doubt that Henry Fairfield Osborn considered these lines, or imagined any popular future for his new discoveries, when he published a conventionally dull, descriptive paper in 1924 on three genera of dinosaurs recently found in Mongolia on the famous Gobi Desert expedition. In this paper, titled "Three New Theropoda, *Protoceratops* Zone, Central Mongolia," Osborn named, and described for the first time, the "skull and jaws, one front claw and adjoining phalanges" of a small but apparently lithe and skillful carnivore. He called his new creature *Velociraptor mongoliensis* to honor these inferred skills, for *Velociraptor* means "quick seizer." *Velociraptor*, Osborn wrote, "seems to have been an alert, swift-moving carnivorous dinosaur." He then describes the teeth as "perfectly adapted to the sudden seizure of . . . swift-moving prey . . . The long rostrum and wide gape of the jaws indicate that the prey was not only living but of considerable size."

Osborn was America's greatest vertebrate paleontologist, but he was also the politically conservative, socially prominent, imperious president of the American Museum of Natural History in New York. He would, I think, have been quite surprised, and not at all amused, to learn that, nearly seventy years later, his creature would win a new, and vastly extended, status as the primary dinosaur hero (or villain, depending on your modes of rooting) in the blockbuster film *Jurassic Park*.

Public fascination has always followed these prehistoric beasts. Just ten years after Richard Owen coined the word *dinosaur* in 1840, sculptor Waterhouse Hawkins was hard at work on a series of full-scale models to display in the Crystal Palace during the Great Exhibition of 1851. (The Crystal Palace burned in 1936, but Hawkins's dinosaurs, recently spruced up with a coat of paint, can still be seen in Sydenham, south of London).

But the popular acclaim of dinosaurs has been fitful and episodic. We saw them in *King Kong* (thanks to Willis O'Brien and his brilliant technique of stop-motion photography using small models, later magnified). We filled our cars under the sign of a giant green *Brontosaurus*, the logo of Sinclair Oil (who also provided a fine exhibit at the 1939 World's Fair in New York). But dinosaurs never became a pervasive cultural icon, and some decades largely ignored the great beasts. I was a "dinosaur nut" as a kid growing up in New York during the late forties and early fifties. Hardly anyone knew or cared about these creatures, and I was viewed as a nerd and misfit on that ultimate field of vocational decision—the school playground at recess. I was called "Fossil Face"; the only other like-minded kid in the school became "Dino" (I am pleased to report that he also became a professional natural historian). The names weren't funny, and they hurt.

During the last twenty years, however, dinosaurs have vaulted to a steady level of culturally pervasive popularity—from gentle Barney who teaches proper values to young children on a PBS television series, to ferocious monsters who can promote films from "G" to "R" ratings. This dinosaurian flooding of popular consciousness guarantees that no paleontologist can ever face a journalist and avoid what seems to be the most pressing question of the nineties: Why are children so fascinated with dinosaurs?

The question may be a commonplace, but it remains poorly formulated in conflating two quite separate issues. The first—the Jungian or archetypal theme, if you will—seeks the universal reason that stirs the soul of childhood (invariably fatuous and speculative, hence my dislike of the question). To this inquiry, I know no better response than the epitome proposed by a psychologist colleague: big, fierce, and extinct—in other words, alluringly scary, but sufficiently safe.

Most questioners stop here, supposing the inquiry resolved when they feel satisfied about archetypal fascination. But this theme cannot touch the heart of current dinomania, culminating in the extraordinary response to *Jurassic Park*, for an obvious, but oddly disregarded, reason: dinosaurs were just as big, as fierce, and as extinct forty years ago, but only a few nerdy kids, and even fewer professional paleontologists, gave a damn about them. We must therefore pose the second question: Why now and not before?

We might propose two solutions to this less general, but more resolvable, question—one that I wish were true (but almost cannot be), and one that I deeply regret (but must surely be correct). As a practicing paleontologist, I would love to believe that current dinomania arose as a direct product of our research, and all the fascinating new ideas that our profession has generated about dinosaurs. The slow, lumbering, stupid, robotic, virtually behaviorless behemoths of my childhood have been replaced by lithe, agile, potentially warm-blooded, adequately smart, and behaviorally complex creatures. The giant sauropods were mired in ponds during my youth, for many paleontologists regarded them as too heavy to hold up their own bodies on land. Now they stride across the plains, necks and tails outstretched. In some reconstructions they even rear up on their hind legs to reach high vegetation, or to scare off predators. (They are so depicted in the first *Brachiosaurus* scene of *Jurassic Park*, and in the full-scale fiberglass model of *Barosaurus* recently installed in the rotunda of the American Museum of Natural History—though most of my colleagues consider such a posture ridiculously unlikely.) When I was a child, ornithopods laid their eggs and then walked away forever. Today, these same creatures are the very models of maternal, caring, politically correct dinosaurs. They watch over their nests, care for their young, form cooperative herds, and bear such lovely, peaceful names as *Maiasaura*, the "earth mother lizard" (in contrast with such earlier monikers as *Pachycephalosaurus*, the "thick bonehead lizard"). Even their extinction now appears in a much more interesting light. They succumbed to vaguely specified types of "climatic change" in my youth; now we have firm evidence for extraterrestrial impact as the trigger for their final removal (see essay 13).

But how can this greening of dinosaurs be the major reason for present faddishness—for if we credit the Jungian theme at all, then the substrate for fascination has always been present, even in the bad old days of dumb and lumbering dinosaurs (who were, after all, still big, fierce, and extinct). What promotes this substrate to overt and pervasive dinomania? To such questions about momentary or periodic fads, one quintessentially American source usually supplies a solution—recognition and exploitation of commercial possibilities.

When I was growing up on the streets of New York City, yo-yo crazes would sweep through kiddie culture every year or two, usually lasting for a month or so. These crazes were not provoked by any technological improvement in the design of yo-yos (just as more-competent dinosaurs do not engender dinomania). Similarly, a Jungian substrate rooted in control over contained circular motion will not explain why every kid needed a yo-yo in July 1951, but not in June 1950 (just as dinosaurs are always available, but only sometimes exploited).

The answer, in short, must lie in commercialization. Every few years, someone figured out how to make yo-yos sell. At some point about twenty years ago, some set of forces discovered how to turn the Jungian substrate into profits from a plethora of products. You just need a little push to kick the positive feedback machine of human herding and copying behavior into its upward spiral (especially powerful in kids with disposable income).

I'd love to know the source of the initial push (a good theme for cultural historians). Should we look to the great expansion of museum gift shops from holes-in-the-wall run by volunteers to glitzy operations crucial to the financial health of their increasingly commercialized parent institutions? Or did some particular product, or character, grip enough youthful imaginations at some point? Should we be looking for an evil genius, or just for an initial chaotic fluctuation, then amplified by cultural loops of positive feedback?

II

Contemporary culture presents no more powerful symbol, or palpable product, of pervasive, coordinated commercialization than the annual release of "blockbuster" films for the summer viewing season. The movies themselves are sufficiently awesome, but when you consider the accompanying publicity machines, and the flood of commercial tie-ins from lunch pails to coffee mugs to T-shirts, the effort looks more like a military Blitzkrieg than an offer of entertainment. Therefore, every American who is not mired in some Paleozoic pit surely knows that dinomania reached its apogee with the release of Steven Spielberg's film version of Michael Crichton's fine novel *Jurassic Park*. As a paleontologist, I could not possibly feel more ambivalent about the result—marveling and cursing, laughing and moaning. One can hardly pay greater tribute to the importance of an event than to proclaim the impossibility of neutrality before it.

John Hammond (an entrepreneur with more than a touch of evil in the book, but kindly and merely overenthusiastic in the film) has built the ultimate theme park (for greedy profits in the book, for mixed but largely honorable motives in the film) by remaking living dinosaurs out of DNA extracted from dinosaur blood preserved within mosquitoes and other biting insects entombed in fossil Mesozoic amber. Kudos to Crichton for developing the most clever and realistic of all scenarios for such an impossible event, for plausibility is the essence of good science fiction. (The idea, as Crichton acknowledges, had been kicking around paleontological labs for quite some time.)

In fact, the amber scenario has been yielding some results—tiny DNA fragments of the entombed insects themselves, not of anyone else's blood within the insects! In the September 25, 1992, issue of *Science*, a group of colleagues, headed by R. De Salle, reported the successful extraction of several DNA fragments (fewer than two hundred base pairs each) from a 25 to 30-million-year-old termite encased in amber. Then, in a publishing event tied to the opening of *Jurassic Park*, the June 10, 1993, issue of the leading British journal *Nature*—same week as the film's premiere—reported results of another group of colleagues, led by R. J. Cano, on the extraction of two slightly larger fragments (315 and 226 base pairs) from a fossil weevil. The amber enclosing this insect is 120–135 million years old[1]—not quite as ancient as Jurassic, but from the next geological period, called Cretaceous and also featuring

1. Since I wrote this review of *Jurassic Park* in 1993, two claims for recovery of dinosaur DNA (teensy fragments, of course, not blueprints of whole organisms) have appeared in the literature—but I am highly skeptical of both claims (and my doubts are widely shared by professional colleagues). Incidentally, I will not be at all surprised (but rather overjoyed) if bits and pieces of genuine dinosaur DNA are recovered in the near future—though I strongly doubt that full DNA programs for entire dinosaurs exist anywhere in the fossil record. I am even more dubious (though I root for the result and hope for its validation) about the May 1995 claim for revivified bacteria from the stomachs of 30-million-year-old bees trapped in amber.

dinosaurs as dominant creatures of the land (most of the *Jurassic Park* dinosaurs are Cretaceous in any case!).[2]

This remarkable blurring of pop and professional domains emphasizes one of the most interesting spinoffs—basically positive in my view—of the *Jurassic Park* phenomenon. When a staid and distinguished British journal uses the premiere of an American blockbuster film to set the sequencing of its own articles, then we have reached an ultimate integration. Museum shops sell the most revolting dinosaur kitsch. Blockbuster films employ the best paleontologists as advisers to heighten the realism of their creatures. Orwell's pigs have become human surrogates walking on two legs—and "already it was impossible to say which was which" (nor do I know anymore who was the pig, and who the person, at the outset—that is, if either category be appropriate).

If all this welcome scientific activity gives people the idea that dinosaurs might actually be re-created by Crichton's scenario, I hasten (with regret) to pour frigid water upon this greatest reverie for any aficionado of ancient life. Aristotle wisely taught us that one swallow doesn't make a summer—nor, his modern acolytes might add, does one gene (or just a fragment thereof) make an organism. Only the most prominent, easily extracted, or multiply copied bits of fossil DNA have been sequenced—and we have no reason to believe that anything approaching the complete genetic program of an organism has been preserved in such ancient rocks. The most comprehensive and rigorous study of fossil DNA—the sequencing of a complete chloroplast gene from a 20 million year old magnolia leaf (see essay 3)—found no nuclear DNA at all, while the recovered gene occurs in numerous copies per cell, with a correspondingly better chance of preservation. More than 90 percent of all attempted extractions in the magnolia study yielded no DNA at all. The amber DNA described above is nuclear, but represents bases coding for the so-called 16S and 18S ribosomal RNA genes—among the most commonly and easily recovered segments of the genetic program.

DNA is not a geologically stable compound. We may recover frag-

2. Pardon some trivial professional carping, but only two of the dinosaurs featured in the film *Jurassic Park* actually lived during the Jurassic period—the giant sauropod *Brachiosaurus*, and the small *Dilophosaurus*. All the others lived during the subsequent Cretaceous period—a perfectly acceptable mixing, given the film's premise that amber of any appropriate age might be scanned for dinosaur blood. Still, the majority might rule in matters of naming, though I suppose that Cretaceous Park just doesn't have the same ring. When I met Michael Crichton (long before the film's completion), I had to ask him this small-minded professional's question: "Why did you place a Cretaceous dinosaur on the cover of *Jurassic Park*?" (for the book's dust jacket—and now the film's logo—features a Cretaceous *Tyrannosaurus rex*). I was delighted with his genuine response: "Oh my God, I never thought of that. We were just fooling around with different cover designs, and this one looked best." Fair enough; he took the issue seriously, and I would ask no more.

ments, or even a whole gene here and there, but no wizardry can make an organism from just a few percent of its codes. *Jurassic Park* acknowledged this limitation when its genetic engineers used modern frog DNA to fill in the missing spaces in their dinosaur programs. But, in so doing, the scientists of *Jurassic Park* commit their worst scientific blunder—the only one that merits censure as a deep philosophical error, rather than a studied and superficial conceit consciously indulged to bolster the drama of science fiction.

An amalgamated code of, say, 80 percent dinosaur DNA and 20 percent frog DNA could never direct the embryological development of a functioning organism. This form of reductionism is simply silly. An animal is an integrated entity, not the summation of its genes, one by one. Fifty percent of your genetic code can't construct a perfectly good half of you; it makes no functioning organism at all. Genetic engineers might get by with a missing dab or two, but large holes cannot be plugged with DNA from a different zoological class. (Moreover, frogs and dinosaurs are not even close evolutionary relatives, for their lines diverged in the Carboniferous period, more than 100 million years before the origin of dinosaurs. I suppose that Crichton used frogs because they conjure up an image of primitivity, and dinosaurs are ancient too. But evolutionary "closeness" involves timing of branching on the tree of life, not external appearances. Jurassic Park's scientists should have used modern birds, the closest living kin of dinosaurs.) The embryological decoding of a DNA program into an organism represents nature's most complex orchestration. You need all the proper instruments and conductors of a unique evolutionary symphony. You cannot throw in 20 percent of a rock band playing its own tunes by its own rules, and hope for harmony.[3]

When a scientist soberly states that something cannot be done, the public has every right to express doubts based on numerous historical precedents for results proclaimed impossible, but later both achieved

3. As my colleague Adam Wilkins said to me, the situation is even more hopeless, and the reductionistic error of *Jurassic Park* more intense. Even 100 percent of dinosaur DNA won't make a functioning organism by itself. A complex newborn animal is not an automatic product of its molecular code—for the code needs to work in, and interact with, the proper environment for embryological growth. After all, the code may act as a chief directing engineer, but not as a mason, carpenter, or plasterer. In dinosaurs, the main material must be present in the maternal egg. Moreover, and most vitally, organisms are not constructed only by their own genes. Certain maternal genes must produce products and chemical signals needed by the fertilized egg nucleus for its early divisions and differentiations. If these maternal transcripts are not present in the dinosaur egg before fertilization, the embryo will not develop. Thus the genetic engineers of *Jurassic Park* must know not only the full code for the dinosaur itself, but also the number, location, and action of maternal genes needed to construct the right environment for development.

and far surpassed. Unfortunately, the implausibility of reconstructing dinosaurs by the amber scenario resides in a different category of stronger argument.

Most proclamations of impossibility only illustrate a scientist's lack of imagination about future discovery—impossible to see the moon's backside because you can't fly there, impossible to see an atom because light microscopy cannot resolve such dimensions. The object was always there: atoms and the moon's far side. We only lacked a technology, possible in principle to attain, but unimagined in practice.

But when we say that a particular historical item—like a dinosaur species—can't be recovered, we are invoking a different and truly ineluctable brand of impossibility. If all information about a historical event has been lost, then the required data just aren't there anymore, and the event cannot be reconstructed. We are not lacking a technology to see something that truly exists; rather, we have lost all information about the thing itself, and no technology can recover an item from the void. Suppose I want to know the name of every soldier who fought in the battle of Marathon. The records, I suspect, simply don't exist, and probably never did. No future technology, no matter how sophisticated, can recover events with crucially missing information. So too, I fear, with dinosaur DNA. We may make gene machines more powerful by orders of magnitude than anything we can now conceive. But if full programs of dinosaur DNA exist nowhere—only the scrap of a gene here and there—then we have permanently lost these particular items of history.

III

I liked the book version of *Jurassic Park*. Crichton not only used the best possible scenario for making dinosaurs, but also based the book's plot upon an interesting invocation of currently fashionable chaos theory. To allay the fears of his creditors, John Hammond brings a set of experts to Jurassic Park, hoping to win their endorsement. His blue-ribbon panel includes two paleontologists and a preachy iconoclast of a mathematician named Ian Malcolm—the novel's intellectual and philosophical center. Malcolm urges—often, colorfully, and at length—a single devastating critique based on his knowledge of chaos and fractals: the park's safety system must collapse because it is too precariously complex in coordinating so many, and such intricate, fail-safe devices. Moreover, the park must fail both unpredictably and spectacularly. Malcolm explains in the book:

> It's chaos theory. But I notice nobody is willing to listen to the consequences of the mathematics. Because they imply very large consequences for human life. Much larger than Heisenberg's principle or Gödel's theorem, which everybody rattles on about . . . Chaos theory concerns everyday life . . . I gave all this information to Hammond before he broke ground on this place. You're going to engineer a bunch of prehistoric animals and set them on an island? Fine. A lovely dream. Charming. But it won't go as planned. It is inherently unpredictable . . . We have soothed ourselves into imagining sudden change as something that happens outside the normal order of things. An accident like a car crash. Or beyond our control, like a fatal illness. We do not conceive of sudden, radical, irrational change as built into the very fabric of existence. Yet it is. And chaos theory teaches us.

Moreover, Malcolm uses this argument—not the usual and vacuous pap about "man treading where God never intended" (Hollywood's only theme in making monster movies, see essay 5)—to urge our self-restraint before such scientific power: "And now chaos theory proves that unpredictability is built into our daily lives. It is as mundane as the rainstorm we cannot predict. And so the grand vision of science, hundreds of years old—the dream of total control—had died, in our century."

This reliance on chaos as a central theme did, however, throw the book's entire story line into a theoretically fatal inconsistency—one that, to my surprise, no reviewer seemed to catch at the time. The book's second half is, basically, a grand old rip-roaring chase novel, with survivors managing to prevail (unscratched no less) through a long sequence of independent and excruciatingly dangerous encounters with dinosaurs. By the same argument that complex sequences cannot proceed as planned—in this case, toward the novelistic necessity of at least some heroes surviving—not a *Homo sapiens* in the park should have been able to move without harm through such a sequence. Malcolm even says so: "Do you have any idea how unlikely it is that you, or any of us, will get off this island alive?" But I do accept the literary convention for bending nature's laws in this case.

I expected to like the film even more. The boy dinosaur enthusiast still dwells within me, and I have seen them all, from *King Kong* to *One Million Years B.C.* to *Godzilla*. The combination of a better story line, with such vast improvement in monster-making technology, seemed to guarantee success at a spectacular new level of achievement.

The dinosaurs themselves certainly delivered. As a practicing paleontologist, I confess to wry amusement at the extended roman-à-clef embedded in the reconstructions. I could recognize nearly every provocative or outré idea of any colleague, every social tie-in now exploited by dinosaurs in their commanding role as cultural icons. The herbivores are so sweet and idyllic. The giant brachiosaurs low to each other like cattle in the peaceable kingdom. They rear up on their hind legs to find the juiciest leaves. Individuals in the smaller species help each other—down to such subtle details as experienced elders keeping young *Gallimimus* in the safer center of the fleeing herd.

Even the carnivores are postmodernists of another sort. The big old fearsome standard, *Tyrannosaurus rex*, presides over Jurassic Park in all her glory (and in the currently fashionable posture, with head down, tail up, and vertebral column nearly parallel to the ground). But the mantle of carnivorous heroism has clearly passed to the much smaller *Velociraptor*, Henry Fairfield Osborn's Mongolian jewel. Downsizing and diversity are in; constrained hugeness has become a tragic flaw. *Velociraptor* is everything that modern corporate life values in a tough competitor—mean, lean, lithe, and intelligent. They hunt in packs, using an old military technique of feinting by one beast in front, followed by attack from the side by a co-conspirator.

Spielberg didn't choose to challenge pop culture's canonical dinosaurs in all details of accuracy and professional speculation; blockbusters must, to some extent, play upon familiarity. Ironically, he found the true size of *Velociraptor*—some six feet in length—too small for the scary effects desired, and he enlarged them to nearly ten feet, thus moving back toward the old stereotype, otherwise so effectively challenged. He experimented, in early plans and models, with bright colors favored by some of my colleagues on the argument that birdlike behavior might imply avian styles of coloration for the smaller dinosaurs. But he eventually opted for conventional reptilian dullness ("your same old, ordinary, dinosaur shit-green," lamented one of my graduate students who had expected more obeisance to modernity).

But let me not carp. The dinosaur scenes are spectacular. Intellectuals too often either pay no attention to such technical wizardry or, even worse, actually disdain special effects with such dismissive epithets as "merely mechanical." I find such small-minded parochialism outrageous. Nothing can be more complex than a living organism, so the technological reconstruction of accurate and believable animals therefore becomes one of the greatest all-time challenges to human ingenuity.

The field has a long and honorable history of continually improving techniques—and who would dare deny this story a place in the annals of human intellectual achievement? An old debate among historians of science asks whether most key technological inventions arise from practical need (more often in war than in any other activity), or from fooling around in maximal freedom from practical pressures. My friend Cyril Smith, the wisest scientist-humanist I ever knew, strongly advocated the centrality of "play domains" as the major field for innovations with immense practical utility down the road. (He argued that the block-and-tackle was invented, or at least substantially improved, in order to lift animals from underground storage pits to the game floor of the Roman Colosseum). Yes, *Jurassic Park* is "just" a movie—but for this very reason, the filmmakers had freedom (and funds) to develop techniques of reconstruction, particularly computer generation, to new heights of astonishing realism. And yes, these advances matter—for immediately aesthetic reasons, and for innumerable practical possibilities in the future.

Spielberg originally felt that computer generation had not yet progressed far enough, and that he would have to film all his dinosaur scenes with the fascinating array of modeling techniques long used, but constantly improving, in Hollywood—stop-motion with small models, people dressed in dinosaur suits, puppetry of various sorts, robotics with hydraulic apparatus moved by people sitting at consoles.

But computer generation improved greatly during the two-year gestation of *Jurassic Park*, and the dinosaurs of the most spectacular scenes are drawn, not modeled—meaning, of course, that performers interacted with empty space during the actual shooting. I learned, after watching the film, that my two favorite dinosaur scenes—the fleeing herd of *Gallimimus*, and the final attack of *Tyrannosaurus* upon the last two *Velociraptors*—were entirely computer generated. (The effect does not always work. The very first dinosaur scene—when paleontologist Grant hops out of his vehicle to encounter a computer-generated *Brachiosaurus*—is the film's worst flop. Grant is clearly not in the same space as the dinosaur, and I could only think of Victor Mature, similarly out of synch with his beasts in *One Million Years B.C.*)

The dinosaurs are wonderful, but they aren't on set enough of the time. (Yes, I know how much more they cost than human actors.) Unfortunately, the plot line for human actors has been reduced to pap and drivel of the worst kind, the very antithesis of the book's grappling with serious themes. I fear that Mammon, and false belief in a need to "dumb down" for mass audiences, has brought us to this impasse of utter inconsistency. How cruel, and how perverse, that we invest the most awesome expertise (and millions of dollars) in the dinosaurs, sparing no knowledge or expense to render every detail, every possible nuance, in the most accurate and realistic manner. I have nothing but praise for the thought and care, the months and years invested in each dinosaur model, the pushing of computer generation into a new realm of utility, the concern for rendering every detail with consummate care, even the tiny bits that few will see and the little sounds that fewer will hear. I think of medieval sculptors who lavished all their skills upon invisible statues on the parapets, for God's view must be best (internal satisfaction based on personal excellence, in modern translation). How ironic that we permit a movie to do all this so superbly well, and then throw away the story because we think that the public will reject, or fail to comprehend, any complexity beyond a Neanderthal "duh," or a brontosaurian bellow.

I simply don't believe that films, to be popular at the box office, must be dumbed down to some least common denominator of universal comprehension. Science fiction, in particular, has a long and honorable tradition for exploring philosophically complex issues about time, history, and the meaning of human life in a cosmos of such vastness. Truly challenging films, like Kubrick and Clarke's 2001, have made money, won friends, and influenced people—and even such truly mass-marketed series as *Star Wars*, *Star Trek*, and *Planet of the Apes* base their themes on the meaty issues traditionally used by the genre as centerpieces of plot lines.

But the film of *Jurassic Park* has gutted Crichton's book and inverted his interesting centerpiece about chaos theory to Hollywood's most conventional and universal pap. We feel this loss most keenly in the reconstruction of Ian Malcolm as the antithesis of his character in the book. He still presents himself as a devotee of chaos theory ("a chaotician"), but he no longer uses the argument to formulate his criticism of the park. Instead he is given the oldest diatribe, the most hackneyed and predictable staple of every Hollywood monster film since *Frankenstein* (see essay 5): that human technology must not disturb the specific and proper course of nature; we must not tinker in God's realm. What dullness and disappointment (and Malcolm, in the film, is a frightful and tendentious bore, obviously so recognized by Spielberg, for he effectively puts Malcolm out of action with a broken leg about halfway through the film).

Not only have we heard this silly argument a hundred times before (can Spielberg really believe that his public could comprehend no other reason for criticizing a dinosaur park?), but Malcolm's invocation of the old chestnut utterly negates his proclaimed persona and cardinal belief, as his cinematic argument invokes the antithesis of chaos, and thereby engenders whopping inconsistency to boot. Two related flubs make the story line entirely incoherent.

First, as Malcolm rails against genetic reconstruction of lost organisms, Hammond asks him if he would really hesitate to bring the California condor back to life (from preserved DNA), should the last bird die. Malcolm answers that he would not object, and would view such an act as benevolent, because the condor's death would have been an accident caused by human malfeasance, not an expression of nature's proper course. But we must not bring back dinosaurs because they disappeared along a natural and intended route. "Dinosaurs," he says, "had their shot and nature selected them for extinction." But such an implied scenario of groups emerging, flourishing, and dying, one after another in an intended and predictable course, negates the primary phenomena of chaos theory and its crucial emphasis on the great, accumulating effect of apparently insignificant perturbations, and on the basic unpredictability of long historical sequences. How can a chaotician talk about nature's proper course at all?

Second, if "nature selected them for extinction," and if later mammals therefore represent such an improvement, why can the dinosaurs of Jurassic Park beat any mammal in the place, including the most arrogant primate of them all? You can't have it both ways. If you take dinosaur revisionism seriously, and portray them as smart and capable creatures able to hold their own with mammals, then you can't argue against reviving them by depicting their extinction as both predictable and appointed, as life ratcheted onward to greater complexity.

Since Malcolm actually preaches the opposite of chaos theory, but presents himself as a chaotician and must therefore talk about the theory, the film's material on chaos becomes a vestigial and irrelevant caricature in the most embarrassing of all scenes—Malcolm's halfhearted courting of the female paleontologist. He grasps her hand, drips water on the top, and uses chaos theory to explain why we can't tell which side the drop will run down! How are the mighty fallen, and the weapons of war perished.

IV

In the film, John Hammond flies his helicopter to the excavation site of Ellie Sattler and Alan Grant, the two paleontologists chosen to "sign off" on his park and satisfy his investors. They say at first that they cannot come, for they are hard at work on the crucial phase of collecting a fossil *Velociraptor*. Hammond promises to fund their research for three years if they will spend one weekend at his site. Grant and Sattler suddenly realize that they would rather be no place else on earth; the *Velociraptor* can wait (little do they know . . .).

This scene epitomizes the ambivalence that I feel as a professional paleontologist about the *Jurassic Park* phenomenon, and about dinomania in general. Natural history is, and has always been, a beggar's game. Our work has never been funded by or for itself. We have always depended upon patrons, and upon other people's perceptions of the utility of our data. We sucked up to princes who wanted to stock their baroque *Wunderkammern* with the most exotic specimens. We sailed on colonial vessels for nations that viewed the cataloging of faunas and floras as one aspect of control (we helped Captain Bligh bring breadfruit from Tahiti to feed slaves in the West Indies). Many, but not all, of these partnerships have been honorable from our point of view, but we have never had the upper hand. Quite the contrary, our hand has always been out.

Few positions are more precarious than that of the little guy in associations based on such unequal sizes and distributions of might. The power brokers need our expertise, but we are so small in comparison, so quickly bedazzled and often silenced by promises (three years as a lifetime's dream for the paleontologists and an insignificant tax write-off for Hammond), so easily swallowed up—if we do not insist on maintaining our island of intact values and concerns in the midst of such a different, and giant, operation. How shall we sing the Lord's song in a strange land?

I do not blame the prince, the captain, or his modern counterparts: the government grantor, the commercial licenser, or the blockbuster filmmaker. These folks know what they want, and they are usually upfront about their needs and bargains. It is our job to stay whole, not to be swallowed in compromise, not to execute a pact of silence, or endorsement, for proffered payoff. The issue is more structural than ethical: we are small, though our ideas may be powerful. If we merge without maintaining our distinctness, we are lost.

Mass commercial culture is so engulfing, so vastly bigger than we can ever be. Mass culture forces compromises, even for the likes of Steven Spielberg. He gets the resources to prepare and film his magnificent special effects; but I cannot believe that he feels comfortable about ballyhooing all the ridiculous kitsch now for sale under the coordinated marketing program of movie tie-ins (from fries in a dinosaur's mouth at McDonald's—sold to kids too young for the movie's scary scenes—to a rush on amber rings at fancy jewelry stores); and I cannot imagine that either he or Michael Crichton is truly satisfied with their gutless and incoherent script as an enjoined substitute for an interesting book. Imagine, then, what compromises the same commercial world forces upon the tiny principality of paleontological research?

As a symbol of our dilemma, consider the plight of natural history museums in the light of commercial dinomania. In the past decade, nearly every major or minor natural history museum has succumbed (not always unwisely) to two great commercial temptations: to sell a plethora of scientifically worthless and often frivolous, or even degrading, dinosaur products by the bushel in their gift shops; and to mount, at high and separate admission charges, special exhibits of colorful robotic dinosaurs that move and growl but (so far as I have ever been able to judge) teach nothing of scientific value about these animals. (Such exhibits could be wonderful educational aids, if properly labeled and integrated with more traditional material; but I have never seen these robots presented for much more than their colors and sound effects [the two aspects of dinosaurs that must, for obvious reasons, remain most in the realm of speculation].)

If you ask my colleagues in museum administration why they have permitted such incursions into their precious and limited spaces, they will reply that these robotic displays bring large crowds into the museum, mostly of people who otherwise never come. These folks can then be led or cajoled into viewing the regular exhibits, and the museum's primary mission of science education receives a giant boost.

I cannot fault the logic of this argument, but I fear that my colleagues are expressing a wish or a hope, not an actual result, and not even an outcome actively pursued by most museums. If the glitzy displays were dispersed among teaching exhibits, if they were used as a springboard for educational programs (sometimes they are), then a proper balance of Mammon and learning might be reached. But, too often, the glitz occupies a separate wing (where the higher admission charges can be monitored), and the real result gets measured in increased body counts and profits. One major museum geared all its fancy fund-raising apparatus for years to the endowment of a new wing—and then filled the space with a massive gift shop, a fancy restaurant, and an Omnimax theater, thus relegating the regular exhibits to neglect and disrepair. Another museum intended the dinosaur robots as a come-on to guide visitors to the permanent exhibits. But they found that the robots wouldn't fit into the regular museum. Did they cancel the show? Not at all. They moved the robots to another building on the extreme opposite end of the campus—and even fewer people visited the regular museum as a result.

I may epitomize my argument in the following way: Institutions have essences—central purposes that define their integrity and being. Dinomania dramatizes a conflict between institutions with disparate essences—museums and theme parks. Museums exist to display authentic objects of nature and culture—yes, they must teach; and yes, they may certainly include all manner of computer graphics and other virtual displays to aid in this worthy effort; but museums must remain wedded to authenticity. Theme parks are gala places of entertainment, committed to using the best displays and devices from the increasingly sophisticated arsenals of virtual reality to titillate, to scare, to thrill, even to teach.

I happen to love theme parks, so I do not speak from a rarefied academic post in a dusty museum office. But theme parks are, in many ways, the antithesis of museums. If each institution respects the other's essence and place, this opposition poses no problem. But theme parks represent the realm of commerce, museums the educational world—and the first, by its power and immensity, must trump the second in any direct encounter. Commerce will swallow museums if educators try to copy the norms of business for immediate financial reward.

Speaking about the economics of major sporting events, George Steinbrenner, the man we all love to hate, once opined that "it's all about getting the fannies into the seats." If we have no other aim than to attract more bodies, and to extract more dollars per fanny, then we might as well convert our museums to theme parks and fill the gift shop with coffee mugs. But then we will be truly lost—necessarily smaller and not as oomphy as Disneyland or Jurassic Park, but endowed with no defining integrity of our own.

Our task is hopeless if museums, in following their essences and respecting authenticity, condemn themselves to marginality, insolvency, and empty corridors. But, fortunately, this need not and should not be our fate. We have an absolutely wonderful product to flog—real objects of nature. We may never entice as many visitors as Jurassic Park, but we can and do attract multitudes for the right reasons. Luckily—and I do not pretend to understand why—authenticity stirs the human soul. The appeal is cerebral and entirely conceptual, not at all visual. Casts and replicas are now sufficiently indistinguishable from the originals that no one but the most seasoned expert can possibly tell the difference. But a cast of the Rosetta Stone is plaster (however intriguing and informative), while the object itself, on display in the British Museum, is magic. A fiberglass *Tyrannosaurus* merits a good look; the real bones send shivers down my spine, for I know that they supported an actual breathing and roaring animal some 70 million years ago. Even the wily John Hammond understood this principle, and awarded museums their garland of ultimate respect. He wanted to build the greatest theme park in the history of the world—but he could do so only by abandoning the virtual reality of most exemplars, and stocking his own park with real, living dinosaurs, reconstructed from authentic dinosaur DNA. (I do appreciate the conscious ironies and recursions embedded in *Jurassic Park*'s own reality—that the best dinosaurs are computer-generated within a movie based on a novel.)

For paleontologists, *Jurassic Park* is both our greatest opportunity and our most oppressive incubus—a spur for unparalleled general interest in our subject, and the source of a commercial flood that may truly extinguish dinosaurs by turning them from sources of awe into clichés and commodities. Will we have strength to stand up in this deluge?

Our success cannot be guaranteed, but we do have one powerful advantage, if we cleave to our essence as guardians of authenticity. Commercial dinosaurs may dominate the moment, but must be ephemeral, for they have no support beyond their immediate profitability. Macbeth, in the soliloquy cited at the outset of this essay, recognized a special problem facing his plans, for he could formulate no justification beyond personal advantage: "I have no spur to prick the sides of my intent, but only vaulting ambition, which o'erleaps itself." This too shall pass, and nothing of human manufacture can possibly challenge the staying power of a dinosaur bone—65 million years (at least) in the making.

18

Cabinet Museums:

Alive, Alive, O!

In Dublin's fair city, at the heart of Georgian elegance near Trinity College and the Old Parliament House, stands an anatomically correct statue of Molly Malone. I do not speak of Molly herself, who may or may not be properly rendered (I didn't particularly notice), but of her legendary wares. She holds two baskets, one full of cockles and the other of mussels—not quite "alive, alive, o!" in their bronzed condition, but clearly sculpted as accurate representatives of the appropriate species. The artist has respected zoological diversity by representing the song's complete natural history. (To comment on diversity of another valued kind, I never understood why the song's third verse includes the only non-rhyming couplet in such a consistent and admirable ditty: "She died of a fever; and no one could save her." But then I realized that these words do rhyme in Ireland—just as "thought" and "note" rhyme in Yorkshire, and therefore in Wordsworth.)

Just a few blocks from Molly and right next to the Dail (the modern Parliament of the Irish Republic) stands the Dublin Museum of Natural History. This museum traces its origin to a private association of fourteen citizens, founded in 1731 as the Dublin Society. The first public exhibit (largely of agricultural implements) opened in 1733 in the basement of the Old Parliament House, mentioned above. George II provided a royal charter in 1749, and Parliamentary grants began in

1761. Growing collections required a new building, and a government grant of five thousand pounds, made in 1853, largely financed the present structure. Lord Carlisle, the Lord Lieutenant and General Governor of Ireland, laid the foundation stone in March 1856. His lordship, speaking in orotund tones suited both to Victorian practice and to the dignity of his official title, expressed a hope "that the building about to arise on this spot . . . may, with its kindred departments, furnish ever-increasing accommodation for the pursuits of useful knowledge and humanizing accomplishments, and open for the coming generations worthy temples of science, art, and learning, at whose shrine they may be taught how most to reverence their creator, and how best to benefit their fellow creatures."

I learned these details of the museum's history in a fine pamphlet by C. E. O'Riordan, titled *The Natural History Museum Dublin*. The museum building, though harmonizing with its earlier Georgian surroundings in exterior design, could not be more quintessentially Victorian within. Two fully mounted, magnificently antlered skeletons of the fossil deer *Megaceros giganteus*—informally, if incorrectly, called the Irish elk—greet visitors at the entrance to the ground floor (while a third skeleton, of an unantlered female, stands just behind). The rest of the ground floor mostly houses representative collections of Irish zoology, phylum by phylum and family by family (a case on the "roundworms of Ireland" or "Irish crabs" certainly conveys an impression of admirable thoroughness in coverage).

The remainder of the museum, a first floor and two galleries above, seems even more frozen into its older style of full and systematic presentation. Cast ironwork and dark wooden cabinets, the mainstays of Victorian exhibition, abound. Copious light enters through the glass ceiling and streams around the shadows made by cabinets and their contents. Heads and horns adorn the walls in profusion, and we wonder for a moment whether we are visiting a museum or a gentleman's trophy room.

The ensemble seems so coherent that we might view the whole display as an embodiment of a blueprint in the head of some Victorian museum worthy under the spell of John Ruskin. In fact, as with any living entity, the exhibits were melded, fused, reordered, and cobbled together over many decades—though these particular decades did end quite some time ago! The horns were not installed until the 1930s, but most of the other exhibits have changed little since Victoria and, later, her son Edward VII ruled this land—or at least since the locals demoted Edward's son, George V, to establish the Irish Free State in 1921.

O'Riordan, who provides a meticulous account of every change in venue for any stuffed bird or seashell, also acknowledges twentieth-century stability. He discusses a massive rearrangement, begun in 1895, to establish the current scheme of Irish specimens on the ground floor, with a run-through of worldwide Linnaean order on the first floor and galleries above. He writes, "The recruitment of extra staff in 1906 enabled work on the invertebrates on the top gallery to proceed quickly and this was completed by 1907. The exhibition on the upper floor and galleries has not radically changed since." He then mentions the addition of several Irish elk skulls to the ground-floor exhibit in 1910, and comments, "Apart from relatively minor alterations in the content and disposition of the exhibits, the overall theme and plan of the exhibition has since remained the same."

We tend—falsely, I shall soon argue—to view such stability as a sure sign of stagnation, if not decrepitude and ruin. Our basic concept of "Victorian" includes images of soot-blackened buildings, cold interior spaces lined with dark wood, chipping paint, peeling wallpaper, and shelves of dusty bric-a-brac. In many towns, the classic late-Victorian mansions are now either funeral homes or lawyers' offices—and neither enterprise seems much beloved of late.

I confess that my first visit to the Dublin Natural History Museum did nothing to dent this stereotype. I spent a good part of 1971, yardstick in hand, measuring the skulls and antlers of Irish elks. I visited the manors of the Marquess of Bath and the Earl of Dunraven, and I measured the mistreated male of commercialized Bunratty Castle (near Shannon Airport), where besotted revelers at the nightly medieval banquet had left the poor fellow with a fat cigar in his jaws, and coffee cups on the tines of his antlers. But the best stash of specimens belongs to the Dublin Museum, where the two skeletons can be supplemented with another fifteen heads and horns, mounted high on the walls of the ground floor, one above each major cabinet.

The same Dr. O'Riordan greeted me warmly and treated me well; his specimens formed the centerpiece of my study (published in the professional journal *Evolution* in 1974, but initially, in a more general version, as my very first article in this series of essays in 1973—and reprinted in my first book, *Ever Since Darwin*). The specimens were terrific, but, oh my, the museum was a dingy place back then. Little light, less comfort, and dust absolutely everywhere. I had to sit on top of the tall cabinets to measure the heads mounted above. There the dust, undisturbed for so many years, had congealed into thick layers of grime. I doubt that any living being had been up there with any sort of cleaning implement since Leopold Bloom met Stephen Dedalus in nighttown (or since Molly Malone last sold the sort of stuff labeled in the ground-floor exhibits as "Mollusca of Ireland").

With such memories, I approached my revisit in September 1993 with some trepidation—for the extrapolated curve of deterioration did not lead to happy expectations. I could not have been more joyously surprised. Not one jot or tittle of any exhibit has been altered, but all the surroundings have been meticulously restored to their original condition—not just accurately, but lovingly as well. An army of brooms has been through the premises (I think of the enormous clone constructed by Mickey Mouse in the "Sorcerer's Apprentice" sequence of *Fantasia*)—and, as my grandmother would surely have said, "You could eat off the floor" (though I never understood why all my older relatives invoked this expression, as I couldn't imagine why anyone would want to try the experiment, however thorough the scrubbing). The glass ceiling has been cleaned, and the light floods through. The dark wood of the cabinets has been repaired and polished, and the glass now shines. The elaborate cast ironwork has been scraped and decorated in colorful patterns reminiscent of the "painted lady" Victorian houses of San Francisco. The ensemble now exudes pride in its own countenance—and I finally understood, viscerally, the coherent and admirable theory behind a classical Victorian "cabinet" museum of natural history.

Two factors—one a prejudice, the other a condition—generally debar us from appreciating the Victorian aesthetic. First, our smugness about progress leads us to view any contrary vision from the past as barbarous. Thus, when modernism espoused simple geometries with unornamented and functional spaces, the Victorian love of busy exuberance became a focus of pity and derision. (We might praise an old Japanese house for anticipating modern simplicity, but what could we do with a shelf of curios?) In a sense, this dismissal might be viewed as payback, for the Victorians aggressively depicted their own times as the pinnacle of progress and often treated the past with condescension. In any case, our knee-jerk dismissal of Victoriana is now fading as the preservationist movement wins more converts, and as postmodernism brings eclecticism and ornament back into architecture and design.

Second, and more important, our image of "Victorian" has not been set by the objects themselves as constructed for their own time, but by their present appearance, usually after a century of neglect and deterioration. The injustice of this situation borders on perversity. I would not, after all, allow my image of "grandfather" to be set by the present state of my Papa Joe's remains at his gravesite. Why, then, do we conceptualize "Victorian" as a ramshackle building with broken steps, creaking floors, and peeling paint—a set fit only for the Addams family, or the Halloween "haunted house" set up by the local Jaycees.

My first, and keenly revealing, experience with Victorian as Victorians knew the style, divested of a century's overlay in deterioration, occurred in 1976, when, to celebrate our nation's two hundredth birthday, the Smithsonian Institution opened a replica of the Philadelphia Centennial Exposition of 1876. This wonderful exhibition included plows, pharmaceuticals, implements for house and farm, and, above all, machines and engines, all spanking new, freshly painted, and entirely in working order, with all their wheels, whistles, and hisses. I particularly remember a case of ax blades—all shiny and sharp. And I realized that I had always pictured Victorian tools as rusted and dull—without ever articulating to myself the obvious point that they must have been gleaming and functional when first made. I am always amazed at the power of a prejudiced assumption (however absurd, and especially when backed by a mental picture) to derail the logical thinking of basically competent people like myself.

I remember Glasgow as the planet's ugliest city upon my first visit in 1961, and as one of the loveliest places I had ever seen upon my return in 1991. The difference: Glasgow is the world's greatest Victorian city in public and commercial architecture. All the major downtown buildings, horribly soot-blackened and decrepit in many other ways in 1961, have now been cleaned and showcased, often by converting traffic orgies into pedestrian malls. I was stunned by the exuberance of these buildings, each different in its curves, ornaments, and filigrees, each vying with all the others, yet somehow forming an integrated cacophony (you have to see them to know why my chosen description is not oxymoronic). I was revolted at my first sight of the Natural History Museum in London, each archway of its elaborate Romanesque entranceway blacker and grimier than the one within—and uplifted by the subtle colors and arching forms of the cleaned building. The Victorian secular glass of Harvard's churchlike Memorial Hall passed beneath my notice for twenty-five years. Now I force these wonderful windows, designed by John La Farge and other great American glassmakers, and resplendent in their newly cleaned state, upon every visitor, for Memorial Hall is stop number one on my personal tour of Harvard.

I now add the Dublin Museum of Natural History to this list of Victorian buildings uplifted from squalor to glory by the simple expedient of restoring them to the original intentions of their architects and designers. Most of all, this splendid restoration taught me something that I had never appreciated about Victorian museum design.

The display of organisms in these museums rests upon concepts strikingly different from modern practice, but fully consonant with Victorian concerns: Today we tend to exhibit one or a few key specimens, surrounded by an odd mixture of extraneous glitz and useful explanation, all in an effort to teach (if the intent be maximally honorable) or simply to dazzle (nothing wrong with this goal either). The Victorians, who viewed their museums as microcosms for national goals of territorial expansion and faith in progress fueled by increasing knowledge, tried to stuff every last specimen into their gloriously crowded cabinets—in order to show the full range and wonder of global diversity. (In my favorite example, Lord Rothschild, the richest and most thorough of all great collectors, displayed zebras and antelopes in kneeling position, or even supine, so that one or two extra rows could be inserted to include all specimens in floor-to-ceiling displays at his museum in Tring.) The standard Victorian cabinet (including many in the Dublin Museum) incorporates several rows of locked wooden drawers beneath the creatures on display under glass—to house all the additional specimens, which may then be shown to professionals and others with specialized interests.

I realize that this tactic of stuffing in every last specimen includes a dubious side in recording the spoils of aggressive and militaristic imperialism, with all the attendant racism and ecological disregard. But do honor and acknowledge the countervailing virtue of displaying such plenitude—as best expressed in the words of Psalm 104: "O Lord, how manifold are thy works! . . . the earth is full of thy riches." You can put one beetle in a cabinet (usually an enlarged model, and not a real specimen), surround it with fancy computer graphics and pushbutton whatzits, and then state that no other group maintains such diversity. Or you can fill the same cabinet with real beetles from each of a thousand different species—all of differing colors, shapes, and sizes—and then state that you have tried to display each kind in the county.

The Victorians preferred this second approach—and I am with them, for nothing thrills me more than the raw diversity of nature. Moreover, the Victorian "cabinet" museum thrives upon an exquisite tension in commingling (not always comfortably, for they truly conflict) two differing traditions from still earlier times: the seventeenth-century Baroque passion for displaying odd, deformed, peculiar, and "prize" (largest, smallest, brightest, ugliest) specimens, the *Wunderkammer* (or "cabinet of curiosities") of older collectors; and the eighteenth-century preference of Linnaeus and the Enlightenment for a systematic display of nature's regular order within a coherent and comprehensive taxonomic scheme.

I have long recognized the theory and aesthetic of such comprehensive display: show everything and incite wonder by sheer variety. But I had never realized how powerfully the decor of a cabinet museum can promote this goal until I saw the Dublin fixtures redone right. Light floods through the glass ceiling, creating a fascinating interplay of brightness and shadow in reflecting off both specimens *and* architectural elements of iron struts, wooden railings, and the dark wood and clear glass of the cabinets themselves. The busy arrangement of cabinets matches the crowding of organisms, while the contrast of dark wood and clear glass reinforces the variegated diversity of creatures within. The regular elements of cast iron and cabinetry echo the order of taxonomic schemes for the allocation of specimens. The exuberance is all of one piece—organic *and* architectural.

I write this essay to offer my warmest congratulations to the Dublin Museum for choosing preservation—a decision not only scientifically right, but also ethically sound and decidedly courageous. The avant-garde is not an exclusive locus of courage; a principled stand within a reconstituted rear unit may call down just as much ridicule and demand equal fortitude. Crowds do not always rush off in admirable or defendable directions.

In choosing to construct a dynamic museum of museums, in asserting the old ideal of displaying nature's full diversity, in restoring their interior space to Victorian intent in harmonizing architecture with organism, the curators of Dublin have stood against most modern trends in museums of science—where fewer specimens, more emphasis on overt pedagogy, and increasing focus on "interactive" display (meaning good and thoughtful rapport of visitor and object when done well, and glitzy, noisy, pushbutton-activated nonsense when done poorly) have become the norm.

Much as I love the cabinet of full variety, I could not defend Dublin's decision if this exhibit in the old style usurped all available space for displaying natural history. After all, we have learned something since the last century, and many of the newer techniques work well, particularly in getting children excited about science. But Dublin has found a lovely solution. They have restored their original housing to one of the world's finest and fullest exhibits in the old and still-stunning cabinet style—not just a room to showcase the past, but an entire building in full integrity. And they have opened a new building on the next street for needed exhibits in a more modern vein (now featuring the great inevitability of our contemporary passions—a display about dinosaurs).

I would not be defending the cabinet style if such museums only honored a worthy past. I support this ideal of fullest possible display as currently vital and exciting, as capable as ever of inspiring interest (as well as awe) in any curious person. I agitate for these old-style museums because they are wonderful today. They provide, first of all, a richness in variety not available elsewhere. When I visited the Dublin Museum, for example, a college course in drawing had convened on the premises—and each student sat in front of a different mammal, sketching at leisure.

But a second reason beyond immediate, practical utility must be embraced if my argument has any power to persuade. This more subtle, and controversial, point was beautifully expressed by Oliver Sacks in two letters written to me, first in December 1990, and then in September 1992:

> My own first love was biology. I spent a great part of my adolescence in the Natural History museum in London (and I still go to the Botanic Garden almost every day, and to the Zoo every Monday). The sense of diversity—of the wonder of innumerable forms of life—has always thrilled me beyond anything else.

> Love of museums was an intense passion for me, for many of us, in adolescence. Erik Korn, Jonathan Miller, and I spent virtually all our spare time in the Natural History Museum, each of us adopting (or being adopted by) different groups—holothuria (Erik), polychaetes (Jonathan), cephalopods (myself). I can still see, with eidetic vividness, the dusty case containing a *Sthenoteuthis carolii* washed up on the Yorkshire coast in 1925. I have no idea whether that case, or any of the dusty cases we were so in love with, still exist—the old museum, the old museum *idea*, has been so swept away. I am all for interactive exhibits, like the San Francisco Exploratorium, but not at the expense of the old cabinet type of museum.

None of these three teenagers grew into a professional zoologist (though others of the same clone and cohort, including me at the New York museum, did)—but all became men of great accomplishment, at least partly because they maintained (and transferred to their chosen profession) a love of detail and diversity, as nurtured by cabinet museums. Erik Korn is England's finest antiquarian book dealer in natural history; Miller's work, in medicine and theater, and Sacks's, in neurology and psychology, are well known to all. Sacks, in particular, has based the passionate humanism of his unique insight into individual personalities—his revival of the old "case study" method in medicine—upon his earlier love for zoological taxonomy. He continued in his letter to me, "I partly see my patients (some of them, at least) as 'forms of life,' and not just as 'damaged,' or 'defective,' or 'abnormal.'" These "old fashioned" museum displays had a profound effect upon the lives of three supremely talented yet remarkably different men.

I must therefore end with a point that may seem outstandingly "politically incorrect," but worthy of strong defense nonetheless. We too often, and tragically, confuse our legitimate dislike of elitism as imposed limitation with an argument for leveling all concentrated excellence to some least common denominator of maximal accessibility. A cabinet museum may never "play" to a majority of children. True majorities, in a TV-dominated and anti-intellectual age, may need sound bites and flashing lights—and I am not against supplying such lures if they draw children into even a transient concern with science. But every classroom has one Sacks, one Korn, or one Miller, usually a lonely child with a passionate curiosity about nature, and a zeal that overcomes pressures for conformity. Do not the one in fifty deserve their institutions as well—magic places, like cabinet museums, that can spark the rare flames of genius?

Elitism is repulsive when based upon external and artificial limitations like race, gender, or social class. Repulsive and utterly false—for that spark of genius is randomly distributed across all the cruel barriers of our social prejudice. We therefore must grant access—and encouragement—to everyone; and we must be unceasingly vigilant, and tirelessly attentive, in providing such opportunities to all children. We will have no justice until this kind of equality can be attained. But if only a small minority respond, and these are our best and brightest of all races, classes, and genders, shall we deny them the pinnacle of their soul's striving because all their colleagues prefer passivity and flashing lights? Let them lift their eyes to hills of books, and at least a few museums that display the full magic of nature's variety. What is wrong with this truly democratic form of elitism?

While in Dublin, I also visited St. Michan's Church, with its beautifully carved organ played by Handel (though some dispute the claim) at the premiere of *Messiah*, first performed in Dublin in 1742—Handel, who wrote four great odes for the coronation of George II; the same King George who then granted a royal charter that eventually led to the Dublin Natural History Museum. And I thought of my favorite chorus (not "Hallelujah!") in part two of *Messiah*, set by Handel with a richly polyphonic beginning and a strong homophonic ending—a lovely analogy, I thought, to the interplay of nature's wondrously variegated diversity with the unity of taxonomic order and evolutionary explanation, contrasting themes so well displayed and intertwined in Dublin's Natural History Museum. And I thought of the words, expressing the most noble mission of teachers: to expand out to the ends of knowledge, and then to gather in—by song, by writing, by instruction, by display. "Great was the company of the preachers . . . Their sound is gone out into all lands, and their words unto the ends of the world."

19

Evolution by Walking

THINK OF THE GREAT DRAMATIC CONFLICTS between women and sinister forces—Fay Wray and King Kong, Sigourney Weaver and the Aliens, or even, if your tastes wander to culture in the canon, Portia and Shylock (even more disturbing, in its anti-Semitism, than gorillahood or parasitic alienation, and unredeemed by Portia's status as the strongest feminist in the group). But I still cast my vote, in this genre, for Raquel Welch and the antibodies in *Fantastic Voyage*. For Ms. Welch, as part of a scientific team reduced to microbial size and injected into a human body, leaves the vehicle for an internal "space walk," and must then struggle with a horde of murderous antibodies who correctly identify her as non-self and merely try to do their appointed job.

I give the nod to Ms. Welch and the antibodies because she embodied a small but distinguished genre of pedagogy that I find particularly effective in using the visceral to grasp the cerebral: scaling a human being up or down in order to illustrate a concept by moving the body directly through a process or phenomenon. I have, for example, been a pawn (literally) in a very large game of chess—and I really did understand the game better after I moved doggedly forward, slipped cleverly on a diagonal to murder a brother of another color, and finally succumbed to the ecclesiastical sweep of a distant bishop along the same diagonal.

Similarly, museum exhibits on the heart may treat each visitor as a blood cell moving through corridors shaped as vena cava and aorta, into rooms modeled as auricles and ventricles. Examples can also become more abstract but shaped to the same end. In a great classic of modern popular science writing *(Mr. Tomkins in Wonderland)*, George Gamow beautifully illustrated the arcana of relativity and quantum theories by granting them control over ordinary human activities at the scales and times of our own bodies, rather than at tiny sizes or great speeds beyond our experience. (His quantum tiger hunt, for example, is a lovely exercise in frustration, as gunners can only fire at a probability spectrum of tiger positions and hope for the best.)

I write this essay to praise the newly opened halls of fossil mammals at the American Museum of Natural History. I admire the new exhibits because they follow the human chessman strategy of pedagogy in teaching us about evolutionary trees by organizing the entire hall as a central trunk and set of branches—and then, so to speak, placing our brains in our feet and letting us learn by walking. Moreover, the chosen geometry of evolutionary organization in this new hall of mammals violates our traditional picture of life's history, thus illustrating, in an unusual scale (large) and mode (visceral), a favorite theme of these essays and an important principle in the history of science: the central role of pictures, graphs, and other forms of visual representation in channeling and constraining our thought. Intellectual innovation often requires, above all else, a new image to embody a novel theory. Primates are visual animals, and we think best in pictorial or geometric terms. Words are an evolutionary afterthought.

The power of pictures, as epitomes or encapsulators of central concepts in our culture, may best be appreciated in studying what I like to call "canonical icons," or standard images that automatically trigger a body of associations connected with an important theory or institution in our lives. The power of canonical icons is greatly enhanced by our highly sophisticated neurological capacity to abstract, process, and distinguish images based on small differences in form. We are particularly good at simple shapes with defining features and clear symmetries—hence the skill of the caricaturist in knowing that we will all recognize a famous figure by exaggeration of one or two key traits.

As I write this essay in December, I think of two simple shapes in bilateral symmetry—the branched menorah as a hemisphere and the Christmas tree as an isosceles triangle—and another pair in radial symmetry—the star of David with six points, and of Bethlehem with five—and consider the power of such simple imagery to provoke immediate recognition and then to evoke emotions of great depth and even danger (for we may as easily be led to tears as to a march into battle). And I remember just how disturbing a departure from canonical iconography can be. For example, as a Jew with no particular stake in the matter, I felt visceral distress when I first saw the beardless Jesus of Byzantine imagery. Distress then triggers thought—another source of iconography's power—and I recalled that we know absolutely nothing about the physical appearance, and for that matter preciously little about the actual existence, of the historical Jesus. Yet a tall, gentle, bearded Caucasian Savior has supported the prayers and intentions of billions.

All scientific theories call upon canonical icons for their illustration (and, often, even for their definition), but my field of evolution and the history of life has been more dependent than most, hence my interest in this theme. My sensitivity has been heightened by the fact that all canonical icons in my profession have presented the same encompassing but fallacious view based on social traditions and psychological hopes rather than the fossil record—namely, the idea of progress as an organizing principle—and that exposure and correction of such iconography therefore assumes special importance for paleontologists.

The canonical icon of life's history presents a series in linear order, starting with something deemed primitive and old, and ending with *Homo sapiens*. These linear arrays may take many forms: a vertical sequence of increasing complexity (the infamous ladder of progress), a horizontal order from stooped ape to upright human (the literal "march of evolutionary advance," always depicted left to right, as we read, although the only Israeli example that I have seen—a recent Pepsi ad—runs right to left), a sequence of exhibits as we walk down a rectangular museum hall, or the order of chapters in a conventional textbook, with protozoans first and mammals last.

As a first, and surely welcome, correction often advocated in these columns, we can switch our icon from ladders to trees and gain accuracy while shedding some prejudices. But this basic topological shift does not solve the problem of progressivist bias for another iconographic reason: trees have conventionally been drawn with their own more subtle set of geometric devices for depicting evolution as continuous advance. The traditional evolutionary bush is a "cone of increasing diversity," an inverted Christmas tree with a central trunk of common ancestry, just a few branches near the base (with each as the primitive precursor of later flowerings), and steady increase in both number of branches and horizontal spread into a verdant canopy at the tree's summit.

Upward growth, according to the captions for most evolutionary trees, only means movement from older to younger periods of geological time, but up and down also stands for good and bad in our mythology (beautiful flowers on thin branches vs. dirty and gnarly roots; the brain vs. the bowels; heaven and hell, Valhalla and Nibelheim)—so *up* on the tree usually becomes conflated with more progressive, and the tree of life, when drawn as a cone of increasing diversity, mimics the ladder of progress to tell the same false tale.

If fire must be fought with fire, then this great fallacy at the core of our canonical icons for evolution can only be corrected by developing and popularizing a more accurate picture. Trees rather than ladders bring us part of the way, but, as my previous paragraph notes, trees may also be drawn restrictively to support the central fallacy of predictable progress. How, then, can we draw a tree differently, to emphasize the genealogical connections that define evolutionary relationship while eschewing the confusion of upward growth with progress, and avoiding the implication that trees always spread out toward more and better as they grow? This search for a new iconography is no minor task, no little frill to an effort that can only be fulfilled in the verbal mode. Icons are primary fashioners of thought, and the search for fundamentally new representation therefore becomes one of the most important efforts a scholar can undertake.

Several biologists have tried to develop new icons to record contingency and unpredictability as central themes in the history of life (see my book *Wonderful Life* for an example). But I write now to praise my colleagues at the American Museum of Natural History for carrying the task further in the different context of an exhibition hall.

Just as paper provides a substrate for a variety of icons to embody the central fallacy of predictable progress—ladders, marches, cones of increasing diversity—so too does the space of a conventional museum exhibit on the history of life record the same biases. Most museum halls are rectangles with a preferred linear flow of visitors in one direction along the major axis. All exhibits on the history of life that I have ever seen in any museum, anywhere in the world, organize the space provided in one of two ways—and both, however unconsciously and often without explicit signage, use the bias of progress as a central principle for arranging organisms.

One favorite scheme simply organizes fossils in temporal order, oldest at one end of the hall, youngest at the other. Such a strategy does not necessarily (and really should not) record the bias of progress, for geologically young does not mean anatomically complex, and bacteria continue to rule the world today, as they have since life's beginnings (and will until the sun explodes). Rather, temporal order expresses the bias because museum exhibits construe arrangement by time in such an oddly restrictive and grossly distorted way. Displays begin with an alcove of invertebrates, followed by a case of fossil fishes, then amphibians, reptiles (including dinosaurs), mammals, and finally human mammals.

All the fossils may be arrayed in temporal order, but what a peculiarly biased set of choices! After all, invertebrates didn't go away, or didn't stop evolving, once fishes arose. And fishes didn't stagnate just because one odd lineage crawled out on land. In fact, the rise of the Teleosti, or so-called "higher bony fishes," must be regarded as the most important event in all vertebrate evolution (at least as recorded by number of modern species)—for more than half of all living vertebrate species are teleosts. But this great group of fishes arose and spread late in geological time, during the reign of dinosaurs on land—so this cardinal event is either not shown at all or relegated to a little space at the "bad" end of the hall.

In other words, temporal order is not construed as a set of representative samples for all animal groups through time, but as a sequential tale of most progressive at any moment, with superseded groups dropped forever once a new "ruler" emerges, even though the old groups may continue to flourish and diversify.

The other major organizing principle presents the same scheme, but doesn't even try to mount a defense in terms of temporal order. This second strategy just announces that life will be shown as a modern morality play of movement from lower to higher. Dinosaurs come before mammals (even though both groups arose at the same time and mammals lived throughout the dinosaurs' reign) because dinosaurs were dumb, lumbering, and primitive; while mammals come last because they are us (more elaborate justifications may be given, but do not doubt this bottom line).

In imagining how an alternative and intellectually challenging iconography might be constructed, I remember the stunning impact that a simple device in a fine book once had upon me. When I studied vertebrate paleontology as a graduate student, we used Ned Colbert's classic textbook, *The Evolution of Vertebrates*. I shall never forget the disturbing sequence of his chapters on mammals. He devoted a section to each of the twenty or so mammalian orders, and he presented them in linear sequence as books must do. But he treated primates, including humans, fifth rather than last. In other words, he discussed the rise of australopithecines, the emergence of *Homo erectus*, the interactions of Cro-Magnon and Neandertal in Europe, *before* he presented the evolution of pigs, elephants, and sea cows.

I was thrown into confusion at first, but then I started to think—and praise be to any scholar who can so provoke a neophyte out of his field's conventions. Why are humans always treated last? After all, although we arose late in geological time, other mammalian groups emerged still later. And we are not, by any stretch of anatomical imagination, particularly far along the sequence of mammalian branches, for we still sport many original mammalian features that later-branching groups lost (including five discrete digits on each hand or foot, all with nails retained).

I then realized what Colbert had done—and his simple device was truly brilliant. He had arranged his chapters *by order of branching,* not by later degree of anatomical "achievement" (usually misinterpreted in human terms). Primates had branched early in the history of mammals, so their chapter belongs near the beginning—no matter what certain peculiar primates later achieved in elaborating one particular organ to an unprecedented degree that, ultimately, may or may not do themselves and the rest of the planet much good.

The fourth floor of the American Museum of Natural History was the shrine, the principal magic place, the sanctum sanctorum of my youth. I first visited with my father at age five, and decided right then to dedicate my life to paleontology. I went back to the dinosaur and fossil mammal halls almost monthly throughout childhood, right to the end of high school. I then left New York for undergraduate studies, but returned to do my Ph.D. at the museum. I loved the old halls, but they had become shabby and outdated. So I did not lament their closure a few years ago for a thorough rethinking and redoing.

The two halls of fossil mammals reopened in 1994. I do not love everything about them. I miss the cases crowded chockablock with titanotheres. But, for the restored murals of Charles R. Knight, with their beautiful bright colors rescued from overlying layers of varnish, grime, and smoke, I can only make comparison with before and after the great cleaning at Michelangelo's Sistine Chapel. But for all my ambivalence at certain points, I offer unstinting praise for the intellectual concept behind the entire presentation. My colleagues have actually done it (writers are trained never to use such indistinct words as "it" in such a context—but this is the big "it," the only "it"). They have ordered all the fossils into an unconventional iconographic tree that fractures the bias of progress. They have created this new icon at gigantic scale, so that we can perambulate along the tree of life and absorb the new scheme viscerally by walking, rather than only conceptually by reading. They have, in short, taken Colbert's radical idea and arranged all the fossils by their branching order, not their later "success" or "advancement." Groups that branch early appear early in the hall, even if they later diversified to a dominating degree (rats and bats) or generated lineages that we regard as particularly complex or advanced (primates). Sea cows and elephants are at the end of the hall, horses in the middle, and primates near the beginning.

Since this essay extols iconography, I must not praise this innovation with words alone. So consider the diagram in the guide pamphlet distributed to visitors. Note that a central dark line indicates the preferred route of passage through the two halls devoted to "mammals and their extinct relatives." Major branching points, numbered one through six, appear along the line in temporal order of events during the history of mammals and their ancestors.

The method used to identify these key branching points follows a theory and philosophy of taxonomy that has motivated most of my colleagues at the museum during the past twenty-five years, stimulating a fine body of important research. (I should say that I am quite agnostic about this theory, called *cladistics*, so I do not write as a shill.) Cladistics classifies organisms in nested hierarchies based exclusively on their order of branching.

The new treelike icon constructed from these branchings is called a cladogram (see the accompanying illustration, also taken from the visitors' pamphlet). The major, sequential branches on the cladogram are defined by traits that arose since the last branching point, and have been held in common by all subsequent lineages on this branch of the tree (such traits are called "shared-derived," or, if jargon attracts you—and this field has invented the most god-awful argot—"synapomorphic," which means the same thing). For example, the earliest mammals did not form a placenta for their embryos, and some modern groups—the egg-laying monotremes (platypuses and echidnas) and the pouch-bearing marsupials—are offshoots of these pre-placentals. The placenta evolved after monotremes and marsupials branched off the main stem, and all subsequent mammals bear placentas. The placenta is therefore a shared-derived character of later mammals, and monotremes and marsupials therefore achieve their early position in the museum's hall, for they must be placed before the branching point marked by acquisition of the placenta (number 3 on the diagram).

To illustrate how sequences of shared-derived characters can be used to build an icon of nested branchings based solely on temporal

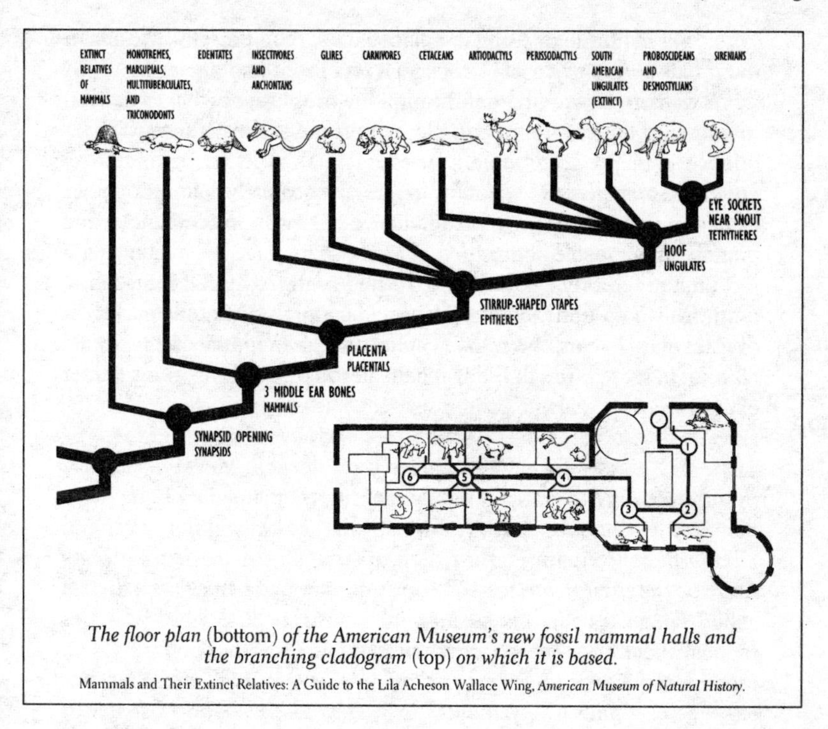

The floor plan (bottom) *of the American Museum's new fossil mammal halls and the branching cladogram* (top) *on which it is based.*

Mammals and Their Extinct Relatives: A Guide to the Lila Acheson Wallace Wing, *American Museum of Natural History.*

order of bifurcations, and not on perceived progress or increasing complexity, let me cite the example presented in the visitors' pamphlet:

> Sharks, salamanders, lizards, kangaroos, and horses all have a backbone composed of vertebrae and belong to a large group called vertebrates. Of the animals mentioned, only salamanders, lizards, kangaroos, and horses have four limbs. So they are more closely related and belong to a group called tetrapods, meaning four-footed. Within tetrapods, lizards, kangaroos, and horses develop in water-tight eggs that are either laid by the mother or are retained inside the mother until the embryo is born. The watertight membrane inside the egg is called the amnion, so lizards, kangaroos, and horses belong to a group called amniotes. Only kangaroos and horses produce milk for their young and have three bones in their ears to conduct sound vibrations. So they are more closely related and belong to a group within amniotes called mammals.

We might continue by using the placenta to group horses with all later mammals and to place kangaroos on a preplacental branch.

A walk in suggested order through the museum's two halls of fossil mammals directs visitors along the main route of the cladogram. A sequence of six key shared-derived characters marks the temporal sequence of branching and establishes the genealogically defined topology of nested groups within more inclusive aggregations. Lineages that branch off before the acquisition of a new shared-derived character appear, in the exhibition hall, before the node defined by that character—as in my example of monotremes and marsupials before the evolution of placentas. The exhibit walks visitors through mammalian history according to six features defining a genealogy of branching, not a ladder of putative advance.

1. The synapsid opening. Late in the Paleozoic era, more than 250 million years ago, a group of reptiles developed an opening in the skull behind the eye socket. This feature apparently evolved but once, and all subsequent creatures with such an opening share a unique genealogical heritage with the common ancestor that first evolved this so-called synapsid opening. All mammals possess this feature (muscles that close the lower jaw attach to the skull around this opening), as do members of ancestral groups formerly placed among the reptiles. For example, the famous pelycosaurs (early tetrapods, like *Dimetrodon*, with sails on their backs, featured in all kiddie plastic "dinosaur" sets from cereal boxes) are synapsids and therefore genealogically closer to mammals than to dinosaurs. By placing pelycosaurs at the foot of the mammal hall, as members of the first branch following acquisition of the synapsid opening, we do not stake a claim for their superior organization or greater complexity (I suspect, in fact, that true dinosaurs were markedly more capable creatures in all important functional ways), but only affirm the genealogical link of pelycosaurs to all other synapsids, including mammals.

2. Middle ear bones. The two bones that articulate the reptilian jaw decreased in size and moved into the middle ear in mammals, where they joined the stapes (or stirrup, the only hearing bone in reptiles) to become the malleus and incus (hammer and anvil) of our middle ear. This highly distinctive feature, ascertainable in the fossil record, defines the branch point separating true mammals from their ancestors. In the new hall, monotremes and marsupials branch off here because they evolved the three ear bones (feature 2), but not the subsequent placenta (feature 3).

3. PLACENTA. As discussed above, the placenta developed after monotremes and marsupials branched off, and all subsequent mammals possess this shared-derived character. In the hall, edentates (sloths, anteaters, and armadillos) branch off here because they have placentas (feature 3) but not feature 4.

4. THE STIRRUP-SHAPED STAPES. In reptiles and early mammals, the stapes is a simple rod. But later on, at this fourth bifurcation, a hole evolved in the stapes (an important blood vessel goes through it), and all subsequent mammals possess this perforated (or stirrup-shaped) stapes. At this key conceptual site in the exhibition (right after room one, at the foot of the second room), all mammals with a stirrup-shaped stapes, but without subsequent feature 5 (the hoof), branch off and thereby secure their location in the sequence. Several major groups diverge at this bifurcation, including carnivores, rodents, bats, and—note the parochial key to a conceptual revolution—primates. Thus, human fossils are properly housed at the beginning of the second hall, in the appropriate position of their genealogical branching, and not at the apex of the icon, or the end of the story.

5. THE HOOF. At a later point, toes coalesced to form hooves in the ancestry of subsequent mammalian groups. The major hoofed orders therefore occupy this large central area in the second hall (horses, rhinoceroses, and tapirs among the perissodactyls, or odd-toed hoofbearers; cows, pigs, sheep, goats, giraffes, deer, antelopes, and many others in the phenomenally diverse group of artiodactyls, or even-toed hoofbearers; a group of distinctive and extinct South American forms; and whales, who evolved from hoof-bearing ancestors, despite their later loss of this feature for obvious reasons of aquatic adaptation).

6. EYE SOCKETS NEAR THE SNOUT. As a final shared-derived character, eye sockets move forward on the skull to a position near the snout—a feature defining the genealogical tie of elephants to sea cows and their relatives.

Consider the extent of the conceptual shift embodied in this new icon based on a different criterion for ordering groups: humans used to occupy the end of the hall, conceived as a pinnacle of progress and complexity, and defined by a criterion of dominion—our exalted brains. Sea cows now cavort at the end because they share, with elephants, a unique feature that evolved late in a sequence defining the major groups of mammals by their order of genealogical branching.

While I strongly praise this new icon for its iconoclasm—literally breaking the conceptual lock imposed by the old canonical drawings, all based on progress as the central feature of life's history—I must remain true to the theme of this essay: all icons embody theories, and are therefore capable of both breaking the conceptual locks of earlier, inadequate icons, and of introducing new (often subtle and unrecognized) biases of their own. Choose your cliché: we are on a two-way street; tit for tat; hoist by one's own petard. This new icon dispels the prejudice of progress, but also introduces some unfortunate restrictions and distortions imposed by strict adherence to cladism as a philosophy of systematics. For cladism is not an accommodating or pluralistic theory, carefully balancing all legitimate concerns, but a zealous advocacy of one admittedly vital criterion: temporal sequences in branching order as the only proper way to depict relationships among organisms.

Two features of evolution, considered terribly important by most professionals (myself included), must be marginalized, if not ignored outright, by cladistic icons: unique traits evolved by single lineages, and trends within groups that do not lead to further branching events. Unique traits, called autapomorphies in the jargon, do not define branching points because they evolve within a single group. Cladists may acknowledge such features, but the new icon grants them no space and no picture. And yet such traits define much of the fascination for evolution among members of the general public, and among specialists as well. We really do want to know what pelycosaurs did with their sails, saber-toothed tigers with their teeth, narwhals with their single tusk, duckbills with their duckbill, armadillos with their armor, and, dare I say so, humans with those damnable brains that keep me up writing essays at four in the morning, five days before Christmas. And we want to understand sustained trends within groups: What about the famous story of fewer toes and higher teeth among horses? What about loss of hind limbs in whales? What about increasing brain size in humans? Again, cladists have little to say because transformations occurring on a single line of a cladogram have no iconographic representation within their system.

Moreover, and especially, such unique features and trends can only be treated in (literally) peripheral positions within the new hall of mammals—for attributes of single groups must be placed off the main line in the corners and alcoves devoted to later change within lineages that have already branched from the central sequence. The iconography that stirred us up now detracts from our legitimate interests by spatially marginalizing some of evolution's most fascinating phenomena.

But let me not carp. What liberating revision, in scholarship, politics, or any human endeavor, ever intruded upon our history without growing pains, or fully formed in optimality? We are, after all, evolutionists, and we do believe in imperfection and change. The followers of Aristotle were called peripatetics because the "master of them that know" valued the linkage of cogitation and ambulation (the covered walk in Aristotle's Lyceum was a *peripatos*). And Emerson made the same connection in his famous plea for a distinctively American excellence: "We will walk on our own feet . . . we will speak our own minds." Let us, then, praise the large scale of this delightful attempt to stir our mental machinery by exercising the historically prior and equally fundamental change that made human evolution possible: upright posture and bipedal locomotion.

The Razumovsky Duet

I LIVE ON A SMALL STREET of some twenty houses, all closely packed together. I presume that most of my neighbors share my mental geography of this terrain—structural divisions into roadway, sidewalks, houses, and gardens, with primary taxonomic separations set by property lines of ownership. But the street also features more cats than people—and I know that these members of the mammalian majority divide the space differently. They clearly have some sense of ownership and territory—for charges, howlings, and spats occur on a daily basis—but their parsings do not match the human separations. Think what we might understand about mammalian mentality if we could ever obtain the cat map of Crescent Street. A potential informant does share my living quarters, but he has been persistently unresponsive and uncooperative (and he still gets fed!).

In a zoologically more restricted framework, this important theme of alternate mappings might give us great insight into differences among human cultures, times, and mentalities—as the French *annales* school of historians has taught us, with their emphasis on changing patterns in ordinary life, featuring working men and women rather than kings and conquerors. In school, I learned the conventional history of dates, nations, and battles. My mental maps of time and geography all follow the usual lines: temporal divisions by kings and presidents, spatial boundaries by nations and languages. But other systems make as much sense, and surely have more relevance to people whose primary activities enjoin different divisions.

I assume that sailors prefer size or function to place of registration as a criterion for classifying vessels (especially when so many ships, for reasons of taxation and licensing, officially carry the Liberian or Panamanian flag). My *Cerion* snails do not recognize a political separation between the Bahamas and the Turks and Caicos Islands, for these places form a unified climatic and geographic province well suited to molluscan lifestyles (yet I have long been frustrated by the difficulty of obtaining maps with both political entities shown in the same way, and to the same scale). This essay emphasizes the artificiality of conventional linguistic and national boundaries in European history—by telling a story about scientists and artists who might have parsed the early 18th century world along different lines of patronage.

So many projects, both little essays like this and lifetimes of effort, begin by accident; you can't, after all, explicitly search for the unexpected. I bought an old book a few months ago, largely because it was so underpriced and the opportunity would therefore not arise again. Its author—Johann Gotthelf Fischer von Waldheim (1771–1853)—cannot claim a place in the generally recognized pantheon of great scientists. But he occupies a small spot in the heart of paleontologists as one member of the pioneering generation who established the basic ordering of life's history when scientists codified the geological time scale. We also owe a debt of thanks to Gotthelf Fischer (to cite the abbreviated moniker that he used in most of his own publications) for popularizing, and perhaps inventing, our modern name of *paleontology*. (Otherwise we might still be stuck with one of the earlier, unpronounceable alternatives, like *oryctology*.)

Fischer was German by birth (from the town of Waldheim in Saxony, which Fischer added to his name when the Russian Czar ennobled him in 1817). He studied with Cuvier, became friendly with Goethe, traveled with the Humboldt brothers, taught in various German universities, and finally, in 1804, moved permanently to Russia as Professor of Natural History and director of the natural history museum at the University of Moscow. Lest his last post seem anomalous, Fischer's emigration followed a common pattern fostered in Russia ever since Peter the Great (1672–1725) sought to leap over Russia's scientific backwardness by importing specialists from other nations, and by purchasing foreign collections. (Peter developed his own great museum of natural history—still partly on view in the building constructed to house it, the Kunstkamera of St. Petersburg—by acquiring two important Dutch collections.)

The growth of Russian universities during the late eighteenth and early nineteenth centuries provided another pathway for importation of foreign scientists and professors—for Russia, lacking universities in its past, could not subsist on home-grown specialists. Fischer's residency in Moscow, beginning in 1804, lies sandwiched between arrivals of the two greatest German-speaking biological emigrés to Russian universities—Peter Simon Pallas in 1768 (see next essay) and Karl Ernst von Baer in 1834. (Both men taught in St. Petersburg. Pallas was born in Berlin, while von Baer belonged to an old Prussian family, then living in Estonia. Von Baer, the greatest embryologist of the nineteenth century, was imperial Russia's greatest academic "catch." He discovered the human ovum in 1827, but abandoned embryological research for an astonishing variety of brilliant studies in ethnology, anthropology, and geomorphology during his Russian years.)

Fischer had a remarkably fruitful and successful career in Russia. He founded three journals (published in French), and wrote nearly two hundred books and articles (also predominantly in French) during his Russian years. His wide-ranging publications spanned all of zoology, but concentrated on pioneering work in describing living Russian insects and fossils of all groups. Fischer was widely recognized by the international scientific community, and became an active or honorary member of nearly ninety institutions and academies, including the American Association of Arts and Sciences in Boston, and the American Philosophical Society in Philadelphia. At his Jubilee in 1847 (to celebrate the fiftieth anniversary of his doctorate), Alexander von Humboldt, the world's most popular living scientist at the time, extolled Fischer as *mein edler, ältester Freund* ("my noble and oldest friend"). Fischer wrote home to his friends in Waldheim, describing the ceremony. He spoke of the six carriages that led the way, five of them drawn by four horses and the last by six. He described the gifts and the encomia, ending with his reaction: *Hier brachen mir die Thränen der Dankbarkeit aus meinen Augen* ("this brought tears of thanks from my eyes").

The book that I bought is not one of Fischer's paleontological works, but a superb example, published in 1813, of a genre that computer technology has driven to extinction—bibliographic compendia of alternative systems for classifying objects, in this case all genera of the animal kingdom. Fischer uses Linnaeus's account of animal genera as a framework, but, in a series of ingenious tables, charts, and lists, displays the correspondences between Linnaeus and all other major systems proposed by leading zoologists, mostly French and German. (The visually attractive result looks like a chart of comparisons among the three synoptic gospels, or among various biblical translations.) The effort may be primarily bibliographic, but we learn much of historical and theoretical interest from Fischer's compilation. For example, historians have often claimed that Lamarck's *Philosophie zoologique*, his major exposition of evolution, written in 1809, was widely ignored as a fatuous and speculative treatise. But one of Fischer's longer charts presents a *Tabula clarissimi Lamarck* ("list of the most celebrated Lamarck"), reproducing Lamarck's chain of being in full evolutionary order, just four years after its publication.

The title of Fischer's book records its intended audience and utility: *Zoognosia: tabulis synopticis illustrata, in usum praelectionum Academiae Imperialis Medico-Chiurgicae Mosquensis* (Zoological knowledge: illustrated with synoptic tables for lectures at the Royal Academy of Medicine and Surgery in Moscow). Such single-volume compendia must have been especially useful for students with little access to extensive libraries and collections. (My own copy went to an even more peripheral location with even more limited primary material, for Fischer has inscribed it "to the Society of Arts and Sciences of Courland"—a duchy on the Baltic Sea, now part of Latvia, that became a Polish fiefdom in the sixteenth century, but passed to the Russian empire in 1795 after the third partition of Poland. Courland enjoyed a seventeenth- century moment of glory, and even held enough power to build a small colonial empire in the West Indies [Tobago] and Africa [Gabon].)

I love to read the dedications of old books written in monarchies—for they invariably honor some (usually insignificant) knight or duke with fulsome words of sycophantic insincerity, praising him as the light of the universe (in hopes, no doubt, for a few ducats to support future work); this old practice makes me feel like such an honest and upright man, by comparison, when I put a positive spin, perhaps ever so slightly exaggerated, on a grant proposal. Fischer's dedication to *Comes Illustrissime, Domine Clementissime*! (Most illustrious count and most compassionate lord!) first struck me as an example of *vin ordinaire* within the genre. But two features of the text sparked my interest and led to this essay. First, Fischer's lament over impediments to his research seemed more fervent and more extreme than usual—for evident reasons most easily inferred. He writes that he is dedicating his book to the illustrious count, and would also like to so dedicate the natural history collections of the museum, but *eheu, omnes perierunt, restant paucissimae.* ("Alas, everything has been destroyed, so very little remains.") He then asks what will happen *post tot calamitates luctuosas casusque tristissimos, quos Musarum cultores Mosquensis experti sumus* ("after so many sorrowful calamities, and so many most dismal events, that we supporters of the muses of Moscow have endured").

I happen to be writing this essay on July 4, so local events in Boston are primed to dramatize the source of Fischer's complaints. Just before sunset, as they do every year, the Boston Pops Orchestra, on the esplanade adjoining the Charles River, will play Tchaikovsky's *1812 Overture*, with glorious fireworks commencing at the booming of cannons (marking the conclusion of this loudest piece in the classical repertory). When I was a child, wallowing in patriotism, I couldn't quite figure out why Tchaikovsky had written an overture to celebrate a little skirmish of an American war that we didn't win. I discovered later that a few other events of passing importance occurred in 1812, most notably, and right in Tchaikovsky's bailiwick, the capture of Moscow by Napoleon, followed by his subsequent retreat and decisive defeat, as the old saying goes, by those greatest of Russian generals, November and December. Napoleon had entered Moscow on September 14, hoping to win quick and favorable peace terms from Czar Alexander. But the Czar would not deal with him and, more important, a massive fire had broken out in Moscow on the day of Napoleon's entry, eventually destroying more than two thirds of the city, and preventing Napoleon from feeding and housing his troops through the winter. The fire may have aided Russia by helping to force Napoleon's withdrawal and subsequent defeat, but the flames also wrecked most of Moscow, including the great libraries and museum collections of the university. Fischer, in other words, had been one of so many victims, largely anonymous to later history, of a signal event in the construction of our modern world.

As a second intriguing feature of Fischer's dedication, I then considered the identity of the illustrious count from whom he sought patronage—and not so subtly—by ending his dedication: *in Te, Comes Illustrissime, spes omnis nostra collocata est.* ("All our hope is placed in you, most illustrious count.") Fischer dedicated his book to Alexis Kirillovich Razumovsky, the Minister of Public Instruction.

Now, speaking of music, and going back a generation or two from Tchaikovsky, we meet a Russian reference known to all lovers of the classics—for Beethoven dedicated three of his most famous string quartets (Opus 59) to a gentleman of the same name. (They are, in fact, called the Razumovsky Quartets, and the first two feature Russian folk melodies in their composition.) I couldn't help wondering about the relationship between Beethoven's actual and Fischer's anticipated patron. Beethoven wrote the quartets in 1806, and Fischer sought help seven years later, so the events are nearly contemporary. The story, as so often happens in our fascinating world of complex interactions, turned out to be well worth reporting.

Beethoven's patron, Andrei Kirillovich Razumovsky, was the brother of Fischer's *Comes Illustrissime*—but the resemblance extends little beyond genealogy. Their story begins two generations further back, with a Ukrainian cossack named Grigor Rozum, who had two remarkable sons. Again, music served as the prod to success. One of his sons, Aleksei Grigorevich Razumovsky (1709–1771), became a singer in the court choir of St. Petersburg. There he attracted the attention, and later the love, of Princess Elizabeth, who became Czarina of Russia in 1741. Aleksei secretly married Elizabeth in 1742. They had no children, and Aleksei took little interest in affairs of state. But he remained the favorite of his secret wife, and became enormously wealthy thanks to her largesse. His brother Kirill Grigorevich (1718–1803, and father of the Razumovskys in our tale of Fischer and Beethoven—as their common patronymic, Kirillovich, indicates) was a more ambitious and accomplished man. He served, for nearly twenty years, as president of the St. Petersburg Academy of Sciences, but held much greater political power (including dominion over more than 100,000 serfs) as the last hetman of Little Russia (ruler of the Ukraine). His two sons therefore inherited all conceivable advantages of wealth and position.

The younger son—Andrei Kirillovich, Beethoven's Razumovsky (1752–1836)—was a warm, generous, and liberal man, and one of the great art patrons of Europe. He spent his professional life as a diplomat in central Europe, wooing women (most notoriously the Queen of Naples) and making deals. He served as Russian ambassador to Vienna from 1790 to 1799 and again from 1801 to 1807. Razumovsky continued to live in Austria for the rest of his life, serving as a leader of the Russian delegation at the Congress of Vienna in 1815 (where Napoleon's spoils were split among the victors, thereby establishing yet another connection to Fischer and the opposite side of our story). For his service, Andrei Razumovsky was promoted from count to prince by the Czar (as His Majesty passed through Vienna).

Music had always been Razumovsky's first love (or perhaps his second, following women). In his initial post as clerk for the Russian Embassy in Vienna, he certainly knew Mozart, and probably met Haydn. He was a more than merely competent violinist and often played in the quartet that he established, largely for Beethoven's use. But he performed his finest service for music as a patron. Paul Nettl's *Beethoven Encyclopedia* states that

> Razumovsky was the most general Maecenas of his time, supporting artists, musicians, and painters. His picture gallery and musical parties were famous throughout Europe. A well-educated, liberal, and generous aristocrat, and brilliant *causeur*, he was one of the most popular and renowned aristocrats of the late eighteenth and early nineteenth centuries.

Razumovsky certainly knew and supported Beethoven as early as 1796, for his name appears on the subscription list to Beethoven's Trios, Opus 1. When Beethoven composed the Razumovsky Quartets ten years later, the Russian nobleman took title to controversial music, not to pretty oompahs—so we must respect his support for artistic license. Many musicians could not grasp or stomach the score's unconventionalities. When one Italian player belligerently asked if Beethoven truly regarded the quartets as music, the master replied, "They aren't for you, but for a later age." Beethoven also dedicated his fifth and sixth symphonies jointly to Razumovsky and to Prince Lobkowitz.

As his finest testimony and favor to Beethoven, Razumovsky established and funded a permanent string quartet, led by Schuppanzigh, in 1808—and placed the players at Beethoven's disposal. A contemporary observer noted,

> Beethoven was . . . cock of the walk in the princely establishment; everything that he composed was rehearsed hot from the griddle and performed to the nicety of a hair, according to his ideas, just as he wanted it and not otherwise, with affectionate interest, obedience and devotion, such as could spring only from such ardent admirers of his lofty genius.

This happy situation persisted until 1816, when Razumovsky held a gigantic New Year's Eve party to celebrate his elevation to princehood after the Council of Vienna. He could not accommodate all seven hundred guests in his capacious palace, so he built an adjacent wooden structure for supplementary space. This addition caught fire, and flames spread to the main palace, consuming the great library and all the works of art, and finally destroying the prince's most famous space—a room full of Canova's statues, reduced to dust and fragments as the ceiling collapsed. Razumovsky, devastated in both spirit and pocketbook, disbanded and pensioned off his quartet.

The older son—Aleksei Kirillovich, Fischer's Razumovsky (1748–1822)—lacked nearly all of his brother's admirable characteristics. Standard sources (not only the *Soviet Encyclopedia*, which might be accused of bias, but others likely to be both more objective and sympathetic) describe him as indolent, dyspeptic, litigious, domineering, and generally miserable. He married the richest heiress in Russia, but sent her packing several years later, after draining most of her wealth. Of his two sons, one was wildly dissolute and the other floridly mad; his two daughters sound quite admirable—one devoted herself to establishing hospitals for the poor—but sexist sources and limitations provide little information about their lives and fates.

Razumovsky hated court life and largely sought to avoid public responsibilities. But he did love botany and natural history, and he established, at his estate in Gorenski, near Moscow, a wonderful botanical garden (specializing in alpine plants) and the most extensive collection of books on natural history in Russia (including the library of P. S. Pallas, which he had purchased). Nonetheless, Razumovsky rarely used those resources to benefit either science or the public. A contemporary observer commented (probably not impartially):

> Count Aleksei Kirillovich has enclosed himself in his property at Gorenski in order to vegetate there with his plants. He is doing with his knowledge the same thing that he does with his immense fortune—that is, it all stays with him and confers no profit upon others.

Razumovsky did, generally after much persuasion, accept some governmental responsibilities, most notably as Minister of Public Instruction, beginning in 1810 (where sources describe him as largely inactive, but by no means incompetent). In this post, he advocated some reforms, notably the suppression of corporal punishment in the schools. He did not, however, act as an advocate of the field he supposedly loved, for he argued that the study of natural history required no place in the training of statesmen and politicians. He adopted a similar hands-off policy in dealing with Russia's universities. In his 1988 book, *The University Reform of Tsar Alexander I*, James T. Flynn writes:

> Razumovsky gave so little active leadership that each university was left to its own devices to work out its foundation and de-

> velopment. This circumstance magnified the importance of each local curator, as well as the particular local situation, and minimized the importance of the statutes and the ministry.

Not necessarily a bad thing.

Razumovsky is best known as the confidant and supporter of the deeply conservative social thinker Joseph de Maistre, perhaps best remembered for his quip that the public executioner is the guardian of social order. Maistre left France during the revolution for a life in Switzerland. He spent many years in St. Petersburg as diplomatic envoy from the King of Sardinia. There he met Razumovsky, directed his letters on public instruction to the Russian minister, and persuaded him to support a variety of conservative doctrines—including increased press censorship and greater integration of religious instruction into the curriculum.

In short, Aleksei K. Razumovsky seems an awfully poor choice as Fischer's hope for active support in rebuilding Moscow's natural history library and collections. Too bad his more affable and public-spirited brother was off playing second fiddle (literally) in a Vienna quartet. Still, I suppose one has to begin with the Minister of Public Instruction in such a dire situation, and hope for the best.

The full extent of Fischer's plight may be appreciated from an extraordinary letter that he wrote on November 20, 1812, to his German-speaking colleague Nikolaus Fuss, secretary of the Royal Academy of Science in St. Petersburg (quoted in the only available biography, by J. W. E. Büttner, titled *Leben und Wirken des Naturforschers Johann Gotthelf Fischer von Waldheim*):

> All scientific institutions are destroyed. Our university has lost so much. Of our library and museum, I was able to save so little—only the best things that I could quickly pack into twenty boxes. What is this against the beautiful totality [*Was ist das gegen das schöne Ganze!*] Your Fischer has lost everything.

Fischer then details the destruction. He only saved five books (they happened to be in his carriage at the time) of five thousand in his library. He lost his most precious items, including a completed manuscript on fossils, and his personal copy of Linnaeus's *Systema naturae* with twenty years' worth of annotations. Nearly all the natural history collections were destroyed, and Fischer particularly lamented his beautiful collection of skulls, insects, and dried plants. He also lost "all my anatomical instruments, my mineralogical apparatus," and even 130 copper plates for engravings, "including the large plate of the mammoth skeleton."

But as he continues the letter to Fuss, a strange alchemy takes hold, and Fischer's zeal and optimism break through:

> We seek to console ourselves with the thought that we are healthy, and that we do not lack bread. I do not know if I will be able to finish my *Zoognosia*, as only nine sheets have been printed [I am delighted that my book—the completed *Zoognosia*—demonstrates the happy negation of this fear]. I am now working on the latest edition of the *Onomasticon Oryctognosiae* [list of names of rocks and fossils]. The work lets me forget all my misfortunes, and makes me most happy.

What more honorable and telling mark of a true scholar—to seek solace in intellectual work, and to rebuild, by painstaking skill, all that misfortune had destroyed. To an old friend, the rector of his school in Waldheim, Fischer wrote, "To be sure, I have lost everything. But I judge myself as so much more lucky than many of my fellow sufferers, for my knowledge remains with me, and with its help, I hope to get everything back again."

And so Fischer appealed to Razumovsky for administrative and financial help. And the indolent minister did nothing. Flynn writes:

> Neither Kutuzov nor Razumovsky helped the university during the calamity of the French invasion in 1812. Razumovsky joined the Committee of Ministers, the army, and several other agencies in issuing orders to the university to close, or stay open; to stay in Moscow or evacuate . . . ; to return to Moscow, or not return, and so on.

Fischer and the other professors therefore turned to the surest of all supports: their own bootstraps and self-help. By importuning local friends for money, gifts of books, and use of buildings; working their butts off; improvising; and even holding the equivalent of yard and bake sales, the professors managed to resume operations in September 1813, with a much-reduced cadre of 129 students. By 1815 the library had reached twelve thousand volumes (compared with twenty thousand lost to the flames). Fischer himself spent the remainder of his career successfully rebuilding the great collections of his museum—with no help from

Razumovsky. As late as 1830 he was still making trips to Germany with explicit plans for replacing parts of the collection destroyed in 1812.

How, then, shall we taxonomize Europe for an optimal understanding of this complicated story about two noble brothers, their temperamental differences, and their beneficiaries, both actual and unrealized. The usual divisions of nations and languages don't seem to help much. Fischer was a German who worked in Moscow and published mostly in French. He wrote a dedication in Latin, hoping that a Russian nobleman would help to rebuild what the fire of a French invasion had destroyed. Meanwhile, this nobleman's brother lived in Austria, where he knew Mozart and then acted for years as Beethoven's most significant patron.

I do not know how students and members of other disciplines—musicians, diplomats, patrons—would choose to slice this particular cake; I can only speak from my own perspective as a scientist. My profession often gets bad press for a variety of sins, both actual and imagined: arrogance, venality, insensitivity to moral issues about the use of knowledge, pandering to sources of funding with insufficient worry about attendant degradation of values. As an advocate for science, I plead "mildly guilty now and then" to all these charges. Scientists are human beings subject to all the foibles and temptations of ordinary life. Some of us are moral rocks; others are reeds. I like to think (though I have no proof) that we are better, on average, than members of many other callings on a variety of issues central to the practice of good science: willingness to alter received opinion in the face of uncomfortable data, dedication to discovering and publicizing our best and most honest account of nature's factuality, judgment of colleagues on the might of their ideas rather than the power of their positions.

But on one issue I do have confidence, based on sufficient personal experience, of adherence to a moral code not always followed in other professions. Science does tend to be international. We share information, try hard to communicate with each other, and deplore the parochialisms that stymie contact. (How, for example, can my field of paleontology prosper if scientists are not free to collect and study the fossils of their expertise wherever they occur?) We all know numerous stories of warm and continued cooperation between scientists in nations dedicated to blasting each other off the face of the earth. We have to work this way, for knowledge is universal.

Jingoism may do no serious harm, when expressed as boycotts of silk products, unofficial bans on German opera at the Met, or campaigns to change the name of a baseball team from the Cincinnati Reds to the

Redlegs (yes, all these happened)—but science cannot and dare not ban a fact because colleagues in hostile nations discovered the information. Fischer published his *Zoognosia* just a year after French aggression provoked the holocaust of his lifetime's work, but he did not exclude the systems of Cuvier and Lamarck from his volume, for these Frenchmen were the world's greatest living taxonomists, and their work transcended the happenstance of their national origin. Fischer lost nearly all his books and specimens in the conflagration of Moscow, and then spent the rest of his life restocking his collection for future generations of students. In so doing, he called upon the generosity of a large network of scientists—who donated, sold, or traded replacements—throughout Europe and the rest of the world. The boundaries of Fischer's world did not lie at national or linguistic borders. His greatest impediment may have been an indolent and arrogant count, a countryman in his adopted land, who did not care, but whose brother in Austria served as midwife to some of the world's most glorious music.

The sciences of natural history have always been strong in international solidarity, and weak in attracting official support. Fischer discovered this fact for himself when Razumovsky wouldn't help, but scientists throughout the world pitched in. Fischer's own name should have taught him the virtues of self-sufficiency. The Austrian Razumovsky first met a great musician who included among his many names the weighty designation of Gottlieb (Theophilus as he was christened in Greek, Amadeus as he later chose to render the name in Latin). They all mean "God's love"—and celestial powers surely adored Wolfgang Amadeus Mozart. The Russian Razumovsky was importuned by a fine scientist named Gotthelf, or "God's help." The count declined, but Fischer persevered and succeeded regardless—as he, no doubt, was meant to do, for Poor Richard told us that God only helps those who help themselves.

21

Four Antelopes of the Apocalypse

GEORGE WASHINGTON DIED on December 14, 1799, after scorning the classic advice of zealous guardians and staying out far too long on horseback in the winter's cold and snow. I have always felt sorry that America's old man missed the 1800s by less than three weeks (though I have no evidence that he cared about such artificial milestones). I also confess to a feeling of privilege and pleasure that most of us will live to witness something even rarer—the inception of a new millennium (see essay 2).

But Washington was not alone in missing this admittedly arbitrary transition by a hair. Half a world away, just east of Capetown in South Africa, a Boer huntsman shot the very last blaauwbock, or blue antelope, thus completing the first extinction of a large-bodied terrestrial mammalian species in historic times. (Perhaps the blue antelope just managed to sneak through, for some sources mention 1800 for the death of the last buck, though most agree on a terminal date of 1799.)

The heads of museum collections are called curators, from a Latin word, *cura*, that has a range of meanings from "care" to "management" to "heartache"—all appropriate at various times (as I should know, for I am, by title, Curator of Invertebrate Paleontology at Harvard's Museum of Comparative Zoology). The job, though often sublime, can be wrenchingly sad when the objects in our charge are remnants of extin-

The blaauwbock, from a 1778 edition of Buffon's Histoire naturelle.
Rijksmuseum van Natuurlijke Historie, Leiden.

guished vigor—scraps, fragments, and artifacts of creatures that should be thriving, but have succumbed to human rapacity. Curators become zealous in such lamentable situations, for we often hold the only palpable remnants of greatness that should still walk among us. We treasure a teenager's notebook in Amsterdam, as we preserve the balcony of a motel room in Memphis, for these are legacies of Anne Frank and Martin Luther King. And we feel triumphant when we discover an object to materialize what had only been a rumor or a memory.

I once tracked down, thanks to the efforts of my colleague Paul Lombardo, the grade-school report card of Vivian Buck, who died at age eight in the midst of the Great Depression. She had figured prominently in the infamous Supreme Court decision of 1927 that established the legality of forced sterilization—see essay 20 in *The Flamingo's Smile*. As a baby she had been judged mentally retarded on evidence of one minute's observation by a Red Cross nurse—the first fruit of a third generation of supposed imbeciles (her mother was subsequently sterilized as an outcome of the case). This humble scrap of paper from her last year of life proved that she had been a competent if not brilliant student—and that the diagnosis had been false. I exulted in the discovery, but then remembered that Vivian Buck should still be alive (for she died of a preventable disease of childhood poverty), enjoying an old age surrounded by a fecund and normal family. I rejoiced in the object (as a curator), but lamented that so little testimony should remain of a life cut so short.

I recently witnessed another example of this principle during a tour through one of the world's finest zoological collections. In the Netherlands, Leiden is to Amsterdam as Boston is to New York—a considerably smaller but older city, lacking the bustling excitement of a commercial center, but rich beyond size in universities and other institutions of learning. Leiden's university, founded in 1575 by William the Silent, Prince of Orange, is among Europe's best. Its natural history museum, formerly under royal patronage as the *Rijksmuseum van Natuurlijke Historie te Leiden*, holds one of the world's great collections, particularly renowned for historical material dating to the dawn of modern zoology. The Netherlands, after all, enjoyed a brief period as the most powerful commercial and shipping nation on earth, and the Dutch East India Company spanned the world in a network of trade (and ultimately left a treasure of zoological specimens to the Leiden museum).

The Leiden collections are housed in a massive warehouse of a building, with a large, winding central staircase that served as the foundation for optical wizardry in some of M. C. Escher's most famous etchings (Escher's brother was a geologist stationed in Leiden). The specimens reside in wooden cabinets, arranged on several floors in tier after vertical tier. The odors are *echt* museum, known to professionals throughout the world—dust, mothballs, and formaldehyde. The visuals would delight any fan of film noir—shafts of light and long shadows, piercing through all tiers at once, for the floors are industrial iron grating, not opaque concrete. The theme of hidden mystery can only be enhanced by knowing that, for several decades, the Leiden museum has presented no public exhibits (this will soon change)—and that these world-class collections have therefore become a private domain of zoologists in the know.

I recently enjoyed the privilege of a tour through this grand monument, conducted by my colleagues Edmund Gittenberger, fellow researcher on mollusks, and Chris Smeenk, specialist in mammals. Smeenk led me up and down the tiers, past tigers, elephants, and narwhals. But he was looking for one specimen, the central jewel of this diadem.

Smeenk found the cabinet and opened the large green door, revealing a slightly mangy (the mothballs do serve a noble purpose), badly faded, mounted specimen of a medium-sized African antelope with annulated horns in a simple, backward-sweeping curve marking the genus *Hippotragus*. This genus includes two living species, the roan antelope (*Hippotragus equinus*) of grasslands from western Ethiopia to South Africa, and the sable antelope (*Hippotragus niger*), with magnificent horns up to five feet in length, of more-wooded areas from Kenya to South Africa. The old specimen in Smeenk's cabinet did not look like either species, though I first thought that I might be looking at a young sable. My face did not therefore register the appropriate surprise and awe; I think that Chris was disappointed.

I had forgotten (if I ever knew) that the genus *Hippotragus* once included, in historic times, a third species—*H. leucophaeus*, the blaauwbock (blue buck) of South Africa. Perhaps my ignorance shall be forgiven, for no large mammal can match the South African blue antelope in obscurity. Western science knew the living blaauwbock for less than half a century, for this species was first mentioned by a traveler in 1719, not formally described until 1766, and then exterminated in 1799. No skeletons exist, and only four mounted specimens grace the collections of Europe. Smeenk had shown me the best and most famous of these four, a precious icon (and a mark of tragedy) in the fraternity of curators.

The first account of the blaauwbock, published in 1719, already included some hints of impending doom. Peter Kolb, a German businessman, worked in Capetown and traveled extensively in South Africa (which he called "Hottentot-Holland") from 1705 to 1712. He noted that blaauwbock meat tasted "good enough . . . but rather dry," and that the

Blaauwbock specimen at the Leiden Museum. Rijksmuseum van Natuurlijke Historie, Leiden.

abundance of more succulent game spared the animal from much hunting for food. But Kolb greatly praised the beauty of the hide, especially for the bluish tinge that had given the animal its name. (Early reports are inconsistent: some attribute the blue color to the hair, others to the underlying skin; some say that the color faded quickly after death, while others disagree. None of the four specimens in modern museums shows any extensive sign of blue color.) "I may truly say," Kolb writes, "that this animal appeared especially beautiful to my eyes because its blue hair is so much like the color of the sky" (*Himmelblau*, or "heaven blue," as he wrote in the original German). Kolb therefore reports that hunters shot the blaauwbock for its hide, noting, as a subsidiary benefit, that the suboptimal meat could then be fed to the dogs. Travelers and naturalists began to record the blaauwbock's decline, virtually from the day of its discovery. In 1774, C. P. Thunberg lamented its increasing rarity, while H. Lichtenstein later reported that the last known animal had been shot in 1799. Regret followed immediately, and Captain

W. Harris, in his *Portrait of the Game and Wild Animals of South Africa*, wrote in 1840, "For a *leucophaea*, I would have willingly given a finger of my right hand."

Only eight years after Linnaeus had codified the binomial system for naming animals (in the tenth edition of his *Systema Naturae* in 1758), the blaauwbock entered the formal roles of science through the work of Peter Simon Pallas (1741–1811), a young German naturalist who was spending several years in the Netherlands, then a virtual Mecca for the best zoological specimens. Pallas never finished the comprehensive treatise on all mammals that he had begun with his description of the blaauwbock and other animals in Dutch collections, for he received an appointment as professor of natural history in St. Petersburg, and soon got diverted to a range of other important projects, including the most significant eighteenth-century treatise on stratigraphic geology and the first observations on frozen mammoths. (Pallas's persona, and brilliant career, well represent the ecumenicism of an admittedly small and ill-funded world of European science at the time. With relatively few places of employment and few sources of publication, and in the last years of Latin as a functioning common language, scientists could move across national boundaries and keep in touch—see the preceding essay.)

Linnaeus himself had given Pallas his imprimatur, describing him as *juvenem acutissimum in Entomologicis, ornithologicis et toto natura versatissimum* ("a most intelligent youth, most competent in the study of entomology, ornithology, and all nature")—not bad credentials for a kid of twenty-five. At least the blaauwbock got its scientific start from the best of a rising generation.

Since South Africa was then being settled by Dutch ancestors of the present-day Afrikaner people, zoological specimens from that part of the world tended to accumulate in collections within the Netherlands. Pallas must have seen several blaauwbock skins, for he began the description of his new species, *Antilope leucophaea*, by writing *Hujus plurimas vidi pelles Promontoriae bonae Spei missas.* ("I have seen several skins of this [species] sent from the Cape of Good Hope.") Pallas described the blaauwbock in his *Miscellanea zoologica*, published in 1766. (In these early days of zoology, when few species had been described, the Linnaean genus had a broader definition, and all antelopes fell into the single genus *Antilope*—thus Pallas's name for the blaauwbock. As more and more antelope species entered the rolls, this genus became unwieldy, and later taxonomists divided *Antilope* into several new genera, genealogically defined. The blaauwbock therefore moved with its closest relatives, the roan and sable antelopes, into the new genus *Hippo-*

tragus. Incidentally, Pallas did not name the blaauwbock for its prominent blue color, but *leucophaea* [meaning dusky white] for a distinctive patch of light color under the eyes.)

Human activity clearly drove the blaauwbock to an accelerated extinction. European hunters delivered the coup de grace, but native African people had introduced domestic sheep into the area as early as 400 A.D., and resulting deterioration of habitat probably set an inevitable course for a species that had, for reasons not involving *Homo sapiens*, already become greatly reduced in numbers and geographic range. If ever a species had been marked for natural death in the ebb and flow of nature, surely the blaauwbock had already received its date with the grim reaper of geological time. No species are more vulnerable than those combining large body size (and consequently small populations—the "more ants than elephants" principle) with highly restricted ecological and geographic ranges. Sources differ on the extent of the blaauwbock when first encountered by Europeans, but most reports fix the range to a tiny area, perhaps no more than sixty miles east-west by forty miles north-south in the province of Swellendam, centered about one hundred miles due east of Capetown. No large mammal with so limited a range can long endure.

As a result of such rapid elimination for such a rare and restricted species, we know almost nothing about the blaauwbock. No reliable data on ecology or behavior were ever recorded in the wild (travelers' accounts are spotty, and often secondhand or contradictory); with the exception of one dubious skull and a few sets of horn cores, we have no skeletal material of blaauwbocks from the time of European settlement (some fossil bones probably belong to this species); the four mounted skins in European museums are too faded to yield knowledge of coloration patterns in life. The blaauwbock passed into western scientific consciousness and then, almost immediately, passed out of existence entirely.

When vitality succumbs, we must make do with the next best evidence—remnants and artifacts. The blaauwbock therefore moves to the world of curators. Preciousness is often defined by rarity. By this criterion, hardly anything in natural history can be more valuable than a scrap of blaauwbock—and blessed be the curator who has one to show. In her monograph on the blaauwbock, Erna Mohr gives a complete accounting of all material in museums—and her list requires but three lines of text.

Only four mounted specimens exist—in the museums of Stockholm, Vienna, Leiden, and Paris. Pairs of horns can be found in Uppsala and London, and perhaps in the Albany Museum of South Africa. The London blitz of 1941 destroyed a lovely skull with horns, once located in the museum of the Royal College of Surgeons. The museum of Glasgow houses a complete skull with horn cores that may (or may not) be a blaauwbock's. No other skins; no other bones. (Pallas's original description cites "several skins sent from the Cape of Good Hope," so Dutch collections may once have held more specimens. But time and wars ravage our works as well as our bodies.) We may easily understand why the Leiden museum treats its blaauwbock with such singular honor.

If my interaction with blaauwbocks had ended with the specimen in its Leiden cabinet, I doubt that I would have written this essay—a good story, yes; but no hook for greater generality or extended contemplation. But Chris Smeenk then told me something else, and gave me two articles to read—and these contributed a fact so poignant and bittersweet that I had to proceed. Smeenk told me that a debate had arisen over which, if any, of the existing specimens had been among those observed by Pallas when he gave the blaauwbock its scientific name. (Such questions should not be viewed as antiquarian indulgences. For legalistic reasons in apportioning scientific names, all species must have a "type" specimen—a single preserved creature that becomes the official name-bearer for the species. We need such types because we often later discover that a named "species" really includes specimens from two or more legitimate species. We must then retain the original name for one of those species, and coin new designations for the others. But which population gets to keep the old name? By the rules of nomenclature, the original name belongs to the type specimen in perpetuity—and its population retains this first designation.)

For more than a century the Leiden curators had assumed that their mounted specimen had been used by Pallas as one of the *plurimas pelles* (several skins) for defining *Hippotragus leucophaeus*. Since none of the other skins can be traced, the Leiden specimen therefore becomes the type—and, by implication, the official and most important specimen of the species (another reason for Smeenk's pride in display). But Erna Mohr had challenged this long-held assumption and disestablished the Leiden blaauwbock to mere mortality as a specimen of uncertain provenance. If anything can be called "fighting words" in the curatorial world, Mohr had launched a haymaker (a real *chmallyeh*, as my grandfather, a true fan of pugilistic arts, would have said), and the curators of Leiden fought back—and won. Smeenk handed me two articles, published by Leiden curators A. M. Husson and L. B. Holthuis in 1969 and 1975 in the museum's house organ, *Zoologische Me-*

dedelingen. The first paper reasserted the specimen's birthright in the title: "On the type of *Antilope leucophaea* Pallas, 1766, preserved in the collection of the Rijksmuseum van Natuurlijke Historie, Leiden."

Husson and Holthuis identify the Leiden blaauwbock as the same specimen figured in the 1778 edition of the standard eighteenth-century work in zoology—Buffon's *Histoire naturelle*. This correspondence already gives the Leiden blaauwbock a special cachet, for Buffon published the only eighteenth-century illustration of the complete animal (two previous sources had depicted horns only, or head and horns, whereas Pallas had included no figure at all in his original description). The 1778 edition of Buffon provides a good account of the history of this specimen: it had first been in the possession of Dr. J. C. Klöckner, a former physician on ships to the East Indies. Klöckner, who had achieved fame for his skill in mounting specimens, lived in Amsterdam between 1764 and 1766, when Pallas published his description. Since we know that Pallas worked in Amsterdam, and certainly tried to visit all major collections, he almost surely saw Klöckner's specimen—making the Leiden blaauwbock one of the *plurimas pelles* of Pallas's original definition, and therefore the type of the species as the only identifiable specimen from the original lot.

Buffon's editor reports that the specimen then went to J. C. Sylvius van Lennep, a young patrician in Haarlem, and thence, upon his death in 1776, to the Holland Society of Sciences in the same city. So far, so good. Buffon's specimen had been seen by Pallas in Amsterdam and had ended up in Haarlem. Buffon's figure certainly seems to be based on the Leiden specimen, but final proof demands a firm link between the last resting place of this specimen in Haarlem and current residence in Leiden. And here the trail ran cold—until Husson and Holthuis made the discovery that clinched the connection.

Mr. P. Tuyn, librarian of the Amsterdam Zoo, found the smoking gun and gave the information to Husson and Holthuis. Tuyn noted an advertisement in the April 5, 1842, edition of the *Opregte Haarlemsche Courant* announcing the sale by auction of part of the Holland Society's zoological collections—and the text of the ad indicated that a blaauwbock had been included among the items. Husson and Holthius then burrowed into the Leiden museum archives and found a stack of bills submitted by the museum to its funding agency, the Ministry of Internal Affairs (the museum then enjoyed royal patronage), on May 31, 1842. The bills prove that Dr. H. Schlegel, then curator of mammals at Leiden, had gone to Haarlem for the sale and had purchased two antelopes, including the blaauwbock now present in the Leiden collections. Of the relevant bills, one particularly caught my eye—a receipt for the passage from Haarlem to Leiden: "For transportation: for the tow-barge f. 2.00, for the Antelopes f. 0.60. For delivery f. 0.85. For consumptions f. 0.75. For purchase of the antelopes f. 47.10. Total f. 51.30."

I don't know why this degree of detail affected me so powerfully. Consider what intimate circumstances we can recover in tracing a case—three quarters of a florin (ancestor of the present Dutch guilder) for beer and sausage to keep the bargemen in good spirits as they brought the blaauwbock from Haarlem to Amsterdam. Yet the blaauwbock itself, as a living, evolving, natural entity, is gone—extinct before we knew its preferred food, the sound of its voice, the color of its hide. When the pith and essence disappear, we are reduced to arguing over scraps and relics. We cry over Anne Frank's notebook; we kneel in reverence, according to our creed, before a scrap of toebone of the Great Buddha, or a vial of the liquefying blood of Saint Januarius (the rationale for New York City's wonderful and annual festival of San Gennaro on Mulberry Street—so we won't ask too many questions about the supposed miracle). We must care about the bargemen's lunch because this beer and sausage has become our link to something beautiful that we lost.

Curators may be guardians of the second best, but when we have lost the living by human malfeasance, what can be more residually noble than faithful preservation and accurate documentation of often pitiable fragments?

This sadness spawns respect for our later compassion and need to preserve—too little and too late for the blaauwbock, but a lesson, perhaps, when we compare the relative worth of a healthy species vs. a receipt 150 years old. An unknown element in the periodic table will one day be made or discovered—for phenomena in such simple and law-bound systems enjoy predictability and repeatability. But when we lose all evidence for a historical item, we cannot reconstruct what we did not preserve—for we have extinguished part of nature's unique richness forever. The blaauwbock came within a hair of total extirpation from memory (as well as from Africa). But we knew this species for a few years during part of a century, and we did save a few scraps and records—miserable, tawdry fragments in one sense, yet magical in another, and therefore of precious rarity, as the only witnesses to a first loss in an accelerating series. They are watching us from Leiden, Paris, Vienna, and Stockholm. Four antelopes of the apocalypse, silently watching to see how many we will bring to their sorry fate.

PART SIX

Disparate Faces of Eugenics

Does the Stoneless Plum Instruct the Thinking Reed?

A DISTINCTIVELY AMERICAN FORM of genius combines personal brilliance with promotional hokum and mountains of hard work. We note varying balances among the three essential ingredients: Barnum as a prototype for hokum, Edison for the diligence (as in his justifiably famous quip that genius includes one percent inspiration and ninety-nine percent perspiration). Luther Burbank (1849–1926) may represent the finest balance of all three factors. He wished to convert American deserts to rangeland by breeding a spineless cactus as cattle fodder. He hybridized two spontaneous mutants—one without prickles on the leaves and the other with no thorns on the young shoots—but he never could eliminate the spines completely. Nevertheless, in public performances he would rub his cheek against a "prop" cactus made so smooth by previous demonstrations that all residual spines had long since worn away.

Yet Burbank's successes made him the greatest plant breeder in American history. An astonishing string of triumphs fostered his reputation as a wizard with a magic wand for transformation—an exaggeration that Burbank, with his taste for publicity, learned to exploit to great advantage. As a young man he developed the Burbank potato in his native Massachusetts and used the proceeds to bankroll his larger, lifetime operation in California. Here he "created" his stoneless plums, white blackberries, the (almost) spineless cactus, the giant Shasta daisy, the Fire poppy, and the Burbank rose.

These triumphs reflected the other two components of American genius. For his brilliance, Burbank had an unparalleled eye for the most minute useful variation. He could scan a field of daisies or an orchard of plums and pick the plant with a slight edge for improving the breed. For his industry, no one worked at Burbank's scale, and with such zealous energy. Despite the hype that enveloped him, Burbank performed no miracles. He evoked nothing new from nature. He just worked harder—far harder than anyone else—with the two tried and trusted techniques of breeding: hybridization and selection. With his uncanny eye and unequalled judgment, he would make novel plants by mixing desired traits of two or more varieties in hybrid offspring. He would then grow field upon field of hybrids, always looking for the rare plant that would emphasize all the favorable characters and suppress the undesirable features. For example, he produced his stoneless plum by crossing productive and richly fruited (but conventionally pitted) lines with a curious variety known in France since the sixteenth century, but treated as useless for its poor and unproductive fruits—the *prune sans noyau* (seedless plum). By repeating cycles of crossing and selection, extending for more than a decade, Burbank managed to produce a large fruit with a stoneless pit.

Burbank fit the mold of populist hero in yet another way: he wrote very little, preferring to nurture his fame by a plethora of proven products and occasional oracular pronouncements. He did, of course, publish catalogs to flog his horticultural achievements, and he did write a few short pieces for agricultural journals. But his longest and most noted article, published by the *The Century* magazine in 1906 and reissued as a small book in 1907, featured the different subject of eugenics. The intriguing title of this little volume had long piqued my interest: *The Training of the Human Plant*. We may rightly suspect that Burbank's "alien" topic for his longest written work did not stray so far from his usual concerns after all.

This essay, while not eschewing the details of spuds and roses, shall be a commentary upon two pitfalls frequently encountered in scientific (and much other) reasoning: misleading taxonomy and false analogy. I shall treat the second subject in analyzing Burbank's article, but consider our usual attitude toward eugenics itself as an example of the former danger—false categorization.

We usually view eugenics as a failed ideology of political conservatives—and for two major reasons. First, eugenics sought to reform society by improving hereditary traits through controls upon breeding (whether forced or voluntary)—and hereditarianism is a conservative strategy par excellence (people get what they deserve by birth; don't look to governmental decrees or handouts for any realignment). Second, the major consumer of eugenics in one extreme form, Mr. Hitler himself, could scarcely be accused of liberal biases (see essay 24).

But the eugenics movement resists cramming into a single, unambiguous political box (and I often wonder how well right and left, liberal and conservative, really work as labels for any opposed pair of boxes). Supporters of eugenics formed a diverse and powerful movement during the early years of our century. Hardly any group owning a single name could have been more motley in all other respects. Politics always makes strange bedfellows, but the range of eugenical support must have generated some legendary pillow fights before "lights out." The movement spanned a full spectrum from hereditarian hardheads who wanted to sterilize the handicapped, the diseased, and even the merely impoverished, to Fabian idealists who hoped to persuade smart and gentle people to have more kids. For anyone with a residual affection for the false equation of eugenics with antediluvian conservatism, consider just one cardinal fact. When the Supreme Court, in the greatest victory of the American eugenics movement, upheld compulsory sterilization of the mentally unfit in 1927, all liberal justices voted aye; the single dissent in this eight-to-one ruling was filed by the court's most conservative member, a catholic who upheld his church's position on reproductive controls.

I mention this political diversity because Luther Burbank's book may be the most influential document of support for a phenomenon that could not exist by the usual false correlations: liberal eugenics. By his choice of title, *The Training of the Human Plant*, Burbank both prepared the ground for his grand, and ultimately false, analogy, and also staked out the "liberal" field by emphasizing nurture through life rather than strict and unalterable hereditary nature.

Burbank's tract is a rigid, elaborate, point-for-point comparison of his vision for improving human society with his methods (as he conceived them) for developing new lines of plants. (Burbank possessed an odd streak of modesty amid his confident self-promotion, for he falsely attributed a key aspect of his own transforming skill to nature's intrinsic mechanisms. We shall see that the error of his eugenic argument lies both in his false analogy with plant breeding and in his misunderstanding of reasons for his own horticultural successes.)

Burbank begins his tract with a recommendation that followed his own first step in producing new lines of plants, but that also must have induced apoplectic rage among his more conservative brethren in the eugenics movement. Burbank achieved his popular reputation as a wizard because people thought he could produce genuine novelty, virtually at will. In fact, he introduced almost every "new" trait by directed immigration—that is, by importing favorable features from other lineages through hybridization. By analogy, Burbank viewed his contemporary America as a land of opportunity for improvement unparalleled in human history. The Statue of Liberty lifted her lamp beside the golden door, and Ellis Island overflowed with a great tide of European immigrants. Burbank titled his chapter one "The Mingling of Races," and began with an explicit analogy:

> I have constantly been impressed with the similarity between the organization and development of plant and human life . . . I have come to find in the crossing of species and in selection, wisely directed, a great and powerful instrument for the transformation of the vegetable kingdom along lines that lead constantly upward . . . Let me now lay emphasis on the opportunity now presented in the United States for observing and, if we are wise, aiding in what I think it fair to say is the grandest opportunity ever presented of developing the finest race the world has ever known out of the vast mingling of races brought here by immigration.

Consider the stunningly radical nature of this proposal in the context of the times. Conservatives, in the eugenics movement and elsewhere, campaigned against immigration as one of their most passionate issues—lest our virile, virtuous, and intelligent "native" stock (meaning "northern European," not truly Native American) be diluted to folly and ineptitude by breeding with the inferior hordes now swamping us from southern and eastern Europe. (My own ancestors all arrived from Hungary, Poland, and Russia during the decade of Burbank's book—so I confess some personal sensitivity to the issue.) These nativists hoped to stem the tide of immigration to a trickle and, above all, to keep those already here from miscegenation. Let the new arrivals work in factories and sweatshops; let the Jew cut my cloth, the Italian my hair, and let the Irishwoman wash my floor; but keep them away from my sons and daughters. Can you imagine how Burbank's proposal for the benefits of interbreeding played in such circles?

Let me not exaggerate. I am not sure how far Burbank meant to go.

He clearly favors mixing of different *European* stocks, but is conspicuously silent about people of other colors (the unstated limit to radicalism, I suspect, in such a racist age). Moreover, Burbank was no egalitarian in a modern sense. He accepted different intrinsic worths and temperaments among European groups (with the traditional contrast of sturdy Nordics and emotional Mediterraneans), but still favored their combination so that the various goods might be sorted into one superior stock:

> Look at the material on which to draw! Here is the North, powerful, virile, aggressive, blended with the luxurious, ease-loving, more impetuous South. Again you have the emerging of a cold phlegmatic temperament with one mercurial and volatile. Still again the union of great native mental strength, developed or undeveloped, with bodily vigor, but with inferior mind.

But mixing can only mark the beginning of human improvement, just as hybridization represents the first stage in producing a new horticultural line. Mixing only creates raw material by mingling formerly separate traits into single lineages. Left alone, with no follow-up by selection, mixing will only create chaos, and make matters worse.

> The mere crossing of species, unaccompanied by selection, wise supervision, intelligent care, and the utmost patience, is not likely to result in marked good, and may result in vast harm. Unorganized effort is often most vicious in its tendencies.

How, then, shall eugenics proceed after this vast and wholesale mixing of European stocks? Since Burbank, by title and intent, wished to improve humans exactly as he made better plants, further progress must follow his horticultural protocol. Burbank outlined a four-step process for improving plant lines, and he transferred this entire apparatus to his vision for human betterment. The poignancy of Burbank's story (also the reason for logical failure of his eugenic argument) lies in our current recognition that he was a far better plant breeder than even he—the consummate self-promoter—realized! For he actually accomplished all his feats with only two of his four steps, while he thought that nature acted as his auxiliary in two additional, but nonexistent processes. Ironically, he rooted the humanistic and "liberal" heart of his eugenics program in the two illusory processes—while his true protocol in horticulture debarred, by his own moral values, any attempted transfer to human improvement, thus eradicating his own analogical argument.

Burbank epitomizes his four-stage method in setting out an argument for transfer to human society:

> There is not a single desirable attribute which, lacking in a plant, may not be bred into it. Choose what improvement you wish in a flower, a fruit, or a tree, and by crossing, selection, cultivation, and persistence you can fix this desired trait irrevocably.

Consider these four steps in order:

1. CROSSING. Burbank, as stated above, performed no miracles of novel evocation. He introduced new traits into old lines by hybridization with different stocks already bearing the desired features. Just as hybridization stirs the raw material for horticultural improvement into a single line, miscegenation of immigrant stocks will produce the world's finest people in America.

2. SELECTION. Crossing only increases the range of variation in offspring, and does not produce any alteration by itself. This net change, called "improvement" when regulated to human advantage, results from selection—the destruction of most plants and the breeding of future generations from just a few that happen to possess a desired set of traits. Darwin, of course, called nature's unconscious version of this process "natural selection." In either mode, natural or horticultural, selection acts with rigor. Many are called, and very few chosen. Rapid improvement demands the exclusion of nearly all from reproduction, with death as the surest means of prevention (though sterilization, celibacy, and other analogues may suffice for humans). The horticulturist uproots; nature generally eliminates. The transfer of this harsh process from nature's amoral realm to human life has been—entirely rightly, in my view—the fatal flaw of all traditional programs in eugenics. Selection is nature's primary path for genetic change in populations, but this process, applied in the ethical realm of human society, must be unacceptably cruel and authoritarian (for someone must play the role of Lord High Uprooter).

3. CULTIVATION. No one doubts the salutary role of cultivation—that is, good environment—in sustaining the health and vigor of an individual plant. But what role can good environment play in genetic improvement, for none of the virtues acquired during life are passed to offspring in Mendel's world? (Proper cultivation keeps a plant vigorous

for reproduction, but this aspect of environment only acts as input to selection and gives nurture no independent status in improvement.) Burbank, however, was a Lamarckian. He believed that characters acquired by good rearing during life could be passed to offspring—and that improvement would accumulate through this transfer, generation by generation and step by constant step.

4. PERSISTENCE. For Lamarckians, persistence enhances evolutionary change as the extension of cultivation. Lamarckian advances come little by little; a breeder (or nature) must keep up the good work for generations, and unremittingly. Mendelian characters are intrinsic; bad environment does not dilute a gene (though extreme malfeasance, of course, may kill the organism and end the experiment). But in Lamarckian theory, traits can be dismembered by generations of neglect just as good rearing had built them sequentially.

With his categories three and four, we arrive at the crux and idiosyncrasy of Burbank's eugenical argument—and its ultimate fallacy. Burbank espoused an incorrect theory about his own efficacy. No Lamarckian component helped his plant breeds along. Nature is Mendelian and does not work in this hopeful manner. In fact, Burbank bred plants on thoroughly Darwinian principles, using only the first two categories of his own list—crossing to mix the Mendelian characters of different lines into hybrid offspring, and selection to collect and accentuate the traits he sought. But he was so damned good at the art of breeding, so much better than anyone else—and this excellence, not fully acknowledged, led Burbank to a false but unshakable conviction that nature must be helping out in some other way! He could not believe that he was doing all this himself, merely by selecting so rigorously and on such a grand scale.

Burbank was surely and spectacularly zealous in the scope of his selection and the heap-size of his hecatomb. He would sacrifice fields of plants to find one worth propagating (and he developed the world's surest eye for finding the best). Let an eyewitness report suffice as testimony to his diligence. The source is particularly interesting, even a bit ironic—Hugo de Vries, the great Dutch botanist and one of the three scientists who rediscovered Mendel's work. (De Vries, who visited California twice, was awestruck by Burbank's practical skills, but dumbstruck by his stubbornness in sticking with Lamarck as Mendelism marched to triumph, with de Vries as a major general in the campaign):

> His methods are hybridization and selection in the broadest sense and on the largest scale. One very illustrative example of

his methods must suffice to convey an idea of the work necessary to produce a new race of superlative excellency. Forty thousand blackberry and raspberry hybrids were produced and grown until the fruit matured. Then from the whole lot a single variety was chosen as the best. It is now known under the name of "Paradox." All others were uprooted with their crop of ripening berries, heaped up into a pile twelve feet wide, fourteen feet high, and twenty-two feet long, and burned. Nothing remains of that expensive and lengthy experiment, except the one parent-plant of the new variety.

Burbank's major emphasis in *Training of the Human Plant*, the source of his "liberal" tilt on eugenics, lies in analogy with the third and fourth phases (alas, the illusory ones) of his supposed horticultural protocol: cultivation and persistence. Burbank makes his plea for healthy childhood environments by his usual comparison with plants:

> If you are cultivating a plant, developing it into something finer and nobler, you must love it, not hate it; be gentle with it, not abusive; be firm, never harsh. I give plants . . . the best possible environment. So should it be with a child, if you want to develop it in right ways. Let the children have music, let them have pictures, let them have laughter.

Burbank's formula for good rearing is strictly Arcadian—a passionately romanticized idyll of clean country living without intellectual pressure (Burbank wanted no formal schooling before age ten, and he continually stresses this single proposal as the beginning, if not the panacea, for all reform):

> Every child should have mud pies, grasshoppers, water-bugs, tadpoles, frogs, mud-turtles, elderberries, wild strawberries, acorns, chestnuts, trees to climb, brooks to wade in, water-lilies, woodchucks, bats, bees, butterflies, various animals to pet, hayfields, pinecones, rocks to roll, sand, snakes, huckleberries and hornets; and any child who has been deprived of these has been deprived of the best part of his education. By being well acquainted with all these, they come into most intimate harmony with nature, whose lessons are, of course, natural and wholesome.

No reformer could fail to advocate rearing for children. But Burbank's book is a eugenic tract, an argument about *genetic* improvement. He must therefore be advocating good rearing as a direct hereditary benefit to future generations—a senseless theme in Mendel's world. But Burbank, as discussed above, advocated Lamarckian inheritance, and he therefore stressed exemplary rearing not only for its obvious value to direct beneficiaries, but also and especially for its Lamarckian transfer to future generations of genetically improved stock. Thus, for Burbank, good nurture now produces better nature for posterity:

> Heredity is not the dark specter which some people have thought—merciless and unchangeable, the embodiment of Fate itself . . . My own studies have led me to be assured that heredity is only the sum of all past environment, in other words environment is the architect of heredity; and I am assured of another fact: acquired characters are transmitted and—even further—that all characters which are transmitted have been acquired.

But what about the bypassed second factor, the bugbear of all "liberal" eugenics, and the actual source of Burbank's major achievements in horticulture: Darwin's force of selection. How can one talk about genetic improvement without a willingness to prevent reproduction of those deemed unfit? And how can one even envisage a humane eugenics if the improvement of people must work like nature or horticulture, where, undeniably, the overwhelming majority of individuals fall into the category of dispensable *and* extirpated?

Burbank skirts around this issue in his tract. In one passage he does admit a preference for laws on prohibition of marriage among the "unfit"—a quite stunning exception to his general defense of civil liberties when you consider the ethically repugnant issues of who gets included and who decides: "It would, if possible, be best absolutely to prohibit in every State in the Union the marriage of the physically, mentally and morally unfit."

But, for the most part, Burbank speaks little about this most powerful force of all. He senses the moral dilemma and, for once, strays from the analogy with nature and horticulture. What of physical weakness? "Shall we, as some have advocated, even from Spartan days, hold that the weaklings should be destroyed? No." He then acknowledges the most difficult question: "But with those who are mentally defective—ah, here is the hardest question of all!—what shall be done with them?" And, for once, decency triumphs over the logic of his own argument. Here Burbank will make an exception and admit the failure of his pervasive analogy:

> In the case of human beings in whom the light of reason does not burn, those who, apparently, can never be other than a burden, shall they be eliminated from the race? Go to the mother of an imbecile child and get your answer. No; here the analogy must cease.

Yet Burbank found a way out—not an ideal solution, but a workable exit that preserved his liberal version of eugenics. Human morality must slow the process of improvement, for we cannot bear the hecatomb that any effective selection imposes. But if heredity is Lamarckian, we may allow the unfit to live, even to reproduce, and still make sure, if slower, progress. For good environment induces improvement, and these benefits are passed on as altered heredity. Nurture the unfit for enough generations and they will eventually become genetically worthy:

> When certain hereditary tendencies are almost indelibly ingrained, environment will have a hard battle to effect a change in the child; but that a change can be wrought by the surroundings we all know. The particular subject may at first be stubborn against these influences, but repeated application of the same modifying forces in succeeding generations will at last accomplish the desired object in the child as it does in the plant. No one shall say what great results for the good of the race may not be attained in the cultivation of abnormal children, transforming them into normal ones.

My exposition may have been lengthy, but the moral of my story is short and unambiguous. Nature, in the wild or in horticulture, works on Darwinian, not Lamarckian, principles. Acquired characters are not inherited, and desired improvement occurs by rigorous selection with elimination of the vast majority from the reproductive stream. Burbank could develop new breeds, but he could not alter the rules. He actually worked by extensive hybridization and uncompromising selection, though his own success fooled him into thinking that nature helped his efforts by Lamarckian inheritance. The Lamarckian theme sets the keystone for Burbank's liberal eugenics, based on the *genetic* effects of good nurturing. The fallacy of Lamarckism marks the utter failure of his argument.

Yet a deeper error pervades the entire enterprise, with Burbank's tract representing but one example in a continuing tradition—the attempt to pattern human ethical conduct by aping nature's way. Burbank was wrong about nature's mode of action. But even if he had been right, his effort would have been just as misguided. Nature's factuality is not, and cannot be, our morality. We must know how nature works in order to understand ourselves and recognize our limits and possibilities in a tough world, but why should a process that regulated three and a half billion years of living creatures without explicit ethical systems provide all the answers for a species that evolved only a geological second ago, and then changed the rules by introducing such new and interesting concepts as justice and righteousness?

In his famous 1893 essay on *Evolution and Ethics*, the classic statement on nature's amorality, Thomas Henry Huxley praised Darwinism for its effectiveness, speaking of "the struggle for existence, which had done such admirable work in cosmic nature." But he quickly added that "the cosmic process has no sort of relation to moral ends." He closes with an argument that human intelligence may choose either to follow nature or, upon deciding that proper ethics requires another course, use the fragile gift of mind as the only available path to transcendence: "In virtue of his intelligence, the dwarf bends the Titan to his will. In every family, in every polity that has been established, the cosmic process in man has been restrained and otherwise modified by law and custom."

Burbank was hard at work in California when Huxley wrote his essay, but Darwin's bulldog did not have plants on his mind when he argued that human intelligence debars nature as the arbiter of ethical questions. Still, I am delighted to report that Huxley, in an unconscious touché to Burbank, used a botanical metaphor—and a grand classic at that—to secure his most important point, a message every bit as vital today:

> The history of civilization details the steps by which men have succeeded in building up an artificial world within the cosmos. Fragile reed as he may be, man, as Pascal says, is a thinking reed: there lies within him a fund of energy, operating intelligently and so far akin to that which pervades the universe, that it is competent to influence and modify the cosmic process.

23

The Smoking Gun of Eugenics

Do Baptist preachers cause public drunkenness? I raise this unlikely inquiry because an old and famous tabulation clearly shows a strong positive correlation between the number of preachers and the frequency of arrests for inebriation during the second half of the nineteenth century in America.

You don't need a Ph.D. in logic to spot the fallacy in my first sentence. Correlation is not causality. The undeniable association of preachers and drunks might mean that hellfire inspires imbibing; but the same correlation could also (and more reasonably) suggest the opposite causal hypothesis that a rise in public drinking promotes the hiring of more preachers. But yet another possibility—almost surely correct in this particular case—holds that preaching and drinking may have no causal relationship, while their simultaneous increase only records a common link to a third, truly determining factor. The steady rise of the American population during the late nineteenth century promoted an increase in thousands of phenomena linked to total numbers, but otherwise unrelated—arrests for drinking and hiring of clergy among them. This tale has long served as the standard textbook example for illustrating the difference between correlation and causality.

But good principles can also be used to buttress bad arguments. I have often stated in these essays that only great thinkers are allowed to fail greatly—meaning that such errors, although large in scope and import, are invariably rich and instructive rather than petty and merely embarrassing. This essay treats the two greatest errors of the twentieth century's patron saint in my profession of evolutionary biology.

Most general readers may not know the name of Sir Ronald Aylmer Fisher (1890–1962), for he wrote nothing for nonprofessional consumption, and the highly mathematical character of his technical work debars access to many full-time naturalists as well. But no scientist is more important as a founder of modern evolutionary theory, particularly for his successful integration of Mendelian genetics with Darwinian natural selection. Fisher's 1930 book, *The Genetical Theory of Natural Selection*, is the keystone for the architecture of modern Darwinism. Fisher built with mathematics, and most biologists will say (though I would disagree in important respects) that the field he founded—population genetics—is the centerpiece of evolutionary theory. Fisher was also one of the world's most distinguished statisticians; he invented a technique called the "analysis of variance"—now about as central to statistics as the alphabet is to orthography. In short, Fisher is the Babe Ruth of statistics and evolutionary theory.

But the Babe also struck out a lot, and Fisher made some major-league errors. Most of my colleagues know about the two key mistakes that I will analyze in this essay, but these errors just aren't discussed in polite, professional company. One is dismissed as an inconsequential foible of Fisher's old age, while the other tends to be bypassed in silence, though it occupies more than one third of Fisher's most important 1930 book.

During the last half-dozen years of his life, Fisher spent considerable time and several publications trying to debunk the idea that smoking can cause lung cancer. Sir Ronald, who enjoyed his pipe, did not deny that a real correlation between smoking and lung cancer had been found. But, following the textbook paradigm of preachers and drunkards, he disputed the claim that causation ran directly from smoke to cancer. He presented the two other logical possibilities, just as the texts always do for Baptists and boozers. First, cancer might cause smoking rather than vice versa. This inherently implausible version seems hard to defend, even as an abstract argument for the sake of conjecture, but Fisher found a way.

As a smoker, Fisher extolled the soothing effects of tobacco. He also recognized that cancers take years to develop and that future sufferers live for several years in a "pre-cancerous state." He supposed that lungs might be chemically irritated during this pre-cancerous phase, and that people so afflicted might be led to increased smoking for psychological relief from an unrecognized physical ailment. A bit strained, but not illogical. Fisher wrote in 1958:

> Is it possible, then, that lung cancer—that is to say, the pre-cancerous condition which must exist and is known to exist for years in those who are going to show overt lung cancer—is one of the causes of smoking cigarettes? I don't think it can be excluded . . . The pre-cancerous condition is one involving a certain amount of slight chronic inflammation . . .
>
> A slight cause of irritation—a slight disappointment, an unexpected delay, some sort of mild rebuff, a frustration—is commonly accompanied by pulling out a cigarette, and getting a little compensation for life's minor ills in that way. And so anyone suffering from chronic inflammation in part of the body (something that does not give rise to conscious pain) is not unlikely to be associated with smoking more frequently, or smoking rather than not smoking . . . To take the poor chap's cigarettes away from him would be rather like taking away his white stick from a blind man.

But Fisher recognized that the second alternative for the correlation of smoking and lung cancer—the association of both, independently, with a truly causal third factor—held much greater plausibility and promise. And Fisher had no doubt about the most likely common factor—genetic predisposition. He wrote: "For my part, I think it is more likely that a common cause supplies the explanation . . . The obvious common cause to think of is the genotype." In other words, genes that make people more susceptible to lung cancer might also lead to behaviors and personalities that encourage smoking. Again, the argument is undeniably logical; genes may have multiple effects, both physical and behavioral. To choose an obvious example, several forms of mental retardation have no causal relationship with correlated physical features. Short stature does not produce retardation (or vice versa) in people with Down's syndrome.

With the hindsight of an additional twenty-five years, we may say conclusively that Fisher was wrong, and tragically so. Smoking is a direct and potent cause of lung cancer—the reason, therefore, for hundreds of thousands of premature deaths in America each year. Yet I cannot fault Fisher on the logic of his argument: correlation is not causality, and the bare fact of correlation does permit the three causal scenarios that Fisher detailed. If Fisher had presented his objections to the indictment of smoking only as a cautionary claim in the absence of conclusive data, then we could not blame him today. (One cannot always be right in our complex world; no dishonor attends an incorrect choice among plausible outcomes drawn from a properly constructed argument.) But in Fisher's case, we have reason to question his motives and his objectivity—and some judgment for his incorrect conclusion may therefore be exacted.

Fisher did present his case with the conventional rhetoric of science. He claimed to be both objective in his weighing of evidence and agnostic about the outcome. He maintained that he raised the issue only in a proper scientific spirit of caution and love of truth. Fisher made three explicit arguments for special and scrupulous care in treating such a socially charged issue, a potential matter of life and death.

1. Millions of people enjoy smoking. We dare not poison the source of their pleasure without conclusive evidence. Fisher pleaded for the psychic health of ordinary smokers in the elitist language of an Oxbridge don (Fisher was the Balfour Professor of Genetics at Cambridge and, at the end of his career, president of Gonville and Caius College):

> After all, a large number of the smokers of the world are not very clever, perhaps not very strong-minded. The habit is an insidious one, difficult to break, and consequently in many, many cases there would be implanted what a psychologist might recognize as a grave conflict . . . Before one interferes with the peace of mind and habits of others, it seems to me that the scientific evidence—the exact weight of the evidence free from emotion—should be rather carefully examined.

Writing more forcefully in a letter to the *British Medical Journal* (July 6, 1957), Fisher compared the claims of anti-smoking forces with the classic case of hysteria-mongering: "Surely the 'yellow peril' of modern times is not the mild and soothing weed but the organized creation of states of frantic alarm."

2. If we make a strident claim for smoking as a cause of cancer, and if we then turn out to be wrong, the entire enterprise of statistics will be discredited. In a further letter to the *British Medical Journal* (August 3, 1957), Fisher pleaded for caution as a protection for science:

> Statistics has gained a place of modest usefulness in medical research. It can deserve and retain this only by complete impartiality . . . I do not relish the prospect of this science being now discredited by a catastrophic and conspicuous howler.

3. Situations of uncertainty require more research above all. Premature conclusions stifle further investigation. In yet another letter, this time to *Nature,* Britain's leading journal for professional scientists, Fisher wrote (August 30, 1958): "Considerable propaganda is now being developed to convince the public that cigarette smoking is dangerous." In his letter of August 1957, Fisher had already specified the perils of such a campaign: "Excessive confidence that the solution has already been found is the main obstacle in the way of more penetrating research."

Fisher's last point about further research backfired strongly upon him—an ironic illustration of its power and truth. Fisher supported his suspicion that smoking does not cause cancer with two poorly documented sets of data—a curious claim that people who inhale develop fewer cancers, for the same amount of smoking, than those who do not inhale; and a puzzling contention that lung cancer had increased faster in men than in women, whereas smoking had risen more rapidly in women.

The data on inhaling came from a very poorly constructed questionnaire. Most respondents may not even have known the meaning of the word "inhale," and may have checked "no" in simple confusion. Later information shows a strongly positive correlation of cancer with inhaling, when all other factors are held constant. As for men and women, Fisher made a sound argument, but the data were wrong. The ever-accelerating incidence of lung cancer in women now ranks among the strongest points of evidence for a causal connection.

The basis for judging Fisher negatively in this sorry incident emerges neither from the logic of his argument (which was sound, despite his false conclusion, based on inadequate data), nor from his proper words of caution, but from a clear inference that he did not live by his own stated strictures. Evidently, Fisher did not approach the issue of smoking and cancer with the open mind that he championed as so necessary for any good science. He maintained an obvious preference for denying that smoking causes cancer—even though he states, again and again, that the raw data of an admitted correlation offers no preference for any of the three potential interpretations. Two aspects of his writing give the game away. First, his language. Consider the small sample quoted above. His words call for argument "free from emotion," and for "complete impartiality." Yet he labels the claim that smoking might cause cancer as "propaganda," probably a "catastrophic and conspicuous howler," and a "frantic alarm" acting as the " 'yellow peril' of modern times."

Second, his treatment of limited data then available. Fisher accepted, virtually without question or criticism, the inadequate but exculpatory data, previously cited, on inhaling and incidence in men vs. women—even though both sets would soon be discredited. Fisher then showcased some even more dubious data supposedly consonant with his favored view that both cancer and smoking arise independently from a common genetic predisposition. Two studies compared the smoking behaviors of identical and fraternal twin pairs. Smoking preferences (either yea or nay) were more often shared by identical than by fraternal twins. Since identicals form from one egg and therefore share the same genetic program, while fraternals develop from two eggs and are no closer genetically than any ordinary pair of siblings, Fisher concluded that the greater similarity of identicals must indicate a strong genetic basis for smoking preferences.

But this inference is both potentially wrong and largely irrelevant to Fisher's argument. First of all, the greater smoking similarity of identicals could, at most, indicate a genetic predisposition for attitudes toward the weed; such data say nothing at all about genetic bases for cancer, or about correlation of the two potential predispositions. Moreover, Fisher's data do not even prove his basic assertion of genetic predisposition for smoking. Fisher's explanation does represent one potential interpretation of the data, but another clearly exists, and he hardly considers this alternative. Identical twins look alike and are frequently raised to emphasize their eerie similarity; they are often dressed alike, learn to act as surrogates one for the other, etc. Perhaps this greater similarity in rearing leads to a stronger likelihood for identical smoking habits.

In any case, Fisher should have considered all these possibilities if he had truly pursued this issue with an open mind. We must conclude, rather, that he entered the fray with a clear preference, even a mission—the debunking of smoking as a cause and a championing of joint genetic predisposition as an alternative explanation. We must therefore probe deeper and ask why Fisher had such a clear preference. Two factors stand out, one immediate and practical, the other long-standing and theoretical.

The immediate reason is easy to state and hard to gainsay: in 1956, Fisher became paid scientific consultant for the Tobacco Manufacturers' Standing Committee. Fisher took great umbrage at any implication that his objectivity might be compromised thereby, arguing that he wouldn't sell his soul for the pittance they paid him. Higher powers must judge the tangled commitments wrought by such employment; I will only observe that we generally, and with good reason, require institutional impartiality as a prerequisite for genuine objectivity of mind.

The long-standing reason is more interesting intellectually, and permits us to work back toward Fisher's first great error, thereby revealing an important continuity in his life and career. Fisher was a strong, lifelong supporter of eugenics, the proposition that human life and culture could be bettered if we implemented strategies for genetic improvement by selective breeding—either encouraging childbearing by those judged genetically more fit (positive eugenics), or preventing procreation by the supposedly unfit (negative eugenics). I must emphasize at the start that I do not single out Fisher for any special opprobrium on this score. The great majority of geneticists advocated some form of eugenics, at least until Hitler showed so graphically how a ruthless program of negative eugenics might operate (see the next essay). Moreover, Fisher's idiosyncratic version was, as we shall see, relatively benign politically, and largely in the positive mode. Eugenics commanded a big and motley group of supporters, including fascists to be sure, but also idealistic socialists and committed democrats (see the preceding essay).

Fisher's strong and lifelong preference for genetic explanations of behavior, the foundation of his eugenical sympathies, surely predisposed him to the argument that both smoking and cancer might be linked to genetic variation among people. The same preference for genetic explanations inspired his much more extensive and encompassing first great error—his general theory of racial decline (and possible eugenic salvation), as presented in his magnum opus of 1930, *The Genetical Theory of Natural Selection.*

Just as most of my colleagues ignore Fisher's late and embarrassing work on smoking, they also pay little or no attention to the eugenical chapters of our profession's Bible. Evolutionists may not know much about Fisher's campaign to exonerate the tobacco industry, but how can they bypass several chapters of a crucial volume present in every professional's library? One leading book on the history of population genetics says this and no more about Fisher's eugenical chapters: "In the concluding five chapters he extended his genetical ideas to human populations."

We don't like to admit flaws in our saints. Perhaps my colleagues are embarrassed that a truly great work, the abstract and theoretical foundation of our field, should include a practical view of society that most of us find both fatally flawed and politically unacceptable. Perhaps we tend to view the eugenical chapters as an unfortunate and discardable appendage to a great work of very different character. But such dismissal cannot be defended. The eugenical chapters are no ending frill; they represent more than one third of the book. Moreover, Fisher explicitly insists that these chapters both follow directly from his general theory and cannot be separated from his more abstract conclusions. He states that he only gathered these chapters together for convenience and might, instead, have scattered the eugenical material throughout the book. Fisher writes, "The deductions respecting man are strictly inseparable from the more general chapters."

A single, if complex, argument runs through the five eugenical chapters: advanced civilizations destroy themselves by "the social promotion of the relatively infertile"—that is, people who rise into the ruling classes (the "better people" so necessary to successful government) tend, alas, to have fewer children for reasons of relative genetic infertility, not mere (and reversible) social choice. The upper classes therefore deplete themselves and society eventually weakens and crumbles. If this assertion seems implausible *a priori*, then follow Fisher's rationale through six steps. Again, as with smoking, the argument is impeccably logical (in the narrow, technical sense of following from premises), but entirely wrong, almost nonsensical, based on the fallacy of these key premises.

1. All great civilizations cycle from initial prosperity to eventual decline and fall. Though conquest may eventually befall a depleted race, the cause of decline is internal and intrinsic. The major reason for ultimate failure must lie with a predictable weakening of the elite classes. Can this decline be stemmed and stability with greatness imparted? Fisher writes: "The fact of the decline of past civilizations is the most patent in history . . . The immediate cause of decay must be the degeneration or depletion of the ruling classes."

2. Fisher now notes, but misinterprets, the long-known and well-documented relationship between family size and social status in modern Western nations. Poorer families have more children, while the elite are relatively infertile because the upper classes marry later, have fewer children after marriage, and contain a higher percentage of permanent bachelors or spinsters. This relative infertility of the upper classes leads to their depletion and eventually, thereby, to a decline of civilization by failure of the most able to replenish themselves. Fisher writes: "The birth

rate is much higher in the poorer than in the more prosperous classes, and this difference has been increasing in recent generations."

Interestingly, and on this basis, Fisher rejects the two most common alternative explanations for racial decay advanced by eugenicists of his time. He denies, first, that the upper classes alone are being vitiated by dangerous inbreeding. The decline in fertility is gradual and pervasive through the social hierarchy, not confined to the ruling elite. Fisher writes: "The deficiency in procreation is not especially characteristic of titled families or of the higher intellects, but is a graded quality extending by a regular declivity from the top to the bottom of the social scale."

Second, Fisher also rejected the common argument with the most unfortunate moral and political consequences—that superior civilizations decay by racial mixture with inferior groups. Fisher's rejection arises directly from his general evolutionary views—and this link supplies the best proof that Fisher's eugenical chapters are integrally connected with his general evolutionary theory, and that the two parts of his book cannot be separated, with the theory exalted and the eugenics ignored in embarrassment. The book's centerpiece is a proposition now known as "Fisher's fundamental theorem of natural selection": "The rate of increase in fitness of any organism at any time is equal to its genetic variance in fitness at that time." Or, roughly, the rate of evolution by natural selection is directly proportional to the amount of usable genetic variation maintained in a population. Or, even more roughly, genetic variation is a good thing if you want to accelerate the rate of evolution. Therefore, since eugenic betterment requires effective evolution, anything that boosts the amount of usable variation should be strongly desirable. In Fisher's view, racial mixing represents a powerful way to increase variation, and Fisher had to acknowledge the potential benefits. (Fisher, following a common prejudice of his time, did not deny the general superiority of some races. Thus, race mixture might lower the average quality of a people. Nonetheless, the *range* of variation would increase, even while the mean declined, and natural selection could produce improvement by favoring rare individuals on the extended upper end.)

3. One might think that the elite have fewer children for purely social reasons (greater access to contraception, postponement of childbearing for work or education, more access to types of leisure enjoyed better without large families); but in fact the cause of the correlation is largely genetic and the elite are less fertile for constitutional reasons.

This statement provides the centerpiece for Fisher's eugenics. He argues that the low fertility of modern elites is a pernicious and recent development, not a permanent state of all societies. In "primitive" social organizations, rulers generally have *more* children. (Fisher discreetly bypasses the major reason for this former positive correlation—concubinage and multiple marriage by males in power—largely, I suspect, because he rejects such practices morally but wishes to think well of elites in any age!) Fisher writes: "The normal destiny of accumulated wealth was to provide for a numerous posterity."

But "advanced" civilization has reversed this old and biologically healthy correlation. The elite now have fewer children, primarily for reasons of relative genetic infertility. How did the tragic reversal occur? This sorry situation could only have arisen, Fisher argues, if tendencies for social promotion of the less fertile predictably arise in "advanced" civilizations, thus flooding the upper classes with the source of their eventual depletion in numbers. But how could such a tendency originate?

4. People who rise from the lower to the upper classes (in a democracy permitting such mobility) do so by virtue of genetic superiority and the advantages thus conferred via intelligence and business acumen. But, unfortunately, these people also tend to be less fertile. Fisher's argument precisely follows the form of his later claim for smoking. High ability does not cause infertility, nor does infertility produce brilliance. Rather, the correlation of high ability and infertility originates because both traits are independently linked to a pernicious circumstance that only arises in "advanced" civilizations. A man has to possess strong, genetically based ability in order to rise at all. (Fisher, following the pervasive sexism of his time, does formulate this argument explicitly for males.) But if a man comes from a large family (and therefore inherits a propensity for high fertility), his chances for rising are diminished because his family will be poorer (more mouths to feed, all other things being equal) and he will have less access to education. But a man with the same high ability, if he comes from a small family (with heritable low fertility), has a better chance to rise. By this noncausal correlation of ability and infertility, the chief reason for declining civilization emerges: the social promotion of the relatively infertile.

This situation breeds tragedy all around. The lower classes decline by loss of their most able members; the upper classes sink by infertility of these upwardly mobile people. Society goes down the tubes. Fisher, at least, tried to do his personal bit to stem the tide by raising a bevy of kids.

5. Fisher now faced a problem in the logic of his argument. If the upper classes are so infertile, shouldn't rising immigrants from lower lev-

els help to replenish the dearth even if these newcomers are less fertile than their compatriots remaining at the bottom? Fisher, following a curious argument first advanced by Francis Galton, argued that men who rise by ability tend to marry particularly infertile upper class women, thus diluting their own capacity for childbearing. Such men, knowing their advantages (and clever enough for exploitation), tend to marry heiresses should the opportunity arise (for these men so desperately need a financial leg-up in order to use their considerable abilities). Now an heiress tends to be particularly infertile because she is so often the only daughter in a family that had no male children. Fisher laments: "This puts in the same class the children of comparatively infertile parents and the men of ability, and their intermarriage has the result of uniting sterility and ability."

When you dissect this preciously absurd argument for its hidden assumptions, sexist and otherwise, you get some sense of Fisher's own background and social biases—and you realize how illusory must be the notion of absolute impartiality, and obedience only to the logic of argument and the dictates of empirical data. Only men rise from lower classes. Infertility is the burden and fault of women. In other words, men advance and women then pull the whole family line down.

6. Fisher summarized the baneful effect of this genetically based inverse correlation between childbearing and social status:

> Whenever, then, the socially lower occupations are the more fertile, we must face a paradox that the biologically successful members of our society are to be found principally among its social failures, and equally that classes of persons who are prosperous and socially successful are, on the whole, the biological failures, the unfit of the struggle for existence, doomed more or less speedily, according to their social distinction, to be eradicated from the human stock.

If the social promotion of infertility is the cause of this destructive inverse correlation, then our only hope for reversal and salvation lies in legislated policies aimed at social *promotion* of the *more fertile*. Fisher advocated some form of payments and bonuses for childbearing, so that lower-class people of both ability *and* fertility would be able to rise—as I said at the outset, a relatively benign form of eugenics.

I need hardly detail the numerous false assumptions that derail Fisher's complex argument. I only note that they represent exactly the same mistake—uncritical acceptance of genetic conjectures—that invalidated his later case for smoking. Why should we assume that people who rise socially do so, in large part, by genetic endowment? And even if this argument is valid, why assume that the well-known negative correlation of childbearing and social status results from differential *genetic* fertility—especially when so many excellent and obvious nongenetic explanations cry out for attention (though Fisher mentions them only in quick derision)—including, as mentioned before, longer years of schooling and delayed marriage, and greater access to contraception and abortion. The first genetic conjecture (a biological basis for social promotion) seems less implausible, though quite unproven; but the second conjecture (a genetic basis for fewer children in the upper classes) seems wildly improbable, even bordering on the absurd. Yet Fisher's case absolutely requires that both genetic conjectures be valid—for if we rise genetically but then have fewer children only for social reasons, then his argument falls apart, for no "social promotion of infertility" would exist.

We may take a kindly view of Fisher's eugenics and say that his genetic conjectures did no harm, for, try as he might in press and before parliament, Fisher's recommendations made no practical headway. But false genetic hypotheses of human behaviors and statuses are politically potent. They represent an ultimate weapon for social conservatives who wish to "blame the victim" for any correctable social ill or inequity. Are workplaces toxic? Screen workers and fire those with genetic predispositions to react badly. Is adequate access available to members of minority races? Argue that these people are inferior by nature and therefore already occupy an appropriate number of slots. The genetic fallacy is generic—and applicable almost anywhere for the all too common and lamentable social aim of preserving an unfair status quo.

We may excuse Fisher's eugenics as relatively harmless, but we cannot be so sanguine about his conceptually similar campaign against a causal link between cancer and smoking. Joan Fisher Box wrote a fine biography of her father, marred only by an understandably hagiographical approach. She depicts Fisher's smoking campaign as rousing good fun for her father, a kind of harmless little game enjoyed by a gadfly against powerful interests. But her last paragraph is chilling, unintentionally so, I suspect:

> In 1958 Fisher was brought into discussion of the evidence in the United States in connection with legal suits expected to be brought to trial against tobacco manufacturers for personal damage caused by their products. Early in 1960 he visited the United

> States at the invitation of a legal firm representing an American tobacco company, whose case was brought to trial in April that year. Other suits were either not brought or were unsuccessful, and the legal pressure on tobacco companies was relieved for a time.

And that, friends, translates into many, many deaths—as pressure to quit and to restrict advertising diminished. Fisher may have been only the tiniest cog in a great machine rolled out by the tobacco industry, but he did contribute. Charles Lamb once wrote a humorous couplet:

> For thy sake, Tobacco I
> Would do anything but die.

Bad and biased arguments can have serious, even deadly, consequences.

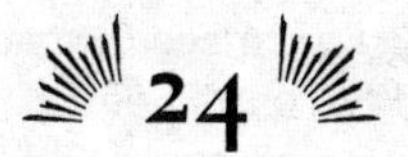

24

The Most Unkindest Cut of All

CONSIDER THIS UNREMARKABLE DESCRIPTION of a perennially pleasant pastime: unwinding after a hard day's work—a smoke and a drink around the fire.

> I remember that at the end of this ——— conference, ——— and I sat very cozily near the stove and then I saw ——— smoke for the first time, and I thought to myself, "——— smoking today"; I'd never seen him do that. "He is drinking brandy"; I hadn't seen him do that for years . . . We all sat together like comrades. Not to talk shop, but to rest after long hours of effort.

Now let's play "fill in the blanks." The man who made such an exception to his usual abstemiousness in order to celebrate his pleasure at such a successful outcome? Reinhard Heydrich, head of the Nazi security police and chief deputy to SS director Heinrich Himmler. The location of the cozy stove? The Wannsee Conference of January 20, 1942, held to prepare a plan for the *Endlösung der Judenfrage*—the "final solution to the Jewish question," the systematic murder of 11 million human beings (by Heydrich's own reckoning) and the genocide of a people. The man who remembered the closing scene quoted above? Adolf

Eichmann, another participant and author of the Wannsee Protocol, the infamous document that summarized the hard day's work.

The maximally chilling content of the Wannsee Protocol is enhanced by its euphemistic and circumlocutory language. Killing and murder are never mentioned directly, and genocide sounds even more evil (if anything can make the ultimately hateful still worse) in its obtuse, but unmistakable, description as a "final solution." But Eichmann recalled at his trial that the verbal discussions could not have been more direct: "What I know is that the gentlemen convened their session, and then in very plain terms—not in the language that I had to use in the minutes, but in absolutely blunt terms—they addressed the issue, with no mincing of words . . . The discussion covered killing, elimination, and annihilation."

In an evasive first half, the Wannsee Protocol enumerates the Jewish population of Europe at some 11 million and then reviews the first two stages of action, now deemed unsuccessful and insufficient. At first, Hitler and company attempted "the expulsion of the Jews from every particular sphere of life of the German people"—read *Kristallnacht*, confiscation and terrorism. The second strategy stressed physical removal: "the expulsion of Jews from the living space *(Lebensraum)* of the German people." But, Eichmann writes in the Wannsee Protocol, emigration spawned too many obstacles and had not worked fast enough: "Financial difficulties, such as the demand for increasing sums of money to be presented at the time of the landing on the part of various foreign governments, lack of shipping space, increasing restriction of entry permits, or canceling of such, extraordinarily increased the difficulties of emigration."

On, then, to a third (and truly ultimate) "solution"—kill them all. (In some sense, as so many others have noted, the most chilling aspect of the Wannsee Conference and Protocol, held to initiate and implement this third strategy, lies neither in Heydrich's ability to conceptualize evil at such grand scale, nor in Eichmann's propensities for composing plans in euphemistic bureaucratese, but in the painstaking and deliberate construction of detailed logistics for such an extensive undertaking—the careful calculation of railroad cars and their volumes, the siting of death camps at the hubs of transportation lines, complex efforts to mask the true intent by depicting genocide as relocation and forced labor.)

Eichmann then introduces the new plan: "Another possible solution of the problem has now taken the place of emigration, i.e., the evacuation of the Jews to the East, provided the Führer agrees to the plan"

(Hitler had, in fact, already ordered such a strategy). Eichmann continues with his masked description of forced transportation to death as emigration for labor:

> Under proper guidance the Jews are now to be allocated for labor to the East in the course of the final solution. Able-bodied Jews will be taken in large labor columns to these districts for work on roads, separated according to sexes, in the course of which action a great part will undoubtedly be eliminated by natural causes . . . In the course of the practical execution of this final settlement of the problem, Europe will be cleaned up from West to East . . . The evacuated Jews will first be sent, group by group, into so-called transit ghettos from which they will be taken to the East . . . It is intended not to evacuate Jews of more than sixty-five years of age but to send them to an old-age ghetto.

Some camps—most notably Auschwitz—served both as killing places and prisons for forced labor (only a slower form of death). But others were gassing sites, pure and simple—Treblinka, Chelmno, Sobibor, Belzec. Architects of the final solution never intended to implement their "cover story" of transportation for labor. We sometimes get a false impression of numbers allocated to immediate death vs. starvation and forced labor at the camps—though little preference can be specified in such a grisly "Sophie's choice." Several thousand survived Auschwitz, and many have told their stories. But the pure death camps are less well known because virtually no one lived to remember. Two people survived at Belzec, three at Chelmno (M. Gilbert, *The Holocaust*; see bibliography). The final solution was always and only about murder—total and unvarnished.

Clearly, I have nothing to add to this ultimate story of human atrocity. I am neither a poet nor historian, and I was not there. Why take up such a subject in a series of essays on natural history and evolutionary theory? The answer lies in the second half of the Wannsee Protocol—the part of the document that is rarely discussed and almost never quoted. My rationale also rests with Hitler's misuse of genetics and evolutionary biology as a centerpiece of plans explicitly stated right from the beginning, at the publication of *Mein Kampf* in 1925.

Insofar as Hitler's evil scheme should be dignified as a "theory" at all, he sunk his argument squarely in paranoia about racial purity and its biological necessity for triumphant peoples in a world of natural selection. The Aryan nation had been great, but its strength had been sapped by racial mixing, encouraged by the insidious and liberal propaganda of parasitic Jews, out to rule the world or at least to feed off the higher morality of Aryan virtue. Hitler wrote in *Mein Kampf*:

> The Aryan gave up the purity of his blood and therefore he also lost his place in the Paradise which he had created for himself. He became submerged in the race-mixture, he gradually lost his cultural ability till . . . he began to resemble more the subjected and aborigines than his ancestors . . . Blood-mixing, with the lowering of the racial level caused by it, is the sole cause of the dying-off of old cultures; for the people do not perish by lost wars [a comment on Germany's defeat in World War I], but by the loss of that force of resistance which is contained only in the pure blood. All that is not race in this world is trash.

(I am quoting from the first complete English translation of *Mein Kampf*, published in America by John Chamberlain and others in 1939 as a warning about the enemy we would soon need to confront. My parents bought this book before my father left to join the battle. Throughout my youth, I stared at this volume on my parents' shelves, taking it down now and again—more to experience the frisson of touching evil than from any desire to read. When my father died a few years ago, and my mother offered me his collection of books, I included this familiar volume, with its blood-red jacket, among the few items that I wanted for my own. To hold the book now, and to quote from it for the first time, gives me an eerie feeling of connectivity with my past, and rekindles my dimmest three-year-old impression of World War II as a fight between my daddy and a bad man named Hitler.)

If the first (and oft-quoted) half of the Wannsee Protocol is a cursory and euphemistic account of genocide at worksites and death camps, the highly specific (and usually neglected) second half is a detailed and specific disquisition on genetics and race-mixing—and for an obvious reason, given Hitler's eugenical ideas. From the Führer's standpoint, killing 11 million people will not solve the Jewish problem if many German citizens still carry tainted blood as a result of partial Jewish ancestry through mixed marriages. A truly final solution demands a set of rules and policies for these *Mischlinge* (literally, mixtures). Kill the others and purify yourselves. The two halves of the Protocol could not be more intimately connected through the sick logic of Hitler's racial doctrines.

Eichmann's transition sentence, written in optimally opaque bureaucratese, states:

> The implementation of the final solution problem is supposed to a certain extent to be based on the Nuremberg Laws [Nazi legislation on eugenic marriage and sterilization], in which connection also the solution of the problems presented by the mixed marriages and the persons of mixed blood is seen to be conditional to an absolutely final clarification of the question.

These pages on racial mixing reveal a form of madness different from Eichmann's first half, on genocide in the east. The first part of the Protocol wallows in the ultimate evil of euphemized mass murder; this second half follows the steely logic of total craziness reasoned right through, with full coverage of all details. Only once before have I experienced such a feeling while reading an official state document: when I studied the annual proceedings of the South African racial classification board (under the old regime, before Nelson Mandela) as they tortured the logic of continuity to find a discrete pigeonhole for each individual under strict apartheid.

Needless to say, any such effort faces an intractable dilemma at the start: people are complexly interbred at all degrees of mixture, and neither pure solution will work. Is everyone with even the slightest trace of Jewish ancestry a Jew (which would condemn most of the nation), or does salvation arise with any Aryan infusion (which would be far too lenient for zealous madmen)? Heydrich, Eichmann, and company therefore invoke the usual trick of argument for breaking a true continuum lacking a compelling point for separation: choose an arbitrary dividing line and then treat your division as a self-evident fact of nature.

The Protocol basically proclaims that half-breeds are Jews (offspring of one pure Jewish and one pure Aryan parent); quarter-breeds are German (offspring of a half-breed and an Aryan). But this tidy little rule required some nuancing at the borders. At one end, half-breeds could avoid their death warrant if they had children by marriage to a person of German blood (but not if such a union had produced no offspring), or "if exemption licenses have been issued by the highest Party or State authorities."

Even this hope for mitigation carried two provisos: first, each exemption must be granted on a case-by-case basis and only for "personal essential merit of the person of mixed blood." Second, anyone so spared must make a little gesture in return: "Any person of mixed blood of the first degree to whom exemption from the evacuation is granted will be sterilized—in order to eliminate the possibility of offspring and to secure a final solution of the problem presented by the persons of mixed blood." But fear not: "The sterilization will take place on a voluntary basis." Consider the alternative, however: "But it will be conditional to a permission to stay in the Reich." One participant at the conference made an astute observation on this score: "SS-Gruppenführer Hofmann advocates the opinion that sterilization must be applied on a large scale; in particular as the person of mixed blood placed before the alternative as whether to be evacuated or to be sterilized would rather submit to sterilization."

At the other end, the supposedly acceptable quarter-breeds could be "demoted" to genocide for any of three reasons all designed to ferret out an unacceptable degree of racial taint: (1) "The person of mixed blood of the second degree is the result of a marriage where both parents are persons of mixed blood" (apparently, a child needs at least one purely Aryan parent to enter the realm of the blessed). (2) "The general appearance of the person of mixed blood of the second degree is racially particularly objectionable so that he already outwardly must be included among the Jews." (3) "The person of mixed blood of the second degree has a particularly bad police and political record sufficient to reveal that he feels and behaves like a Jew."

In the gray area between, marriages of two people with mixed blood sealed everyone's doom, parents and children alike. Even a normally exempt quarter-breed must die (with all children) if such a person marries a half-breed: "Marriage between persons of mixed blood of the first degree and persons of mixed blood of the second degree: Both partners will be evacuated, regardless of whether or not they have children . . . Since as a rule these children will racially reveal the admixture of Jewish blood more strongly than persons of mixed blood of the second degree." What can be more insane than madness that constructs its own byzantine taxonomy—or are we just witnessing the orderly mind of the petty bureaucrat applied to human lives rather than office files?

The Protocol's most stunning misuse of evolutionary biology, however, appears not in this tedious taxonomy of genetic nonsense, but right in the heart of the document's chief operational paragraph. I quoted the words above, but inserted ellipses to signify an omission that I now wish to restore. Eichmann speaks of deportation to the east, and of hard labor on the roads, leading to the death of most evacuees. He then continues (to flesh out my ellipses): "The possible final remnant will, as it must undoubtedly consist of the toughest, have to be treated accordingly . . . " Let me now switch to the original German: ". . . *da dieser, eine natürliche Auslese darstellend, bei Freilassung als Keimzelle eines neuen jüdischen Aufbaues anzusprechen ist.*" Or: ". . . as it is the *product of natural selection,* and would, if liberated, act as a bud cell of a Jewish reconstruction."

Perhaps you do not see the special horror of this line (embedded, as it is, in such maximal evil). But what can be more wrenching than the violation of one's own child, or the perversion for vicious purpose of the most noble item in a person's world? I am an evolutionary biologist by original training and a quarter-century of practice. Charles Darwin is the resident hero in my realm—and few professions can name a man so brilliant, so admirable, and so genial as both founder and continuing inspiration. Darwin, of course, gave a distinctive name to his theory of evolutionary change: natural selection. This theory has a history of misuse almost as long as its proper pedigree. Claptrap and bogus Darwinian formulations have been used to justify every form of social exploitation—rich over poor, technologically complex over traditional, imperialist over aborigine, conqueror over defeated in war. Every evolutionist knows this history only too well, and we bear some measure of collective responsibility for the uncritical fascination that many of us have shown for such unjustified extensions. But most false expropriations of our chief phrase have been undertaken without our knowledge and against our will.

I have known this story all my professional life. I can rattle off lists of such misuses, collectively called "social Darwinism." (See my book *The Mismeasure of Man.*) But until the fiftieth anniversary of the Wannsee Conference piqued my curiosity and led me to read Eichmann's Protocol for the first time, I had not known about the absolute ultimate in all conceivable misappropriation—and the discovery hit me as a sudden, visceral haymaker, especially since I had steeled myself to supposed unshockability before reading the document. *Natürliche Auslese* is the standard German translation of Darwin's "natural selection." To think that the key phrase of my professional world lies so perversely violated in the very heart of the chief operative paragraph in the most evil document ever written! What symbol of misuse could possibly be more powerful? Surely this is the literary equivalent of imagining one's daughter shackled in a dungeon operated by sadistic rapists.

It scarcely matters that the phrase is so ludicrously inverted in its misapplication—for we dare hardly dignify Eichmann's argument with space for refutation. Natural selection is nature's process of differential reproductive success, however that advantage be attained (sometimes by dominating one's fellows, to be sure, but also by cooperation for mutual benefit). What could be more unnatural, more irrelevant to Darwin's process, than the intricately planned murder and starvation of several million people by human technology?

I could simply end this essay here, washing my profession's hands of all conceivable responsibility. After all, the Wannsee Protocol would have proclaimed and implemented its horrors even if Eichmann had not employed Darwin's phrase to justify and embellish his major recommendation. And Eichmann's use, in any case, ranks as a perverse misinterpretation of Darwin on both key grounds—the false application of a natural principle to human moral conduct, and the distortion of a statement about differential reproductive success into a bogus validation of mass murder as natural. Of the second mistake, Darwin wrote in a key passage in the *Origin of Species*:

> I should presume that I use the term Struggle for Existence in a large and metaphorical sense, including dependence of one being on another, and including (which is more important) not only the life of the individual, but success in leaving progeny. Two canine animals in a time of dearth, may be truly said to struggle with each other which shall get food and live. But a plant on the edge of a desert is said to struggle for life against the drought.

But resolutions are never so clean or simple. Science, as a profession, does have a little something to answer for, or at least something to think about. Darwin may have been explicit in labeling the struggle for existence as metaphorical, but most nineteenth-century versions (including Darwin's own illustrations, most of the time) stressed overt competition and victory by death—surely a more congenial image than peaceful cooperation for an age of aggressive expansion and conquest, both ethnographic and industrial.

Hitler didn't invent the mistaken translation to human affairs. Claptrap Darwinism had served as an official rationale for German military conquest in World War I (while our side often used the same argument, though less zealously and systematically). In fact, William Jennings Bryan (see essay 28 in my book *Bully for Brontosaurus*) first decided to oppose evolution when he mistook Darwin's actual formulation for the egregious German misuse that so deeply disturbed him.

Many scientists consistently opposed this misapplication, but others, probably the majority, remained silent (many enjoying the prestige, even if falsely won), while a few actively abetted the cooptation of their field for a variety of motives, including misplaced patriotism and immediate personal reward. Several English and American eugenicists offered initial praise for Hitler's laws on restriction of marriage and enforced sterilization—before they realized what the Führer really intended. The text of the German legislation borrowed heavily from eugenical sterilization statutes then on the books of several American states, and upheld by the Supreme Court in 1927. German evolutionists did not raise a chorus of protest against Hitler's misuse of natural selection, dating to *Mein Kampf* in 1925. Wannsee is the logical extension of the following fulmination in *Mein Kampf*, with its final explicit misanalogy to nature, and its call for elimination and sterilization of the supposedly unfit:

> The fight for daily bread [in Nature] makes all those succumb who are weak, sickly and less determined . . . The fight is always a means for the promotion of the species' health and force of resistance, and thus a cause for its development to a higher level. If it were different, every further development towards higher levels would stop, and rather the contrary would happen. For, since according to numbers, the inferior element always outweighs the superior element, under the same preservation of life and under the same propagating possibilities, the inferior element would increase so much more rapidly that finally the best element would be forced to step into the background, if no correction of this condition were carried out. But just this is done by Nature, by subjecting the weaker part to such difficult living conditions that even by this the number is restricted, and finally by preventing the remainder, without choice, from increasing . . . [Did Eichmann have this passage in mind when he advocated the same two-step process for the final solution: starve to a remnant by "difficult living conditions"; then kill the rest.] Man, by trying to resist this iron logic of Nature, becomes entangled in a fight against the principles to which alone he, too, owes his existence as a human being. Thus his attack is bound to lead to his own doom.

A scientist's best defense against such misappropriation lies in a combination that may seem to mix two disparate traits, vigilance and humility: vigilance in combating misuses that threaten effectiveness (you don't have to refute every kook who writes a letter to the local newspaper—time doesn't permit—but how can we distinguish the young Hitler from "just another nut"?); humility in recognizing that science does not, and cannot in principle, find answers to moral questions. A popular linguistic construction of late would have us believe that morality may be measured as a kind of "fiber"—as though one might pour ethics out of a cereal box prominently labeled (by exacting laboratory standards) with precise content and contribution to minimal daily requirements!

Science can supply information as input to a moral decision, but the ethical realm of "oughts" cannot be logically specified by the factual "is" of the natural world—the only aspect of reality that science can adjudicate. As a scientist, I can refute the stated genetic rationale for Nazi evil and nonsense. But when I stand against Nazi policy, I must do so as everyman—as a human being. For I win my right to engage moral issues by my membership in *Homo sapiens*—a right vested in absolutely every human being who has ever graced this earth, and a responsibility for all who are able.

If we ever grasped this deepest sense of a truly universal community—the equal worth of all as members of a single entity, the species *Homo sapiens*, whatever our individual misfortunes or disabilities—then Isaiah's vision could be realized, and our human wolves would dwell in peace with lambs, for "they shall not hurt nor destroy in all my holy mountain." We are freighted by heritage, both biological and cultural, granting us capacity both for infinite sweetness and unspeakable evil. What is morality but the struggle to harness the first and suppress the second? Darwin's soulmate, the great American born on the very same day, said much the same in the famous words of his first inaugural address, in March 1861, when he still hoped to spare the nation from the horrors of civil war. Lincoln asked us to remember the former unity of North and South, and to avoid destruction by applying our better nature to this memory. Let us extend his hope to all bodies of the single human species:

> The mystic chords of memory, stretching from every battlefield and patriot grave to every living heart and hearthstone all over this broad land, will yet swell the chorus of the Union when again touched, as surely they will be, by the better angels of our nature.

EPILOGUE

I doubt that I have ever written a more thoroughly serious essay, without even an attempt at passing humor. What else can one do with this most totally tragic of all conceivable subjects? Still, an old literary tradition holds that some lightness must be introduced—for the relief and variety that our emotional natures crave—especially in such situations of sustained sadness. Thus, Hamlet jests with the gravedigger about poor Yorick, while Puccini introduces three courtiers, Ping, Pang, and Pong, as comic relief between Turandot's beheading of several sequential suitors.

Therefore, with such a historically distinguished excuse, I have decided to cite in full a remarkable letter received by the editor of *Natural History* in response to the original version of this essay:

> As a long-time faculty member concerned about correct usage of the English language, I was disturbed by the title of Stephen Jay Gould's latest article. Then I decided something in the article would explain why the double superlative was used. When that supposition proved groundless, I could only conclude that someone intended it as a joke or as a means to provoke correspondence from readers like myself. Am I correct? I will continue to enjoy your periodical, but hope that you will not continue to provide shocks of this kind.

I replied that I grasped both her point and her distress, and that, while I hate to pass the buck and believe that a man must bear the consequences of his own actions, she really would have to air her complaint with Mr. Shakespeare!

> If you have tears, prepare to shed them now . . .
> Look! in this place ran Cassius' dagger through:
> See what a rent the envious Casca made:
> Through this the well-beloved Brutus stabb'd . . .
> Judge O you gods, how dearly Caesar loved him!
> This was the most unkindest cut of all.
>
> —Mark Antony in *Julius Caesar*

PART SEVEN

Evolutionary Theory, Evolutionary Stories

THEORY

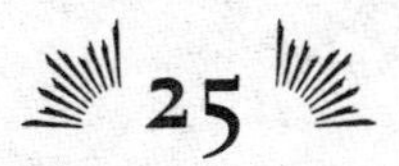

Can We Complete Darwin's Revolution?

IN A WONDERFULLY WISE and frequently cited statement, Sigmund Freud identified the common component of all major scientific revolutions: "Humanity has . . . had to endure . . . great outrages upon its naive self-love." In other words, great revolutions smash pedestals—the previous props for our cosmic arrogance. Freud then identified the two most significant fracturings: first, the cosmological shift from a geocentric to a heliocentric universe, "when [humanity] realized that our earth was not the center of the universe, but only a speck in a world-system of a magnitude hardly conceivable"; second, the Darwinian discovery of evolution, which "robbed man of his particular privilege of having been specially created, and relegated him to descent from the animal world." Freud then hinted that the discovery and elucidation of the unconscious, in large part his own work, might smash a third pedestal in setting aside our convictions about mental rationality.

This statement suggests a criterion for judging the completion of scientific revolutions—namely, pedestal-smashing itself. Revolutions are not consummated when people accept the physical reconstruction of the universe thus implied, but when they grasp the meaning of this reconstruction for the demotion of human status in the cosmos. The two phenomena—realignment of the physical universe and reassessment of human status—are truly distinct, a separation best understood by invoking an old mental strategy that has received a striking new name in contemporary culture: spin doctoring.

In spin doctoring, an art practiced best by politicians from time immemorial, one accepts a sorry fact, but provides an interpretation based entirely on the silver lining said to accompany all dark clouds. For example, Dr. Pangloss of Voltaire's *Candide*, surely the greatest spin doctor in Western literature, stated that syphilis, inadvertently transmitted from the New World to Europe, might be unpleasant but that, on balance, all must be for the best because the Americas had also provided such wonderful products as chocolate.

We may say, I think, that Freud's first revolution is complete in his pedestal-smashing sense. All thinking people accept that we live on a peripheral hunk of rock on the edge of one galaxy among gezillions—and no one seems enveloped by cosmic *Angst*, or despairing about the meaning of human life, on these grounds. (Perhaps we have come to terms through the passage of centuries, for the new cosmology did not always seem so unthreatening, and we have not forgotten Galileo's torment. Many early versions of heliocentrism retained the pedestal by placing our own personal star, the sun, in the center of a limited universe.)

But, having spent a professional lifetime explicating and defending evolution in both public and technical fora, I feel certain that Freud's second revolution has not been able to surmount a mental roadblock. Evolution still floats in the limbo of our unwillingness to face the implications of Darwinism for the cosmic estate of *Homo sapiens*. Physical reconstruction, the first step in a Freudian revolution, has been accomplished: all thinking people accept the biological fact of our "descent from the animal world." But the second stage, mental accommodation toward pedestal-smashing, has scarcely begun. Public perception of evolution has been so spin doctored that we have managed to retain an interpretation of human importance scarcely different, in many crucial respects, from the exalted state we occupied as the supposed products of direct creation in God's image. (I am not even discussing the sociologically significant fact that millions of Americans, but no large numbers in any other Western nation, don't accept evolution at all, and continue to espouse the literal reading of Genesis for a creation of all life in a few days of twenty-four hours each. The observation that some people cannot even take the first Freudian step only emphasizes the particular fear and reluctance that this revolution raises in us.)

We need no great philosophical or cultural acumen to recognize why the Darwinian revolution has been most difficult to accept, and therefore remains least complete in the Freudian sense. I don't think that any other ideological revolution in the history of science has ever so strongly or directly impacted our view of our own meaning and purpose. (Some scientific revolutions, though equally portentous and revisionary in their physical reconstruction, just don't pack as much *oomph* for the human soul. For example, plate tectonics has thoroughly changed our view of the earth's history and dynamics, but few people have staked much about the meaning of their lives upon the issue of whether or not Europe and America were once physically connected, and whether or not continents reside within thin plates floating over the earth's surface as new seafloor arises at oceanic ridges.)

I like to summarize what I regard as the pedestal-smashing messages of Darwin's revolution in the following statement, which might be chanted several times a day, like a Hare Krishna mantra, to encourage penetration into the soul: Humans are not the end result of predictable evolutionary progress, but rather a fortuitous cosmic afterthought, a tiny little twig on the enormously arborescent bush of life, which, if replanted from seed, would almost surely not grow this twig again, or perhaps any twig with any property that we would care to call consciousness.

All the classic forms of evolutionary spin doctoring are designed to avoid the radical and unwanted consequences of this mantra. Spin doctoring centers on two different subjects: the *process* of evolution as a theory and mechanism; and the *pathway* of evolution as a description of life's history. Spin doctoring for the process tries to depict evolution as inherently progressive, and as working toward some "higher" good in acting "for" the benefit of such groups as species or communities (not just for advantages of individual organisms), thereby producing such desired ends as harmonious ecosystems and well designed organisms. Spin doctoring for the pathway reads the history of life as continuous flux with sensible directionality toward more complex and more brainy beings, thereby allowing us to view the late evolution of *Homo sapiens* as the highest stage, so far realized, of a predictable progress.

How best can we illustrate that spin doctoring has had this pervasive and baleful effect, and that public understanding of evolution lies immured in biases preventing the completion of Darwin's revolution in Freud's crucial sense of pedestal smashing? Well, I grew up in New York City, and I remain a quintessentially partisan New Yorker. I still root for the Yankees after twenty-five years in Boston, and my mental map of the United States matches the celebrated Steinberg *New Yorker* cover, with Fifth Avenue essentially dividing the nation and the Hudson River near the Nevada-California border. The United States is too diverse to have a canonical media source for identifying the pulse of an educated culture—as the BBC might do in Britain, or *l'Osservatore Romano* in the Vatican. But grant to this parochial New Yorker that *The New York Times* comes as close to such a status as any American publication.

I therefore suggest that a compendium of commentary from the *Times* might give us some insight into the spin doctoring of Darwin's revolution. I have been struck by three items that appeared in *The New York Times* during the past year, for each represents a primary component of the spin-doctored view, yet all are stated with complete assurance that evolution *must* work in such a manner. I therefore found these three items, in their collectivity, particularly compelling as an illustration of our deep miring in the spin doctored view of evolution as sensible and predictable progress, continuously moving toward desired ends by working for the good of groups and communities.

1. EVOLUTION FOR THE COLLECTIVE GOOD. On June 4, 1944, the allied armies launched a great attack that, without cynicism, may be regarded as one of history's finest efforts for global human good. On the fiftieth anniversary of D-Day, June 6, 1994, *The New York Times* lauded the invasion in many front-page articles, and reprinted both General Eisenhower's announcement of the landings and their own editorial of praise. On the same day, the letters column of the *Times* printed this more general commentary on working for the collective good—in response to an earlier article in the Tuesday "Science Times" section that asked how sexual reproduction might benefit the evolutionary success of individuals.

Evolution Benefits Species as a Whole

> To the Editor:
>
> . . . The question why sexual reproduction has evolved, should be asked not from the standpoint of individuals . . . but from the standpoint of the species itself. While sexual reproduction continually introduces mutations that can damage individuals of the species, the advantages of this continual introduction of new genetic material into the gene pool is an evolutionary plus for that species . . . You miss the point. Evolution is not about a good deal for individual females or individual males, but about a good deal for the species.

I regret to inform the writer that he has missed the point. Darwin's central theory of natural selection is about advantages ("good deals," if you will) that accrue to individuals, explicitly not to species. In fact, this counterintuitive proposal—that individual bodies, not "higher" groups like species, act as units and targets of natural selection—lies at the heart of Darwin's radicalism, and explains a large part of our difficulty in grasping and owning his powerful idea. Natural selection may lead to benefits for species, but these "higher" advantages can only arise as sequelae or side consequences of natural selection's causal mechanism: differential reproductive success *of individuals.*

Warm and fuzzy ideas about direct action for the good of species represent a classical strategy in spin doctoring, one that has precluded proper understanding of natural selection for more than a century. If evolution worked explicitly for species, then we could soften the blow of Darwin's radicalism. The transition from God's overt beneficence *toward* species to evolution's direct operation *on* species permits a soft landing in transferring allegiance from creationism to evolution—for the central focus on "higher" good as raison d'être remains unchanged.

But Darwin's real theory of natural selection is uncompromising in kicking this prop away. Natural selection is a theory of ultimate individualism. Darwin's mechanism works through the differential reproductive success of individuals who, by fortuitous possession of features rendering them more successful in changing local environments, leave more surviving offspring. Benefits accrue thereby to species in the same paradoxical and indirect sense that Adam Smith's economic theory of laissez faire may lead to an ordered economy by freeing individuals to struggle for personal profit alone—no accident in overlap, because Darwin partly derived his theory of natural selection as a creative intellectual transfer from Smith's ideas.

If we free individual businesses to act for their own benefit, Smith argues, then the most efficient firms both drive out incompetent competitors, and balance each other to provide an orderly economy. But such order only arises as a side consequence, through the action of an "invisible hand," in Smith's unforgettable phrase; all direct causality resides in the struggle among individuals. Similarly, in Darwin's world, natural selection acts only for the benefit of individuals in reproductive success (firms in profit, by analogy to Smith); well designed organisms and balanced ecosystems emerge as side consequences.

We can say all we want about the beauty and radicalism of Darwin's central notion, but how do we know that his idea is true? How can we tell that nature is Darwinian, rather than shaped by some other set of evolutionary forces?

A convincing set of proofs lies around us, though this simple and powerful point has rarely been articulated in popular writing, and therefore remains largely unappreciated. We begin with something of a paradox—the proof of Darwinian nature does not reside in the best and classic cases of organic design for optimal biomechanical function: the aerodynamic perfection of the bird's wing, or the hydrodynamic shaping of a fish's body. Darwin's natural selection, working for the reproductive success of individuals, might build such excellence in design, but another kind of evolutionary force that worked for the good of species might yield the same result. Excellent wings are both good for species and good for individual birds. To show that nature is Darwinian, we need a set of phenomena that can only be built by forces working for the benefit of individuals, and *not* for species.

Such phenomena exist in abundance: organs and devices that aid individuals in sexual combat for mates and matings against other individuals of the same species. Such organs cannot be beneficial for species, since they only aid individuals in struggle against others of the same population, and cannot aid the species in competition against other species. Moreover, these organs are often elaborate and devilishly clever; they represent enormous "investments" of evolutionary energy, not mere superficial frills. Much of evolution's causal effort must therefore be devoted to building such organs for individual benefit.

The peacock's tail provides a classic case. This gaudy and brilliant, but cumbersome, structure does the bird no good in a biomechanical sense (and probably acts as a positive disadvantage in this regard). But peacocks use their showy tail to compete with other peacocks for attention of the peahen in the essential Darwinian activity of passing more genes along to future generations. Showier tails help individual males in competition with other males; they do not benefit the species. In fact, fancy tails probably injure the species's prospects for extended geological longevity—and can therefore only arise if evolution works for advantages of individuals.

But even this classic case is indirect, for the tail doesn't raise reproductive success by itself, but only by impressing females or intimidating other males. A host of more direct adaptations work explicitly for individuals in the reproductive act itself. For example, males may hold on to females for weeks or months, thus assuring that no other sperm but their own can fertilize the eggs. (This odd phenomenon, called amplexus in frogs, does the species no good, but surely boosts the reproductive success of amplexing males.)

The ever-diverse world of insects yields thousands of stunning examples (see W. G. Eberhard's remarkable book, *Sexual Selection and Animal Genitalia*, for the details). Males of many species, for example, will reach into the female's vulva and pull out the sperm from any previous matings before depositing their own. Others, after mating, secrete a genital plug (nature's chastity belt after the fact) into the vulva—a rocky substance that blocks copulation with any other male. These examples of "sperm competition" (as professionals label the subject) can only evolve if natural selection works for advantages of individuals, not of species.

2. SENSIBLE DIRECTIONALITY. Another letter to the *Times* (January 8, 1995), again commenting upon a previous report from the "Science Times" section, subtly illustrates a major theme in spin doctoring the pathway of evolution (rather than the process). The correspondent objects to a sentence from an article upholding the theory that dinosaurs disappeared in a cosmic catastrophe triggered by impact of a large extraterrestrial body (see essay 12):

Dinosaurs and Destiny

> To the Editor:
>
> In a Jan. 3 "Science Times" article you report on a theory that dinosaurs died out after an asteroid hit sulfur-rich rock in what is now the Yucatán Peninsula of Mexico, producing a haze of sulfuric acid that blocked sunlight for decades. Had the rock not been rich in sulfur, you say, "the dinosaurs might well have survived the impact, thereby changing the course of evolution."
>
> Actually, it was the demise of the dinosaurs that changed the course of evolution. Had the dinosaurs not been wiped out, evolution would have continued on the same path it had been following for at least 150 million years.

While I will not defend the *Times*'s fuzzy language about "the course of evolution," the writer of the letter labors under the false impression that life's history follows a definite path, and that catastrophic episodes can only be read as disruptors of sensible continuity. I see nothing amiss in what the *Times* stated on January 3. If the impact hadn't occurred, dinosaurs would probably have survived and evolution would then have proceeded differently from the pathway actually followed during the past 65 million years (an alternate route, I hasten to add, that would almost surely have kept mammals as small creatures in the interstices of a dinosaurian world, thus preventing the origin of a peculiar group of large mammals with consciousness, and the eventual invention of *The New York Times*).

The error of the letter-writer lies in an assumption that evolution, if not disrupted somehow, follows a path that will sensibly continue into an indefinite future. But no such road exists. The course of evolution is only the summation of fortuitous contingencies, not a pathway with predictable directions. What is the supposed route that evolution had followed for 150 million years before the disruption at the end of the Cretaceous period? For starters, this 150 million year interval included a mass extinction just as intense (and perhaps just as catastrophically triggered) as the later event that wiped out the dinosaurs—the mass dying at the end of the Triassic period. More basically, evolution's unpredictability is fractal and present at all scales. We can trace, in retrospect, what happened during those 150 million years, and we may be able to explain the results in evolutionary terms. But we could not have predicted the outcome at the outset, any more than we could have looked out from Concord Bridge on April 19, 1775, and known that Eisenhower's forces would defeat Nazi Germany 170 years later. Evolution has no pathway that goes forward in sameness if not disrupted by externalities.

3. CONTINUOUS FLUX. Since my first two examples involved letters mistakenly critical of articles in the "Science Times" section, let me strive for journalistic balance by exposing a spin-doctored fallacy in a "Science Times" article of March 14, 1995.

Judging from the dozen or so requests that I later received for interviews and comments based upon this article, the piece obviously inspired a great deal of interest and struck most readers as strange, fascinating, and unexpected. I declined all the interviews because, as I explained, the article was correct and had expressed something important about evolution—but the result described was entirely expected and orthodox, not at all surprising unless one has adopted a spin-doctored view of evolution.

The article, by William K. Stevens, bore the title, "Evolution of Humans May at Last Be Faltering." It opened with the following lead sentence: "Natural evolutionary forces are losing much of their power to shape the human species, scientists say, and the realization is raising tantalizing questions about where humanity will go from here. Is human evolution ending, ushering in a long maturity in which *Homo sapiens* persists pretty much unchanged?" (Oh, how I love that universal and anonymous appeal to authority—"scientists say"!) The article then gave an accurate account of the fact that human anatomy has not altered substantially for the past 100,000 years or so. The Cro-Magnon people who painted the great caves of Europe some twenty to thirty thousand years ago were indistinguishable from us.

Interesting fallacies are often subtle, often based upon hidden assumptions, unstated and probably unconsciously held. As a professional evolutionist, I find nothing whatever surprising about human stability over 100,000 years (see essays 10 and 11). This interval, while not quite so short as an evolutionary eyeblink, represents a pretty damned small unit of geological time. Most species are stable during most of their geological duration. Large, successful, well adapted, mobile, geographically widespread species are particularly prone to stability—because evolutionary events are concentrated in episodes of branching speciation within small, isolated populations. *Homo sapiens* possesses all these attributes for stability, so why should we be surprised at the reported results? And why should Stevens's article have elicited such a strong response of virtual astonishment?

I can only conclude that the spin doctored view of life's history conceives of evolution within species as a continuous flux of improvement and adaptation. We are particularly prone to expect such a result for our own species. After all, we evolved from small-brained ancestors, and we have achieved our exalted status by cranial enlargement. Shouldn't this process, as intrinsic, be continuing during our period of maximal spread and success? Therefore, if we have truly stabilized, isn't something funny going on, and mustn't that something be an imposition of our cultural discoveries upon our biological estate? No, no, a thousand times no. Our stability is orthodox—at least in a fully revolutionary Darwinism with smashed pedestals.

Correct the three errors, and we may grasp evolution as a process causally driven by struggle among individuals for reproductive success, and not by any principle working bountifully for the good of species or any other "higher" entity in nature. We may then view life's history as an unpredictable set of largely fortuitous, and eminently interruptible, excursions down highly contingent pathways. And we will understand successful species as islands of temporary stability, not as striving entities in a flux of constant improvement.

Just as the first error appeared in the *Times* on the fiftieth anniversary of D-Day, the last occurred on March 14, 1995, the date of the *Times*'s 50,000th issue. The editor marked the occasion with the restrained fanfare typical of a newspaper that still refuses to publish the funnies. Arthur Ochs Sulzberger, chairman of the New York Times Company, sent a memorandum to his staff: "The best way we can celebrate is by insuring that our 50,001st edition is the best newspaper we can possibly produce." Bravo, Mr. Sulzberger—and how like evolution devoid of the spin doctoring that has so sadly prevented the completion of Freud's revolution. Not the saccharine motto of faith cures for the past hundred years: Every day, in every way, I'm getting better and better. But the toughness and true heroism of a player up against a house with infinite resources: Hang in there, as best you can, for as long as you can. No ignobility, but only enlightenment, attends our reduction to appropriate size. For when we smash pedestals, we do grant a ray of freedom to our very own defining evolutionary peculiarity: the human mind. I don't know if the truth can make us free, but I do believe that our unique mentality thrives on this form of soul food, whatever the pain of lost illusions.

26

A Humongous Fungus Among Us

WHEN AN ANIMAL ACHIEVES disembodied immortality by becoming a verb, human speakers usually honor its behavior: we hawk our wares, gull or buffalo our naive competitors, hound our adversaries, and clam up in the face of adversity; we have also been known to man the barricades and kid around with our companions. But plants and other rooted creatures do not feature so great a range of overt actions, and our botanically based verbs therefore tout growth and appearance as sources of metaphor. Consider the two most prominent examples, citing comparable phenomena but with such different meaning—for one usually expresses our joy and the other our fear. Art and prosperity *flower*; taxes and urban violence *mushroom*. The burden of difference reflects an obvious source in our culture and legends. We love the bright flowers of "higher" plants, either radiant in the sunlight or jewellike in the quiet darkness of a forest. We loathe the spongy fruiting bodies of "lowly" fungi, growing in dank dampness, sprouting in cancerous formlessness from rotting logs. (Even a colorful mushroom usually strikes us as sinister rather than lovely.) I well remember a common schoolyard taunt of my youth, often cruelly directed at unloved classmates: "There's a fungus among us"—a cry that always inspired the ritualistic retort, "Kill it before it multiplies."

Combine this image of uncontrolled and noxious fungal growth with our general fascination for superlatives—the biggest and most persistent in this case—and we can easily understand the flurry of press commentary that greeted a technical article by M. L. Smith, J. N. Bruhn, and J. P. Anderson in the April 2, 1992, issue of *Nature*: "The fungus *Armillaria bulbosa* is among the largest and oldest living organisms." On the same day, *The New York Times* reported the discovery in a front-page story with a more sprightly headline: "Thirty-Acre Fungus Called World's Largest Organism."

We divide multicellular life into three great kingdoms: animals, plants, and fungi. How does this new *Armillaria* stack up against champions of the other domains? Blue whales, by far the largest animals that ever lived (edging out both *Ultrasaurus* and *Supersaurus* from the world of dinosaurs), may reach one hundred feet and as many tons. (Incidentally, since female whales exceed males in size, the largest individual animal of all time was undoubtedly a female.) Sequoia trees are bigger than whales by far, sometimes exceeding one thousand tons (though mostly in nonliving wood), not to mention heights measured in hundreds of feet and years in thousands.

Armillaria bulbosa lives in and around tree roots in European and eastern North American mixed hardwood forests. Clones begin as a single fertilized spore and then spread out by vegetative growth. The basic unit of spread is a hypha, or threadlike filament that forms the structural unit of growth in many fungi. In *Armillaria*, the hyphae are then bundled into cordlike accumulations called rhizomorphs. Since these rhizomorphs grow and extend underground (in and around tree roots), a human observer sees nothing of this interwoven subterranean mat except for the occasional and spatially discontinuous mushrooms that poke through the forest floor.

Since we mistake an individual mushroom body for a discrete item of life, we might look at the area of this *Armillaria* clone and, seeing nothing of the underground continuity, view the species as consisting of a few widely scattered, entirely separate individuals—a population of several insignificant organisms. But Smith and his colleagues, working in a forest near Crystal Falls on the Upper Peninsula of Michigan, found a region of some thirty acres underlain by the interconnected rhizomorphs of a single *Armillaria* clone. Extrapolating from best assessments of rhizomorph growth rates, they infer a minimal age of 1,500 years—probably a substantial underestimate, since neighboring clones have inhibited further spread for some time, while the calculation assumes continuous expansion. The authors estimate the weight of rhizomorphs in the soil at about ten tons, but that figure excludes much of the clone's bulk, growing in places harder to assess—for example, within tree roots and below fifteen centimeters deep in the soil; and also, perhaps more important, the mass of fine hyphae extending from the rhizomorphs into surrounding soil and wood. Smith and colleagues estimate that the weight of the entire clone may exceed their figure by ten times or more, yielding a total closer to one hundred tons.

Thus, this champion *Armillaria* clone may weigh as much as a blue whale (though substantially less than a large sequoia), and probably lies in the ballpark of the oldest tree for age. Its claim to "firstest with the mostest" must rest upon the additional criterion of lateral spread—in other words, of mushrooming. Thirty acres, however thinly distributed in spots, amounts to quite a creature. Even *The Blob*, of B-movie fame, will have to take a back seat.

Since haste does make waste, the first flurry of commentary on scientific findings often emphasizes some superficial features, while the truly interesting conceptual issues fall into the background. I can get as het up as the next guy about claims for maximal age, size, and weight. I know that Robert Wadlow nearly reached nine feet, that Robert Earl Hughes exceeded one thousand pounds, and that Mr. Izumi of Japan lived past his 120th birthday. The Michigan clone of *Armillaria*, while falling short in weight, and perhaps in age, does enter my mental list as the biggest organic spread. But the deeper fascination of this tale lies elsewhere—in the striking way that this underground fungal mat forces us to wrestle with the vital biological (and philosophical) question of proper definitions for individuality.

First of all, the novelty of Smith, Bruhn, and Anderson's study does not lie in the discovery of something so big, but in the establishment of criteria to test the clone's status as an individual. Large and continuous underground mats of fungal rhizomorphs have long been recognized. Heretofore, however, we had no way of ascertaining whether such a mat grew from a single source. Fungal spores pervade our spaces (as we all know from experiences with bread mold). A single mushroom can produce several million fertile spores per hour over several days. With so many spores, representing so many different clones, falling all over the forest floor, why suspect that a large mat of rhizomorphs might represent growth from a single initiating spore, and therefore count as one individual by genealogy? Why not propose that such gigantic mats of rhizomorphs form as congeries, or aggregations made of products grown from several founding spores (representing many different parents), all twisted and matted together—in other words, a heap rather than a person?

Spores are certainly ubiquitous, but a constant rain does not guarantee a cooperative matting to form massive aggregations—for organisms have immunological mechanisms for distinguishing self from non-self and rejecting amalgamation (abutting coral colonies do not generally fuse on growing reefs), while Darwinian theory views competition among different genetic units within a species as a major thrust of life's game. Thus, single vs. multiple sources for large fungal mats has been viewed as an open and interesting question for some time.

Genetic tests are now available for resolving such an issue. Smith and colleagues first sampled several portions of the *Armillaria* mat both for genes that determine compatibility in interbreeding (called mating-type alleles) and for several fragments of mitochondrial DNA. Both the mating-type and the mitochondrial genes are highly variable within the species *Armillaria bulbosa*, but all samples of the Michigan mat yielded the same array. This identity demonstrates close genetic relationship, but does not prove individuality, for how can we know whether the mat derives from many highly inbred siblings or from a single source of origin? An additional genetic test then indicated a single founder. Repeated close inbreeding leads to a marked reduction of genetic variability among offspring—the presumed "deep reason" behind our various laws and taboos against incestuous marriage. Genes that are variable within an individual (called "heterozygous" because the maternal and paternal copies differ) will tend to become uniform in offspring produced by continual mating, over many generations, among very close relatives. Smith and colleagues traced markers of several DNA sequences from heterozygous genes throughout the mat, and found no reduction in their variability—a sign that the entire mat has grown from a single source and does not represent an incestuous amalgamation of numerous sibling subclones.

But this elegant demonstration that the Michigan mat formed from a single source only opens the more interesting and portentous issue of defining individuality—*the* central question, as we shall see, for applying Darwinian theory to nature. The Michigan *Armillaria* mat grew vegetatively from a single source, but does such a thirty-acre spread qualify as an individual under our usual vernacular definitions? Clive Brasier addressed this question in a commentary that accompanied Smith, Bruhn, and Anderson's original article in the April 2 issue of *Nature:*

> The suggestion of Smith *et al.* that [the Michigan mat] deserves recognition as one of the largest living organisms, rivalling the blue whale or the giant redwood, invites closer scrutiny. The

> blue whale and redwood exhibit relatively determinate growth within a defined boundary, whereas fungal mycelia do not.

In other words, a whale is a whale, with flippers and tail, but the Michigan mat just spreads. Moreover, we cannot be sure that the Michigan mat remains truly continuous throughout. Pieces may break off and become physically separated from the main mass; but a whale's flipper, if broken off, is dead meat, not a miniwhale. Brasier concludes, with justice:

> So although [the Michigan mat's] reputation as a champion genotype [discrete genetic system] may yet be secure, its status as a champion organism depends upon one's interpretation of the rules.

What, then, shall we accept as a definition of individuality, and why is this question important to biological theory, not merely a verbal game? A single genetic origin followed by growth in physical continuity might provide a good criterion as a first stab, but the Michigan mat raises a cardinal problem also illustrated by many other creatures including grass blades and bamboo stalks (although the problem may only record an unfortunate parochialism based upon our unfair extrapolation from the character of our own bodies to a false criterion for all life).

Our canonical individuals are bounded entities with definable form—a whale, a tree, a cockroach, a human being. By analogy, we wish to label a grass blade, a bamboo stalk, or a mushroom as an individual. But consider the proposed criterion of contiguity and unique genetic origin. A bamboo stalk looks like the entity we call an entire plant in other circumstances, but each stalk in a field may arise from a common system of underground runners, all united and all formed by vegetative growth from a single seed. Isn't the single bamboo stalk, then, just a part of this larger individual, as the mushrooms of *Armillaria* are only the visible fruiting bodies of an underground totality? (Bamboo stalks are giant grass blades, so the same argument applies to our lawns.)

Botanists, more often than zoologists, encounter this problem of apparent parts that look like individuals (though a colony of coral animals raises exactly the same issue). Botanists have therefore devised a special terminology to treat these ambiguous cases of parts that look like entire organisms of vernacular usage, but act as organs of a larger totality by the genetic definition. Botanists speak of the morphologically well-defined part (the grass blade, bamboo stalk, or mushroom) as a *ramet*, and the entire interconnected system (the underground runners and stalks, the mat of rhizomorphs with occasional mushroom buds) as a *genet*. In other words, the vernacular individual is a ramet, and the genetic individual a genet. This terminology does not solve the conceptual problem of defining individuality, but merely devises names to acknowledge the classical case of inherent ambiguity.

The vernacular and genetic definitions are driven even further apart when we recognize that even the chief feature of connectedness can fail when we advocate the genetic criterion. In 1977, zoologist Dan Janzen wrote a provocative article on this subject (see bibliography), including the subheading "What is an aphid?" Many aphid species breed sexually just once a year. Females born from this sexual union produce subsequent generations by parthenogenesis—that is, without fertilization. The parthenogenetic offspring are all identical (and all female), both to their sisters and their mothers (except for rare new mutations). In other words, the asexually produced offspring of one original mother form a clone of genetically identical bodies—and billions of aphids may ultimately arise from one founding mother. (Eventually the females begin to produce some male offspring, and fertilization can then proceed—but that is a different story.)

Now what shall we call all these aphid bodies in a single parthenogenetic clone? By any vernacular criterion, they are individuals. They look like any other unambiguously individual insect—like a cockroach, a ladybeetle, or a cricket. They have six legs and mouths, just like any ordinary insect. Yet they are parts of one genetic system, all formed by an analogue of vegetative growth from a single starting point (the founding mother). The entire clone of separated aphid bodies is (or are) just like the stalks of a bamboo field or the mushrooms of an *Armillaria* mat, but without the underground connections. In other words, the aphid bodies are disconnected ramets of a single genet. Why, Janzen argues, should we not label the aphid bodies as parts, and the entire clone as a single EI, or "evolutionary individual"? This redefinition yields some startling consequences, as Janzen notes (you may need to read this quotation twice to put yourself in its unconventional framework):

> Each EI grows rapidly by parthenogenesis, with occasional pieces (aphids) being bitten out of it by parasites (in conventional discussions these would be called . . . predators). Only very rarely is an EI preyed upon (i.e., all of it eaten), since part of its growth pattern is to spread itself very thinly over the sur-

face of the plants in its habitat, so thinly that a potential predator is very unlikely to find all of it at once.

Before this sequence of increasing ambiguity becomes any more maddening, let me propose a different approach, and a potential solution. Terms are best defined within the context of explanatory theories. Gravity may hold a variety of vernacular meanings, but changing technical definitions in the successive formulations of Newton and Einstein define the character of gravity as a scientific concept. Similarly, the term *individual* holds a central role and technical meaning within Darwinian theory, our ruling paradigm for understanding nature. Shouldn't we accept this theoretical definition as our primary biological meaning (the vernacular may stray, but science cannot and should not control all ordinary usage)?

Darwin's central postulate states that natural selection works upon *individuals* engaged in a struggle (metaphorical and without conscious intent, to be sure) for reproductive success. *Individuals* leaving more surviving offspring obtain a Darwinian edge, and populations change thereby. Fine, but how shall we define the "individual" engaged in such a struggle? Darwin gives us a clear answer: individuals are organisms—that is, conventional bodies (with some nuancing for such ambiguous cases as mushrooms and aphids). Natural selection works on creatures—on individual lions fighting for a limited supply of zebras; on individual trees struggling with others for access to light.

This emphasis on ordinary vernacular organisms played a central role in Darwin's radical reformulation of nature (see the preceding essay), for he consciously sought to overthrow the comforting and conventional idea of nature's intrinsic benevolence, with a creator fashioning good organic design and harmonious ecosystems directly. How delicious to contemplate that these "benevolent" results arise only as side consequences of a mechanism operating "below" divine superintendence, and pursuing no "goal" but the selfish propagation of individuals—that is, organisms struggling for personal reproductive success, and nothing else.

We continue to accept Darwin's abstract formulation today—individuals struggling for personal reproductive success—but a substantial rethinking has enlarged Darwin's concept of individuality. For Darwin, only organisms are individuals, or "units of selection." But what properties must an entity possess in order to operate as a Darwinian individual—and are organisms the only such entities in nature? We can specify five such properties: an *individual* must have a clear beginning (or birth) point, a clear ending (or death) point, and sufficient stability between to be recognized as an entity. These first three properties suffice to define an "individual" in the most abstract sense. But an entity requires two further properties to enter a Darwinian process of reproductive competition: a Darwinian individual must bear children, and these offspring must be produced by a principle of inheritance that makes children resemble parents, with the possibility of some differences.

Darwin was surely right in holding that ordinary organisms (also *Armillaria* mats and aphid clones) possess these five properties: they are born and die at definable points; they are stable enough during their lifetimes; they have children; and offspring resemble parents with the possibility of difference. Organisms can therefore be units of selection.

But what about entities either more or less inclusive than organisms? What about genes "below" organisms, and species "above"? We usually think of genes as *parts* and species as *collectivities*, but perhaps this conventional view only represents the bias of restricted focus on our personal lives. Maybe genes and species are just as good Darwinian individuals as bodies. After all, species are born when a population becomes isolated and branches off from the parental stock. Species die unambiguously at their extinction. Most species are quite stable throughout their geological duration. Genes also possess the five key properties of birth, death, stability, reproduction, and inheritance with the possibility of difference.

Thus, individuality extends beyond *Armillaria* mats and aphid clones to encompass different levels of biological organization—so different that we have usually called them parts or collectivities under the parochial assumption that only organisms can be units of selection. Genes and species are also Darwinian individuals, and selection can operate upon these larger and smaller entities as well. Natural selection can work simultaneously at several levels of a genealogical hierarchy—on genes and cell lineages "below" organisms, and on populations and species "above" organisms. All these levels produce legitimate Darwinian individuals—and this hierarchical definition gives us the large, inclusive, and proper biological meaning of the term *individual*.

Evolutionary theory does not operate as Darwin proposed when selection acts simultaneously upon several kinds and levels of individuals. Balances and feedbacks, rather than adaptive perfection, become a source of temporary stabilities. Fancy feathers are wonderful for an individual peacock, but ultimately harmful (in geological time) for the individual species *Pavo cristatus.* Jose Canseco's salary is terrific for him and his family line, but of dubious value for long-term persistence of the individual called Major League Baseball.

Nature is not an intrinsic harmony of clearly defined units. Nature exists in multiple levels, interacting with fuzziness at their borders. We cannot even formulate an unambiguous definition of "individual" at the single level of organic bodies—as *Armillaria* mats and aphid clones demonstrate. Furthermore, in Darwinian terms, legitimate individuals exist and operate at several levels of a genealogical hierarchy—genes and species, as well as organisms. But what a fascination when this maelstrom of differing individuals builds its meshwork of interaction to produce life's history by Darwinian evolution. Does nature herself then sing Walt Whitman's "Song of Myself"?

> Do I contradict myself?
> Very well then I contradict myself,
> (I am large, I contain multitudes.)

27

Speaking of Snails and Scales

THE PSALMIST ASKED: "What is man, that thou art mindful of him?" We have searched, no doubt vainly, for a nature or essence of humanity ever since we evolved enough cognition to ask. Just consider the variety of classical responses, each emphasizing a central part of the human totality. "A reasoning animal," said Seneca, honoring an aspect of our mental life. "A political animal," proclaimed Aristotle, focusing upon our social instincts. "One name belonging to every nation . . . one soul through many tongues," wrote Tertullian, signaling our unity in diversity. We should also not forget the famous definition, attributed to Plato and based on overt appearance in contrast with other vertebrates: "a featherless biped." (According to legend, Diogenes the Cynic plucked a cock and brought it to the Academy, proclaiming, "Here is Plato's man." The old definition then received an addendum: "with broad flat nails.") All in all, however, I favor the celebrated and much later definition of Blaise Pascal, for he emphasized our sublime weakness: "a thinking reed."

No one-liner can ever be optimal, but my preferred characterization at least has the virtue of combining, in one descriptor, all the elements cited above—our social needs, cognitive abilities, uniqueness among animals, and unity amid variety. Human beings are storytellers, spinners of tales.

We gather the complexities of our world into stories; we give order to the confusion of our lives, and to the apparent senselessness or cruelty of our surroundings, by constructing narratives that imbue the totality with meaning. This propensity to tell stories grants us resolution, but also spells danger in avenues thereby opened for distortion and misreading. For our favored stories unroll along definite and limited pathways (we call them epics, myths, and sagas, and they often show eerie similarity across disparate cultures)—and we often try to channel a much more varied nature along these familiar and edifying routes.

Since all discovery emerges from an interaction of mind and nature, thoughtful scientists must scrutinize the many biases that record our socialization, our moment in political and geographic history, even the limitations (if we can hope to comprehend them from within) imposed by a mental machinery jury-rigged in the immensity of evolution.

We are most attuned to obvious biases of a social or political character. We can easily grasp how racism has distorted our view of human diversity, or how creationism once precluded any adequate understanding of life's history. We have been less able to recognize the subtler, but equally constraining, prejudices that arise from more universal properties easily hidden by their lack of evident variation across cultures and classes. Into this less visible category I would place our tendency to order complex reality into stories with restricted themes and outcomes. I call this propensity "literary bias" (see essay 16 in *Bully for Brontosaurus*). Anton Chekhov wrote that "one must not put a loaded rifle on stage if no one is thinking of firing it." Good drama requires spare and purposive action, sensible linking of potential causes with realized effects. Life is much messier; nothing happens most of the time (see essay 10). Millions of Americans (many hotheaded) own rifles (many loaded), but the great majority, thank God, do not go off most of the time. We spend most of real life waiting for Godot, not charging once more unto the breach.

A particular class of stories holds special power to distort by combining the standard form of sociopolitical prejudice with subtler literary bias. Our conventional explanations of historical sequences tend to be regulated by the primary sociopolitical theme of Western life since the late seventeenth century: the idea of progress, with corollaries of movement from small to large, simple to complex, primitive to advanced—an ideal of perpetual growth and expansion. When we add our more general storytelling preferences, our literary biases for narrative continuity between stages and causal unity of transforming forces, we obtain the standard format for historical stories: purposeful, directional, and sensible change. Given the failure and inadequacy of so many tales in this mode—the pageant of prehistoric life from monad to man; Marx's theory of historical stages toward a communist ideal—we must begin to wonder how often nature deigns to venture even close to our hopes for her constitution.

In any case, our standard historical stories of meaningful progress do not stand as mere abstractions; they are heuristic devices that strongly encourage scientists to proceed in a certain way. In particular, such stories virtually dictate that complex systems can best be understood by searching for a simpler state, an earlier stage, or a more primitive version to serve as a comprehensible model for a more intricate reality presently beyond our grasp (study my pea patch and you might eventually fathom agrobusiness). Darwin scrutinized a few hundred years of pigeon breeding to model the history of life over millions; earlier generations of anthropologists, in their more than mildly racist language, scoured the earth for "primitive" peoples who, in the simplicity of their commerce and social relations, might stand as surrogates for early stages of Western urbanity.

These differently complex, non-Western cultures are fascinating in themselves; so too, believe it or not, are pigeons. Events of small scale or short times should be studied in their own right. But the interpretation of such events as early stages in a narrative of rising complexity may backfire because causal continuity often fails. Small and short are often just different from big and long, not little brothers gliding toward a more intricate manhood.

To illustrate this theme of misconstruing the less complex as the primitive precursor, I want to tell a story about a scientific misjudgment of my own early career. In 1969, I first visited the island of Curaçao (in the Netherlands Antilles, off the coast of Venezuela) to study the land snail *Cerion uva* (I will mention later why I chose this peripheral island in the general distribution of *Cerion*). In 1994, exactly twenty-five years after my first visit, I traveled to Curaçao again (I will also recount the purpose of this visit later—essays are stories, and I am therefore permitted the literary devices of foreshadowing and mild mystery).

Curaçao is a land of mixtures and contrasts. As a Dutch island off Spanish South America, a cactus-filled desert in the Caribbean tropics, Curaçao has melded its disparate parts into a distinctive culture. Consider the anomalous mixture of oil and sun. Curaçao's geology and geography have set its destiny. The periphery of the island is built of hard, solid limestone (primarily of reef coral, as tectonic forces raised Curaçao from the sea); the interior is covered by soft and crumbly volcanic rock.

Consequently, Curaçao features many large and strategic harbors—for narrow piercings of the hard limestone, easily defended by forts (or even, in early times, by heavy chains stretched from shore to nearby shore), open into ample basins eroded from the soft volcanics. The discovery of oil in nearby Venezuela set Curaçao's current economy as a safe place for refining and transshipment (on a politically stable Dutch island with magnificent harbors). But think of the anomalous mixture, given the island's other economic staple, tourism. Visitors can easily smell the contrast, as the easterly trade winds blow the effluvia from refineries along the Schottegat over the fancy new hotels of Piscaderabaai.

The people of Curaçao form a grand polyglot as well, recording the realities and evils of such a Caribbean mixing point. The Caucasian component derives mainly from Dutch officials and businessmen, Spanish and Portuguese planters, and Jewish merchants. The Jewish community of Curaçao, the oldest in our hemisphere, dates to the seventeenth century, when the Portuguese Inquisition reached Brazil and expelled a vigorous community of Sephardic Jews, who then fled to havens under liberal Dutch control—some to New York (then Dutch New Amsterdam) to become the first Jews in the future United States, but most to nearby Curaçao. Congregation Mikve Israel still worships in an elegant synagogue built in 1732, the oldest in continuous use in our hemisphere (Curaçao Jews helped to establish the oldest American temple, the Touro synagogue of Newport, Rhode Island, in the 1760s). I felt privileged, and more than a little awestruck, to attend the Friday night service and to think that people of my heritage have been saying the same prayers in the same spot for more than 250 years within this New World of constant change.

The larger African component reached Curaçao involuntarily as chattels in the brutal system of plantation slavery. From this great and disparate mixture of people, a local language emerged, a creole called Papiamentu, spoken only by the few hundred thousand people of Curaçao and the neighboring islands of Aruba and Bonaire. In this tongue, *papiamentu* means "speaking"—so the language's name recalls the common practice of cultures who choose, as a designation for themselves, the native word for "people" in general. Several features of this fascinating language provide my first of two examples from Curaçao for the fallacy of misequating less complex with primitive or historically prior.

Plantation owners, in part for reasons of availability, but also as a conscious strategy for preventing a solidarity that might lead to insurrection, built their retinue of slaves from people of different linguistic backgrounds, so that no secret system of communication could emerge in an "unknown" native language. These slaves heard no common tongue but the language of their owners and overseers. They consequently built new languages upon this European base, but influenced by African sources and imbued with features of syntax derived more from universals of human grammar (asserting themselves in this caldron of invention) than from any immediate historical background of the originators. Such languages, called creoles, have arisen all over the world during the past five hundred years; Derek Bickerton, a specialist on creoles (especially the Hawaiian version that emerged so quickly after expansion of the sugar industry led to importation of workers from such varied places as China, Japan, Korea, Portugal, and the Philippines) wrote in his celebrated book *Roots of Language* (I also used his companion volume, *Language and Species*, as a source for this essay):

> Creole languages arose as a direct result of European colonial expansion. Between 1500 and 1900, there came into existence . . . small, autocratic, rigidly stratified societies, most engaged in monoculture (usually of sugar), which consisted of a ruling minority from some European nation and a large mass of (mainly non-European) laborers, drawn in most cases from many different language groups . . . It is generally assumed that speakers of different languages at first evolved some form of auxiliary contact-language, native to none of them (known as a *pidgin*), and that this language, suitably expanded, eventually became the native (or *creole*) language of the community which exists today. These creoles were in most cases different enough from any of the languages of the original contact situation to be considered "new" languages.

In *Language and Species*, Bickerton defines the difference between an original pidgin and a subsequent creole. "The gulf between a pidgin and its associated creole, in terms of formal structure, is immense. A pidgin . . . is structureless, whereas a creole exhibits the same type of structure as any other natural human language."

I became fascinated with Papiamentu when I first visited Curaçao, and have tried to learn something of its structure (my thanks to many local people, and especially to E. R. Goilo's *Papiamentu Textbook*, the source of most following observations). Papiamentu has a Spanish and Portuguese base, with a strong admixture of Dutch, already quite an amalgamation of Romance and Germanic sources, as noted in such phrases as *Danki Dios* ("Thank God," with Germanic gratitude to a Romance deity).

Beyond these overt amalgamations, the most obvious and distinctive feature of Papiamentu lies in the stripped-down logic of its grammar and syntax. For one who has struggled so many years with the complex cases, conjugations, declensions, pluralizings, and genderings of nouns and verbs in most European languages, the simplified, barebones structure of Papiamentu provides a distinct pleasure. The stems of verbs, for example, never change their form, either in different tenses or agreements. *Bai* is "going" in the infinitive, the imperative, the past, present, or future, and whether the motion involves you, we, I, or they. Past and future tenses just modify this universal form with an adverb. The future takes *lo*, from a Portuguese word meaning "later on"—*lo mi bai* ("I shall go") literally means "later on I go." The past definite takes *a*. Goilo traces this form to a grammatical holdover from the Spanish auxiliary *ha* of the past tense, but Bickerton argues that, as for the future tense, *a* is an adverb appended to the universal form—from the Portuguese *ja* for "already," so that *mi a bai* ("I went") literally means "I already go." Interestingly, the one major exception to the rule of no change for the verb stem lies in the infrequently used gerund, where the Spanish *ando* and *endo* terminations have been retained (as in *sabiendo* for "knowing"). Even more interestingly, and showing the vestigial power retained by grammatic structure of a source, the few verbs of Dutch origin do not take this Spanish gerund and have no distinctive ending in this form, but do—and uniquely—alter the verb's stem in the past tense by adding the Dutch *ge* (*skop* is "to kick," but a former kick is *geskop*). What a fascinating mixture of novel logical simplicity and past historical distinctiveness!

Nouns have neither gender nor plural forms. *Buki* is either one or a hundred books. But if the needed distinction can't be drawn from context, you add *nan* (the personal pronoun for "they") to designate a plural. Thus, *e buki* is "a book," and *dies buki* is "ten books"—but "the books" can be *e bukinan* (literally "the book they"). Sometimes the logic is truly compelling. Papiamentu has no word for "son" or "daughter," but *yiu* is a child of the family (while *mucha* is a child in general)—so *yiu homber* (child man) is "your son," and *yiu muhé* (child woman) "your daughter." Similarly, *ruman* is a sibling, and *ruman homber* "your brother," with *ruman muhé* "your sister."

In the past, constrained by stories of progress and a sociopolitical legacy of racism, Western scholars tended to view non-Western languages of nonliterate people as primitive stages (or degenerated reversions) of an evolutionary sequence leading to modern Indoeuropean tongues. Consider, for example, a standard nineteenth-century work inspired by the Darwinian revolution—William Dwight Whitney's *Life and Growth of Language*, first published in 1875 (Whitney was professor of Sanskrit at Yale; even the title of his work proclaims the new developmental paradigm, with overt analogy to the progressive and programmed growth of children). Whitney exposes his sociopolitical bias in ranking modern "primitive" languages as intermediate between a conjectured first human tongue and the maximal modern complexity of, say, his own English:

> If we hold him [primitive man] to have gradually developed them [the elements of civilization] out of scanty beginnings . . . there is no reason why we should not hold the same view in respect to language . . . Even in existing languages the differences of degree are great, as in existing stages of culture in general. An infinity of things can be said in English which cannot be said in Fijian or Hottentot; a vast deal, doubtless, can be said in Fijian or Hottentot which could not be said in the first human languages.

But consider also Whitney's more abstract and largely literary or storytelling biases for a slow succession of upward steps ("that the process was a slow one, all our knowledge of the history of later speech gives us reason to believe"), and for causal continuity in general ("no account of the origin of language is scientific which does not join directly on to the later history of language without a break, being of one piece with that history").

Given this legacy, we should not be surprised that linguists once viewed creoles either with contempt ("despised and neglected for centuries by mainstream linguists as the degenerate products of illiterate nonwhites," as the jacket of Bickerton's *Roots of Language* notes) or with paternalistic interest as a primitive lingo to expose the evolutionary roots of speech ("those [theories] which sought to derive creoles from the babyish imitations of Europeans' condescending simplifications").

But once we drop these biases and rephrase our questions, we can conceptualize creoles in a new light—not as something simple and prior, a primitive stage of a progressive sequence, but as something unique and different, a potential source of insight into the structural nature of language in general. I cannot judge the validity of Bickerton's theory for the origin of creoles (and I recognize that his opinions are controversial among linguistic scholars), but his hypothesis does illustrate the important principle that less overtly complex can connote difference with a general message, rather than primitivity to ennoble our current status under our storytelling biases.

Pidgins, according to Bickerton, are rough and ready accommodations to necessary communication and have little linguistic structure—while the creoles that derive from pidgins develop all the formal complexity of any true human language (without all the overt frills of conjugation, declension, etc.). So how do creoles originate?—especially since they often arise so quickly, usually in a single generation. (This surprising claim for maximal rapidity is well documented, particularly in Hawaii, where polyglot immigration, and its resulting pidgin, only began in 1876, when revision of U.S. tariff laws permitted free importation of Hawaiian sugar, while creolization occurred between 1910 and 1920.)

Bickerton, in short, argues for invention of creoles by children, as they hear the surrounding pidgin and enrich this base by adding the universal grammar that, according to Chomsky's generative theory, all human beings inherit as a product of the evolutionary development of our brains. If Bickerton is correct, then creoles—as known, traceable, novel languages, replete with full structure but often stripped to the bare bones of universality—provide our best possible evidence for both the existence and the nature of this most precious and defining of human universals, the core for any meaningful concept of human nature. If children create creoles in one pass, and if their parents speak only the unstructured pidgin, then the formal properties of language must emerge from within as shared properties of all people. "Out of the mouths of babes . . ." Bickerton writes in *Language and Species:*

> What happened in Hawaii was a jump from protolanguage to language in a single generation. Moreover, the grammar of the language that resulted bore the closest resemblance not to grammars of the languages of Hawaii's immigrants; nor to that of Hawaiian, the indigenous language; nor to that of English, the politically dominant language; but rather to the grammars of other creole languages that had come into existence in other parts of the world. This fact argues that creole languages form an unusually direct expression of a species-specific biological

> characteristic, a capacity to recreate language in the absence of any specific model from which the properties of language could be "learned" in the ways we normally learn things.

Bickerton then reminds us that we could lose this precious insight if, by hewing too closely to a politically correct notion that "all languages are equal" (in complexity or vocabulary), we neglected the genuine way in which creoles are simpler. Less complex need not connote primitive in a pejorative sense; less complex can also mean "stripped to bare essentials" (a Japanese house compared with a high Victorian cottage), so that the universal properties of construction emerge for all to see. All human languages, including all creoles, possess the entire complexity of universal grammar, but some languages have more bric-a-brac, more geegaws, more tchotchkes than others—and the "cleaner" creoles may therefore make the underlying universals more visible. "Creoles," Bickerton writes, "far from being 'primitive' in anything but the sense of 'primary,' give us access to the essential bedrock on which our humanity is founded."

As I restudied Papiamentu during my recent visit to Curaçao, and as I developed my thoughts about the error of storytelling that automatically equates less complex with prior or primitive, I realized that Curaçao could provide me with two examples—one from the local language, and one (I recognized with some embarrassment) from my own studies of the local land snails. I am proud of the work I did on Curaçao, but I now realize that I undertook my investigations for the wrong reason. I went to Curaçao because I thought that comprehension of a simple system would unlock the greater complexities of other West Indian islands. I found, on Curaçao, many phenomena of interest in themselves, and I did resolve an old debate in a way that satisfied me deeply and brought rigorous sense to prior confusion. But I never discovered a key to the greater complexity elsewhere—and I now attribute such a hope to the wiles of Scheherazade rather than the messages of nature.

The landsnail *Cerion* lives throughout the northern West Indies, often in fantastic variety. Hundreds of species have been named on Cuba and the Bahamas (though only a small percentage are valid). But *Cerion*'s geographic outliers tend to house restricted diversity—only one species at the northern limit of the Florida keys *(Cerion incanum)*, and only one at the eastern edge of the Virgin Islands *(Cerion striatellum)*. The three islands of Aruba, Bonaire, and Curaçao form a distant and isolated southern limit, and also feature but a single species, *Cerion uva.*

In fact, since the Dutch were early colonizers and assiduous collectors, *Cerion uva* is the so-called type (first designated species) of the entire genus—and the name, given by Linnaeus himself, could not have a more honored source (see essay 32). Consequently, *Cerion uva* has been long known and much studied. Two major investigations of the twentieth century, based on exhaustive collections on all three islands, had reached opposite conclusions. In 1924 the American zoologist H. B. Baker argued, from rudimentary statistical analysis grafted onto the subtle expertise of sustained, if subjective, observation, that populations on the three islands could be recognized by minor but consistent differences in the size and shape of shells. He also claimed that shells from western and eastern Curaçao differed in significant ways (Curaçao, shaped like a dumbbell, split into two separate islands during past times of higher sea level). But, in 1940, the Dutch zoologist P. W. Hummelinck reached an opposite conclusion based on measurements of hundreds of samples, and his student W. de Vries then verified these results in 1974 with an even more comprehensive study. Hummelinck and de Vries found no consistent differences among islands or regions, but great amounts of local variation tied to immediate conditions of soil, wind, sun, and rainfall.

When I began my work in the late 1960s, tools had just been invented, thanks to that greatest of modern midwives, the computer, for more adequate statistical analysis based on simultaneous consideration of many measurements for each specimen (a family of techniques known as multivariate analysis). Baker, Hummelinck, and de Vries, by force of habit and limits of technology, had treated their variables one by one, or at most in pairs (bivariate analysis). Since a shell is an integrated structure with a multiplicity of measurable parts, I reasoned that multivariate analysis might resolve the debate between sensible regional differentiation (Baker) and crazy-quilt local variation without general pattern (Hummelinck and de Vries).

Again, I visited all three islands and collected hundreds of samples. I measured (or, rather, I must confess, paid a very careful assistant to measure) nineteen variables on twenty snails from each of 135 samples—for a total of more than fifty thousand measurements, all then submitted to multivariate analysis of various types (my best technical paper on the subject appeared in *Systematic Zoology* during 1984). I was able to resolve the debate in an interesting way, thanks to the power of the new statistical techniques. Both sides had seen part of a fascinating totality. I could discern the large-scale variation—exactly as Baker had specified, with four areas of Aruba, Bonaire, western and eastern Curaçao. But I also found the basis for the smaller-scale variation within regions, linked to immediate environments, that Hummelinck and de Vries had identified.

Interestingly, these two components of variation could be separated on orthogonal axes (*orthogonal* means "at right angles," and orthogonal axes are both literally and mathematically independent of each other). Nearly all diversity within this highly varied species could be captured by two independent factors, each attributable to a different biological source, and each identified by a set of shell features recording a pattern of growth through life. I took pride in discovering such an elegantly simple and sensible scheme; I do not think that I have ever been so satisfied with an empirical study, so pleased to extract an underlying pattern from overt complexity that had stymied all prior researchers.

But my study was an utter failure in terms of my original intent. I had gone to Curaçao because I believed that a scientist had to grasp the simple version before tackling the maximally complex manifestation. I went to Curaçao because I wanted to resolve the riotous diversity of *Cerion* in the Bahamas, where hundreds of species have been named. I saw Curaçao as a baby Bahama. If I could resolve the variation on one island with one species, then surely I would know how to handle a hundred species on a score of islands.

Well, I think I did resolve Curaçao in my intended terms. But this effort hardly aided my Bahamian quest at all. Curaçao is not an embryonic Nassau. Variation within a species doesn't tell you how to treat interactions between species; the phenomena are disparate and exist at different scales. The contrast between me and my neighbor doesn't explain, by simple extension, the gap between me and a chimpanzee. Recent creoles are not baby versions of old and established languages; variation within *Cerion uva* is not a model for interaction among several *Cerion* species in larger regions. Causal continuity does not unite all levels; the small does not always aggregate smoothly into the large. But the small may be interesting in itself and may teach us much about our world in general; Papiamentu and other creoles may lay bare the universals of human language, and *Cerion uva* may permit us to tease apart the sources of evolutionary variation within stable species.

Conventional stories about progressive complexity and causal continuity led earlier linguists to misinterpret (or to ignore) creole tongues, and led me to study *Cerion uva* for the wrong reason. But the solution to these errors does not lie in avoiding stories, for we do not have this option, given our essence (called "human nature"). We must, instead, become more aware of the stories that underlie our methods and choices of topics for research; and we must learn to recognize the constraints and prejudices that any particular story must specify. We should, above all, enlarge our range of potential stories, for a choice among a thousand and one nights provides so much more scope than Cinderella (or another progressive tale of rags to riches) told every night.

I made my recent Curaçao visit to attend a conference held to honor the seventieth birthday of mathematician Benoit Mandelbrot, the inventor of fractal geometry. Ironically, in the light of my two tales about mistaken effort in Curaçao, fractals tell a different story that might have provided an exit from the trap of equating small with simple or primitive. Fractal curves are self-similar—that is, invariant across scales. In the paradigm for a natural fractal, the coastline of North America has no absolutely ascertainable length—for what you measure depends upon the chosen scale, and all scales may show the same basic form. If I try to measure around every sand grain, the length will be effectively infinite. If I measure only around the appropriate wiggles at a scale that depicts Maine to Florida on a piece of paper one or two feet long, then the total length may be exactly the same as a similar-sized map of Acadia National Park measured around each promontory, or a chart of a few feet on one of Acadia's beaches measured around each pebble. In such a fractal world, no scale has any special standing as higher or lower. The beach is not simpler than the whole coastline, and the beach does not build the coastline by degrees.

I began with a famous line from the eighth psalm (and fleetingly quoted another bit in noting the power of children to create creoles). I should end by citing the lead-in to my first statement: "When I consider thy heavens, the work of thy fingers, the moon and the stars, which thou hast ordained; What is man, that thou art mindful of him?"

Who am I to criticize good King David, who slew the giant, and paid so grievously for dubious actions in losing his friend Jonathan and his son Absalom? But isn't David falling into the same trap by wondering why God should bother with little humanity when he made the stars and all the heavens? For if people occupy just one level in a predominantly fractal universe, then we are like the galaxies—but the amoeba that we swallowed in that last glass of water and the mites crawling on our eyebrows are also like us.

Stories

28

Hooking Leviathan by Its Past

THE LANDSCAPE of every career contains a few crevasses, and usually a more extensive valley or two—for every Ruth's bat a Buckner's legs; for every lopsided victory at Agincourt, a bloodbath at Antietam. Darwin's *Origin of Species* contains some wonderful insights and magnificent lines, but this masterpiece also includes a few notable clunkers. Darwin experienced most embarrassment from the following passage, curtailed and largely expunged from later editions of his book:

> In North America the black bear was seen by Hearne swimming for hours with widely open mouth, thus catching, like a whale, insects in the water. Even in so extreme a case as this, if the supply of insects were constant, and if better adapted competitors did not already exist in the country, I can see no difficulty in a race of bears being rendered, by natural selection, more aquatic in their structure and habits, with larger and larger mouths, till a creature was produced as monstrous as a whale.

Why did Darwin become so chagrined about this passage? His hypothetical tale may be pure speculation and conjecture, but the scenario is not entirely absurd. Darwin's discomfort arose, I think, from his failure to follow a scientific norm of a more sociocultural nature. Scientific conclusions supposedly rest upon facts and information. Speculation is not entirely taboo, and may sometimes be necessary *faute de mieux*. But when scientists propose truly novel and comprehensive theories—as Darwin tried to do in advancing natural selection as the primary mechanism of evolution—they need particularly good support, and invented hypothetical cases just don't supply sufficient confidence for crucial conclusions.

Natural selection (or the human analogue of differential breeding) clearly worked at small scale—in the production of dog breeds and strains of wheat, for example. But could such a process account for the transitions of greater scope that set our concept of evolution in the fullness of time—the passage of reptilian lineages to birds and mammals; the origin of humans from an ancestral stock of apes? For these larger changes, Darwin could provide little direct evidence, for a set of well-known and much-lamented reasons based on the extreme spottiness of the fossil record.

Some splendid cases began to accumulate in years following the *Origin of Species*, most notably the discovery of *Archaeopteryx*, an initial bird chock-full of reptilian features, in 1861; and the first findings of human fossils late in the nineteenth century. But Darwin had little to present in his first edition of 1859, and he tried to fill this factual gap with hypothetical fables about swimming bears eventually turning into whales—a fancy that yielded far more trouble in easy ridicule than aid in useful illustration. Just two years after penning his bear-to-whale tale, Darwin lamented to a friend (letter to James Lamont, February 25, 1861), "It is laughable how often I have been attacked and misrepresented about this bear."

The supposed lack of intermediary forms in the fossil record remains the fundamental canard of current antievolutionism. Such transitional forms are sparse, to be sure, and for two sets of good reasons—geological (the gappiness of the fossil record) and biological (the episodic nature of evolutionary change, including patterns of punctuated equilibrium, and transition within small populations of limited geographic extent). But paleontologists have discovered several superb examples of intermediary forms and sequences, more than enough to convince any fair-minded skeptic about the reality of life's physical genealogy.

The first "terrestrial" vertebrates retained six to eight digits on each limb (more like a fish paddle than a hand), a persistent tailfin, and a lateral-line system for sensing sound vibrations underwater. The anatomical transition from reptiles to mammals is particularly well documented in the key anatomical change of jaw articulation to hearing bones. Only one bone, called the dentary, builds the mammalian jaw, while reptiles retain several small bones in the rear portion of the jaw. We can trace, through a lovely sequence of intermediates, the reduction of these small reptilian bones, and their eventual disappearance or exclusion from the jaw, including the remarkable passage of the reptilian articulation bones into the mammalian middle ear (where they became our malleus and incus, or hammer and anvil). We have even found the transitional form that creationists often proclaim inconceivable in theory—for how can jawbones become ear bones if intermediaries must live with an unhinged jaw before the new joint forms? The transitional species maintains a double jaw joint, with both the old articulation of reptiles (quadrate to articular bones) and the new connection of mammals (squamosal to dentary) already in place! Thus, one joint could be lost, with passage of its bones into the ear, while the other articulation continued to guarantee a properly hinged jaw.

Still, our creationist incubi, who would never let facts spoil a favorite argument, refuse to yield, and continue to assert the absence of *all* transitional forms by ignoring those that have been found, and continuing to taunt us with admittedly frequent examples of absence. Darwin's old case for the origin of whales remains a perennial favorite, for if Darwin had to invent a fanciful swimming bear, and if paleontologists haven't come to the rescue by discovering an intermediary form with functional legs and potential motion on land, then Jonah's scourge may gobble up the evolutionary heathens as well. God's taunt to Job might be sounded again: "Canst thou draw out leviathan with an hook?" (The biblical Leviathan is usually interpreted as a crocodile, but many alternate readings favor whales.)

Every creationist book on my shelf cites the actual absence and inherent inconceivability of transitional forms between terrestrial mammals and whales. Alan Haywood, for example, writes in his *Creation and Evolution* (see bibliography):

> Darwinists rarely mention the whale because it presents them with one of their most insoluble problems. They believe that somehow a whale must have evolved from an ordinary land-dwelling animal, which took to the sea and lost its legs . . . A land mammal that was in process of becoming a whale would fall between two stools—it would not be fitted for life on land or at sea, and would have no hope of survival.

Duane Gish, creationism's most ardent debater, makes the same argument in his more colorful style (*Evolution: The Challenge of the Fossil Record*):

> There simply are no transitional forms in the fossil record between the marine mammals and their supposed land mammal ancestors . . . It is quite entertaining, starting with cows, pigs, or buffaloes, to attempt to visualize what the intermediates may have looked like. Starting with a cow, one could even imagine one line of descent which prematurely became extinct, due to what might be called an "udder failure."

The most "sophisticated" (I should really say "glossy") of creationist texts, *Of Pandas and People* by P. Davis, D. H. Kenyon, and C. B. Thaxton says much the same, but more in the lingo of academese:

> The absence of unambiguous transitional fossils is strikingly illustrated by the fossil record of whales . . . If whales did have land mammal ancestors, we should expect to find some transitional fossils. Why? Because the anatomical differences between whales and terrestrial mammals are so great that innumerable in-between stages must have paddled and swam the ancient seas before a whale as we know it appeared. So far these transitional forms have not been found.

Three major groups of mammals have returned to the ways of distant ancestors in their seafaring modes of life (while smaller lineages within several other mammalian orders have become at least semiaquatic, often to a remarkable degree, as in river and sea otters): the suborder Pinnepedia (seals, sea lions, and walruses) within the order Carnivora (dogs, cats, and Darwin's bears among others); and two entire orders—the Sirenia (dugongs and manatees) and Cetacea (whales and dolphins). I confess that I have never quite grasped the creationists' point about inconceivability of transition—for a good structural (though admittedly not a phylogenetic) series of intermediate anatomies may be extracted from these groups. Otters have remarkable aquatic abilities, but retain fully functional limbs for land. Sea lions are clearly adapted for water, but can still flop about on land with sufficient dexterity to negotiate ice floes, breeding grounds, and circus rings.

But I admit, of course, that the transition to manatees and whales represents no trivial extension, for these fully aquatic mammals propel themselves by powerful, horizontal tail flukes and have no visible hind limbs at all—and how can a lineage both develop a flat propulsive tail from the standard mammalian length of rope, and then forfeit the usual equipment of back feet so completely? (Sirenians have lost every vestige of back legs; whales often retain tiny, splintlike pelvic and leg bones, but no foot or finger bones, embedded in musculature of the body wall, but with no visible expression in external anatomy.)

The loss of back legs, and the development of flukes, fins, and flippers by whales, therefore stands as a classic case of a supposed cardinal problem in evolutionary theory—the failure to find intermediary fossils for major anatomical transitions, or even to imagine how such a bridging form might look or work. Darwin acknowledged the issue by constructing a much-criticized fable about swimming bears, instead of presenting any direct evidence at all, when he tried to conceptualize the evolution of whales. Modern creationists continue to use this example and stress the absence of intermediary forms in this supposed (they would say impossible) transition from land to sea.

Goethe told us to "love those who yearn for the impossible." But Pliny the Elder, before dying of curiosity by straying too close to Mount Vesuvius at the worst of all possible moments, urged us to treat impossibility as a relative claim: "How many things, too, are looked upon as quite impossible until they have been actually effected." Armed with such wisdom of human ages, I am absolutely delighted to report that our usually recalcitrant fossil record has come through in exemplary fashion. During the past fifteen years, new discoveries in Africa and Pakistan have greatly added to our paleontological knowledge of the earliest history of whales. The embarrassment of past absence has been replaced by a bounty of new evidence—and by the sweetest series of transitional fossils an evolutionist could ever hope to find. Truly, we have met the enemy and he is now ours. Moreover, to add blessed insult to the creationists' injury, these discoveries have arrived in a gradual and sequential fashion—a little bit at a time, step by step, from a tentative hint fifteen years ago to a remarkable smoking gun early in 1994. Intellectual history has matched life's genealogy by spanning the gaps in sequential steps. Consider the four main events in chronological order.

CASE ONE: *Discovery of the oldest whale.* Paleontologists have been fairly confident, since Leigh Van Valen's demonstration in 1966, that whales descended from mesonychids, an early group of primarily carnivorous running mammals that spanned a great range of sizes and habits from eating fishes at river edges to crushing bones of carrion.

Whales must have evolved during the Eocene epoch, some 50 million years ago, because Late Eocene and Oligocene rocks already contain fully marine cetaceans, well past any point of intermediacy.

In 1983, my colleague Phil Gingerich from the University of Michigan, along with N. A. Wells, D. E. Russell, and S. M. Ibrahim Shah, reported their discovery of the oldest whale, named *Pakicetus* to honor its country of present residence, from Middle Eocene sediments some 52 million years old in Pakistan. In terms of intermediacy, one could hardly have hoped for more from the limited material available, for only the skull of *Pakicetus* has been found. The teeth strongly resemble those of terrestrial mesonychids, as anticipated, but the skull, in feature after feature, clearly belongs to the developing lineage of whales.

Both the anatomy of the skull, particularly in the ear region, and the inferred habitat of the animal in life, testify to transitional status. The ears of modern whales contain modified bones and passageways that permit directional hearing in the dense medium of water. Modern whales have also evolved enlarged sinuses that can be filled with blood to maintain pressure during diving. The skull of *Pakicetus* lacks both these features, and this first whale could neither dive deeply nor hear directionally with any efficiency in water.

In 1993, J. G. M. Thewissen and S. T. Hussain affirmed these conclusions and added more details on the intermediacy of skull architecture in *Pakicetus*. Modern whales achieve much of their hearing through their jaws, as sound vibrations pass through the jaw to a "fat pad" (the technical literature, for once, invents no jargon and employs the good old English vernacular in naming this structure), and thence to the middle ear. Terrestrial mammals, by contrast, detect most sound through the ear hole (called the "external auditory meatus," which means the same thing in more refined language). Since *Pakicetus* lacked the enlarged jaw hole that holds the fat pad, this first whale probably continued to hear through the pathways of its terrestrial ancestors. Gingerich concluded that "the auditory mechanism of *Pakicetus* appears more similar to that of land mammals than it is to any group of extant marine mammals."

As for place of discovery, Gingerich and colleagues found *Pakicetus* in river sediments bordering an ancient sea—an ideal habitat for the first stages of such an evolutionary transition (and a good explanation for lack of diving specializations if *Pakicetus* inhabited the mouths of rivers and adjacent shallow seas). My colleagues judged *Pakicetus* as "an amphibious stage in the gradual evolutionary transition of primitive whales from land to sea . . . *Pakicetus* was well equipped to feed on fishes in the surface waters of shallow seas, but it lacked auditory adaptations necessary for a fully marine existence."

Verdict: In terms of intermediacy, one could hardly hope for more from the limited material of skull bones alone. But the limit remains severe, and the results therefore inconclusive. We know nothing of the limbs, tail, or body form of *Pakicetus*, and therefore cannot judge transitional status in these key features of anyone's ordinary conception of a whale.

CASE TWO. *Discovery of the first complete hind limb in a fossil whale.* In the most famous mistake of early American paleontology, Thomas Jefferson, while not engaged in other pursuits usually judged more important, misidentified the claw of a fossil ground sloth as a lion. My prize for second worst error must go to R. Harlan, who, in 1834, named a marine fossil vertebrate *Basilosaurus* in the *Transactions of the American Philosophical Society*. *Basilosaurus* means "king lizard," but Harlan's creature is an early whale. Richard Owen, England's greatest anatomist, corrected Mr. Harlan before the decade's end, but the name sticks—and must be retained by the official rules of zoological nomenclature. (The Linnaean naming system is a device for information retrieval, not a guarantor of appropriateness. The rules require that each species have a distinctive name, so that data can be associated unambiguously with a stable tag. Often, and inevitably, the names originally given become literally inappropriate for the unsurprising reason that scientists make frequent mistakes, and that new discoveries modify old conceptions. If we had to change names every time our ideas about a species altered, taxonomy would devolve into chaos. So *Basilosaurus* will always be *Basilosaurus* because Harlan followed the rules when he gave the name. And we do not change ourselves to *Homo horribilis* after Auschwitz, or to *Homo ridiculosis* after Tonya Harding—but remain, however dubiously, *Homo sapiens*, now and into whatever forever we allow ourselves.)

Basilosaurus, represented by two species, one from the United States and the other from Egypt, is the "standard" and best-known early whale. A few fragments of pelvic and leg bones had been found before, but not enough to know whether *Basilosaurus* bore working hind legs—the crucial feature for our usual concept of a satisfying intermediate form in both anatomical and functional senses.

In 1990, Phil Gingerich, B. H. Smith, and E. L. Simons reported their excavation and study of several hundred partial skeletons of the Egyptian species *Basilosaurus isis*, which lived some 5 to 10 million years after *Pakicetus*. In an exciting discovery, they reported the first complete hind limb skeleton found in any whale—a lovely and elegant structure (put together from several partial specimens), including all pelvic bones, all leg bones (femur, tibia, fibula, and even the patella, or kneecap), and nearly all foot and finger bones, right down to the phalanges (finger bones) of the three preserved digits.

This remarkable find might seem to clinch our proof of intermediacy, but for one small problem. The limbs are elegant but tiny (see the accompanying illustration), a mere 3 percent of the animal's total length. They are anatomically complete, and they did project from the body wall (unlike the truly vestigial hind limbs of modern whales), but these miniature legs could not have made any important contribution to locomotion—the real functional test of intermediacy. Gingerich et al. write: "Hind limbs of *Basilosaurus* appear to have been too small relative to body size to have assisted in swimming, and they could not possibly have supported the body on land." The authors strive bravely to invent some potential function for these minuscule limbs, and end up speculating that they may have served as "guides during copulation, which may otherwise have been difficult in a serpentine aquatic mammal." (I regard such guesswork as unnecessary, if not ill-conceived. We need not justify the existence of a structure by inventing some putative Darwinian function. All bodies contain vestigial features of little, if any, utility. Structures of lost usefulness in genealogical transitions do not disappear in an evolutionary overnight.)

Verdict: Terrific and exciting, but no cigar, and no bag-packer for creationists. The limbs, though complete, are too small to work as true intermediates must (if these particular limbs worked at all)—that is, for locomotion on both land and sea. I intend no criticism of *Basilosaurus,* but merely point out that this creature had already crossed the bridge (while retaining a most informative remnant of the other side). We must search for an earlier inhabitant of the bridge itself.

CASE THREE. *Hind limb bones of appropriate size. Indocetus ramani* is an early whale, found in shallow-water marine deposits of India and Pakistan, and intermediate in age between the *Pakicetus* skull and the *Basilosaurus* hind legs (cases one and two above). In 1993, P. D. Gingerich, S. M. Raza, M. Arif, M. Anwar, and X. Zhou reported the discovery of leg bones of substantial size from this species.

Gingerich and colleagues found pelvic bones and the ends of both femur and tibia, but no foot bones, and insufficient evidence for reconstructing the full limb and its articulations. The leg bones are large and presumably functional on both land and sea (the tibia, in particular, differs little in size and complexity from the same bone in the re-

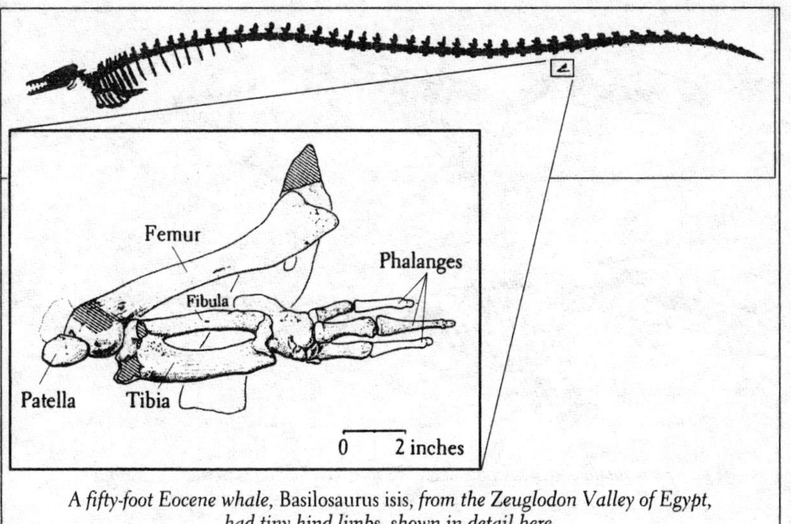

A fifty-foot Eocene whale, Basilosaurus isis, *from the Zeuglodon Valley of Egypt, had tiny hind limbs, shown in detail here.*

Adapted from Science, *vol.* 249, 13 *July* 1990.

lated and fully terrestrial mesonychid *Pachyaena ossifraga*). The authors conclude: "The pelvis has a large and deep acetabulum [the socket for articulation of the femur, or thighbone], the proximal femur is robust, the tibia is long . . . All these features, taken together, indicate the *Indocetus* was probably able to support its weight on land, and it was almost certainly amphibious, as early Eocene *Pakicetus* is interpreted to have been . . . We speculate that *Indocetus,* like *Pakicetus,* entered the sea to feed on fish, but returned to land to rest and to birth and raise its young."

Verdict: Almost there, but not quite. We need better material. All the right features are now in place—primarily leg bones of sufficient size and complexity—but we need more and better-preserved fossils.

CASE FOUR: *Large, complete, and functional hind legs for land and sea—finding the smoking gun.* The first three cases, all discovered within ten years, surely indicate an increasingly successful paleontological assault upon an old and classic problem. Once you know where to look, and once high interest spurs great attention, full satisfaction often follows in short order. I was therefore delighted to read, in the January 14, 1994, issue of *Science,* an article by J. G. M. Thewissen, S. T. Hussain, and M. Arif, titled "Fossil evidence for the origin of aquatic locomotion in archaeocete whales."

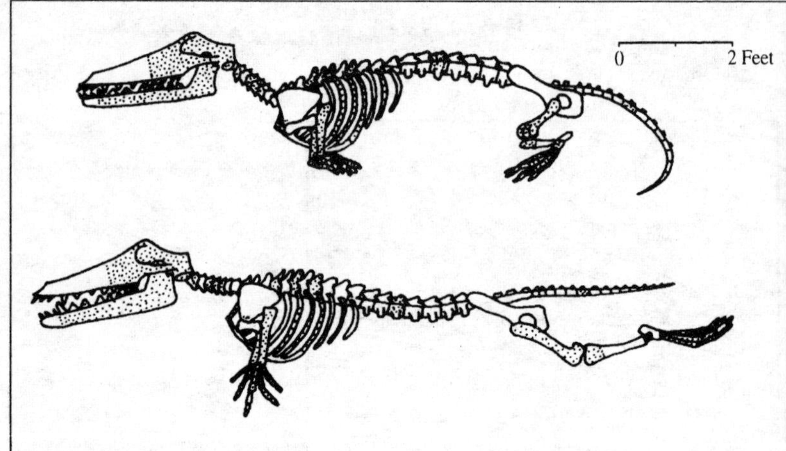

Two reconstructions show Ambulocetus, *a fossil whale from Pakistan, standing* (top) *and at the end of a swimming stroke* (bottom).

Adapted from Science, *vol.* 263, 14 *January* 1994.

In Pakistan, in sediments 120 meters above the beds that yielded *Pakicetus* (and therefore a bit younger in age), Thewissen and colleagues collected a remarkable skeleton of a new whale—not complete, but far better preserved than anything previously found of this age, and with crucial parts in place to illustrate a truly transitional status between land and sea. The chosen name, *Ambulocetus natans* (literally, the swimming walking-whale) advertises the excitement of this discovery.

Ambulocetus natans weighed some 650 pounds, the size of a hefty sea lion. The preserved tail vertebra is elongated, indicating that *Ambulocetus* still retained the long, thin mammalian tail, and had not yet transmuted this structure to a locomotory blade (as modern whales do in shortening the tail and evolving a prominent horizontal fluke as the animal's major means of propulsion). Unfortunately, no pelvic bones have been found, but most elements of a large and powerful hind leg were recovered—including a complete femur, parts of the tibia and fibula, an astragalus (ankle bone), three metatarsals (foot bones), and several phalanges (finger bones). To quote the authors: "The feet are enormous." The fourth metatarsal, for example, is nearly six inches long, and the associated toe almost seven inches in length. Interestingly, the last phalanx of each toe ends in a small hoof, as in terrestrial mesonychid ancestors.

Moreover, this new bounty of information allows us to infer not only the form of this transitional whale, but also, with good confidence, an intermediary style of locomotion and mode of life (an impossibility with the first three cases, for *Pakicetus* is only a skull, *Basilosaurus* had already crossed the bridge, and *Indocetus* is too fragmentary). The forelimbs were smaller than the hind, and limited in motion; these front legs were, to quote the authors, "probably used in maneuvering and steering while swimming, as in extant cetaceans ["modern whales" in ordinary language], and they lacked a major propulsive force in water."

Modern whales move through the water by powerful beats of their horizontal tail flukes—a motion made possible by strong undulation of a flexible rear spinal column. *Ambulocetus* had not yet evolved a tail fluke, but the spine had requisite flexibility. Thewissen et al. write: "*Ambulocetus* swam by means of dorsoventral [back-to-belly] undulations of its vertebral column, as evidenced by the shape of the lumbar [lower back] vertebra." These undulations then functioned with (and powered) the paddling of *Ambulocetus*'s large feet—and these feet provided the major propulsive force for swimming. Thewissen et al. conclude their article by writing: "Like modern cetaceans—it swam by moving its spine up and down, but like seals, the main propulsive surface was provided by its feet. As such, *Ambulocetus* represents a critical intermediate between land mammals and marine cetaceans."

Ambulocetus was no ballet dancer on land, but we have no reason to judge this creature as any less efficient than modern sea lions, which do manage, however inelegantly. Forelimbs may have extended out to the sides, largely for stability, with forward motion mostly supplied by extension of the back and consequent flexing of the hind limbs—again, rather like sea lions.

Verdict: Greedy paleontologists, used to working with fragments in reconstructing wholes, always want more (some pelvic bones would be nice, for starters), but if you had given me both a blank sheet of paper and a blank check, I could not have drawn you a theoretical intermediate any better or more convincing than *Ambulocetus*. Those dogmatists who can make white black, and black white, by verbal trickery will never be convinced by anything, but *Ambulocetus* is the very animal that creationists proclaimed impossible in theory.

Some discoveries in science are exciting because they revise or reverse previous expectations, others because they affirm with elegance something well suspected, but previously undocumented. Our four-case story, culminating in *Ambulocetus*, falls into this second category. This sequential discovery of picture-perfect intermediacy in the evolution of whales stands as a triumph in the history of paleontology. I cannot imagine a better tale for popular presentation of science, or a more satisfying, and intellectually based, political victory over lingering creationist opposition. As such, I present the story in this series of essays with both delight and relish.

Still, I must confess that this part of the tale does not intrigue me most as a scientist and evolutionary biologist. I don't mean to sound jaded or dogmatic, but *Ambulocetus* is so close to our expectation for a transitional form that its discovery could not provide a professional paleontologist with the greatest of all pleasures in science—surprise. As a public illustration and sociopolitical victory, transitional whales may provide the story of the decade, but paleontologists didn't doubt their existence or feel that a central theory would collapse if their absence continued. We love to place flesh upon our expectations (or put bones under them, to be more precise), but this kind of delight takes second place to the intellectual jolting of surprise.

I therefore find myself far more intrigued by another aspect of *Ambulocetus* that has not received much attention, either in technical or popular reports. For the anatomy of this transitional form illustrates a vital principle in evolutionary theory—one rarely discussed, or even explicitly formulated, but central to any understanding of nature's fascinating historical complexity.

In our Darwinian traditions, we focus too narrowly on the adaptive nature of organic form, and too little on the quirks and oddities encoded into every animal by history. We are so overwhelmed—as well we should be—by the intricacy of aerodynamic optimality of a bird's wing, or by the uncannily precise mimicry of a dead leaf by a butterfly. We do not ask often enough why natural selection had homed in upon this *particular* optimum—and not another among a set of unrealized alternatives. In other words, we are dazzled by good design and therefore stop our inquiry too soon when we have answered, "How does this feature work so well?"—when we should also be asking the historian's questions: "Why *this* and not *that*?" or "Why *this* over here, and *that* in a related creature living elsewhere?"

To give the cardinal example from seagoing mammals: The two fully marine orders, Sirenia and Cetacea, both swim by beating horizontal tail flukes up and down. Since these two orders arose separately from terrestrial ancestors, the horizontal tail fluke evolved twice independently. Many hydrodynamic studies have documented both the mode and the excellence of such underwater locomotion, but researchers too often stop at an expression of engineering wonder, and do not ask the equally intriguing historian's question. Fishes swim in a truly opposite manner—also by propulsion from the rear, but with vertical tail flukes that beat from side to side (seals also hold their rear feet vertically and move them from side to side while swimming).

Both systems work equally well; both may be "optimal." But why should ancestral fishes favor one system, and returning mammals the orthogonal alternative? We do not wish to throw up our hands, and simply say "six of one, half a dozen of the other." Either way will do, and the manner chosen by evolution is effectively random in any individual case. "Random" is a deep and profound concept of great positive utility and value, but some vernacular meanings amount to pure cop-out, as in this case. It may not matter in the "grand scheme of things" whether optimality be achieved vertically or horizontally, but one or the other solution occurs for a reason in any particular case. The reasons may be unique to an individual lineage, and historically bound—that is, not related to any grand concept of pattern or predictability in the overall history of life—but local reasons do exist and should be ascertainable.

This subject, when discussed at all in evolutionary theory, goes by the name of "multiple adaptive peaks." We have developed some standard examples, but few with any real documentation; most are hypothetical, with no paleontological backup. (For example, my colleague Dick Lewontin loves to present the following case in our joint introductory course in evolutionary biology: some rhinoceros species have two horns, others one horn. The two alternatives may work equally well for whatever rhinos do with their horns, and the pathway chosen may not matter. Two and one may be comparable solutions, or multiple adaptive peaks. Lewontin then points out that a reason must exist for two or one in any case, but that the explanation probably resides in happenstances of history, rather than in abstract predictions based on universal optimality. So far, so good. History's quirkiness, by populating the earth with a *variety* of *unpredictable* but sensible and well-working anatomical designs, does constitute the main fascination of evolution as a subject. But we can go no further with rhinos, for we have no data for understanding the particular pathway chosen in any individual case.)

I love the story of *Ambulocetus* because this transitional whale has provided hard data on reasons for a chosen pathway in one of our best examples of multiple adaptive peaks. Why did both orders of fully marine mammals choose the solution of horizontal tail flukes? Previous discussions have made the plausible argument that particular legacies of terrestrial mammalian ancestry established an anatomical predisposition. In particular, many mammals (but not other terrestrial vertebrates), especially among agile and fast-moving carnivores, run by flexing the spinal column up and down (conjure up a running tiger in your mind, and picture the undulating back). Mammals that are not particularly comfortable in water—dogs dog-paddling, for example—may keep their backs rigid and move only by flailing their legs. But semiaquatic mammals that swim for a living—notably the river otter *(Lutra)* and the sea otter *(Enhydra)*—move in water by powerful vertical bending of the spinal column in the rear part of the body. This vertical bending propels the body forward both by itself (and by driving the tail up and down), and by sweeping the hind limbs back and forth in paddling as the body undulates.

Thus, horizontal tail flukes may evolve in fully marine mammals because inherited spinal flexibility for movement up and down (rather than side to side) directed this pathway from a terrestrial past. This scenario has only been a good story up to now, with limited symbolic support from living otters, but no direct evidence at all from the ancestry of whales or sirenians. *Ambulocetus* provides this direct evidence in a most elegant manner—for all pieces of the puzzle lie within the recovered fossil skeleton.

We may infer from a tail vertebra that *Ambulocetus* retained a long and thin mammalian tail, and had not yet evolved the horizontal fluke. We know from the spinal column that this transitional whale retained its mammalian signature of flexibility for up and down movement—and from the large hind legs that undulation of the back must have supplied propulsion to powerful paddling feet, as in modern otters.

Thewissen and colleagues draw the proper evolutionary conclusion from these facts, thus supplying beautiful evidence to nail down a classic case of multiple adaptive peaks with paleontological data: "*Ambulocetus* shows that spinal undulation evolved before the tail fluke . . . Cetaceans have gone through a stage that combined hindlimb paddling and spinal undulation, resembling the aquatic locomotion of fast swimming otters." The horizontal tail fluke, in other words, evolved because whales carried their terrestrial system of spinal motion to the water.

History channels a pathway among numerous theoretical alternatives. In his last play, Shakespeare noted that "what's past is prologue; what to come, in yours and my discharge." But present moments build no such wall of separation between a past that molds us and a future under our control. The hand of the past reaches forward right through us and into an uncertain future that we cannot fully specify.

Epilogue

I wrote this essay in a flush of excitement during the week that Thewissen and colleagues published their discovery of the definitive intermediate whale *Ambulocetus*, in January 1994. With my lead time of three months from composition to the first publication of these essays in *Natural History* magazine, "Hooking Leviathan by Its Past" appeared in April 1994—complete with central theme of a chronologically developing story in four stages.

I think of the old spiritual: "Sometimes I get discouraged, and think my work's in vain. But then the Holy Spirit revives my soul again." I'm actually a fairly cheerful soul, but we all need replenishment now and then. If "there is a balm in Gilead" (the song's title) for scientists, that elixir, that infusion of the holy spirit, takes the form of new discoveries. On the very week of my essay's publication, Phil Gingerich and colleagues (see bibliography) published their description of yet another intermediate fossil whale, a fifth tale for this gorgeous sequence of evolutionary and paleontological affirmation. (I did feel a bit funny about the superannuation of my essay on the day of its birth, but all exciting science must be obsolescent from inception—and I knew I could write this epilogue for my next book!)

Gingerich and colleagues discovered and named a new fossil Eocene whale from Pakistan, *Rodhocetus kasrani* (*Rodho* for the local name of the region, *kasrani* for the group of Baluchi people living in the area. *Rodhocetus*, estimated at some ten feet in length, lived about 46.5 million years ago. This new whale is thus about 3 million years younger than the "smoking gun" *Ambulocetus* (Case Four and the key story in the main essay), and about the same age as *Indocetus* (stage three in the main essay). No forelimb bones have been found, and the spinal column lacks tail vertebrae, but much of the skull has been recovered with, perhaps more important, a nearly complete vertebral column from the neck all the way back to the beginning of the tail. Most of the pelvis has also been found and, crucial to evidence about intermediacy, a complete femur (but no other elements of the hind limb).

We may summarize the importance of *Rodhocetus*, and its gratifying extension of our story about "hard" evidence for intermediacy in the evolution of whales from terrestrial ancestors, by summarizing evidence in the three great categories of paleontological data: form (anatomy), habitat (environment), and function.

FORM. I was most struck by two features of *Rodhocetus*'s anatomy. First, the excellent preservation of the vertebral column provides good evidence of intermediacy in a mixture of features retained from a terrestrial past with others newly acquired for an aquatic present. The high neural spines (upward projections) of the anterior thoracic vertebrae (just behind the neck) support muscles that help to hold up the head in terrestrial animals (not a functional necessity in the buoyancy of marine environments; whales evolved from a terrestrial group, the mesonychids, with particularly large heads). Direct articulation of the pelvis with the sacrum (the adjacent region of the vertebral column) also characterizes both *Rodhocetus* and terrestrial mammals (where gravity requires this extra strength), but does not occur in modern whales. Gingerich and colleagues conclude: "These are primitive characteristics of mammals that support their weight on land, and both suggest that *Rodhocetus* or an immediate predecessor was still partly terrestrial."

But other features of the spinal column indicate adaptation for swimming: short cervical (neck) vertebrae, implying rigidity for the front end of the body (good for cutting through the water as the rear parts of the animal provide propulsion); and, especially, the seamless flexibility of posterior vertebrae (sacral vertebrae are fused together in most large terrestrial mammals, but unfused in both modern whales and *Rodhocetus*), an important configuration for providing forward thrust in swimming. Gingerich and colleagues conclude: "These are derived characteristics of later archaeocetes [ancient whales] and modern whales associated with aquatic locomotion."

Second, and even more striking for this essay's case of graded intermediacy, sequentially discovered during the past twenty years, *Rodhocetus* is about 3 million years younger than the "smoking gun" *Ambulocetus* (a marine whale with limbs large enough for movement on land as well), and a good deal older than later whales that had already crossed the bridge to fully marine life (*Basilosaurus*, my Case Two, with well-formed but tiny hind limbs that could not have functioned on land, and probably didn't do much in water either). In the most exciting discovery of this new Case Five, the femur of *Rodhocetus* is about two thirds as long as the same bone in the older *Ambulocetus*—still functional on land (probably), but already further reduced after 3 million additional years of evolution.

HABITAT. *Rodhocetus* is the oldest whale from fully and fairly deep marine waters. The oldest of all whales, *Pakicetus* of Case One, lived around the mouths of rivers; *Ambulocetus* and *Indocetus* of Cases Three and Four inhabited very shallow marine waters. Interestingly, the more fully marine habitat of *Rodhocetus* correlates with greater reduction of the hind limb, for *Indocetus* is a contemporary of *Rodhocetus*, yet grew a larger femur comparable in length with the earlier *Ambulocetus*. (All three creatures had about the same body size). Thus, admittedly on limited evidence, limbs decreased in size over time and became smaller faster in whales from more fully marine environments. (Perhaps *Rodhocetus* had already ceased making excursions on land, while the earlier *Ambulocetus*, with a larger femur, almost surely inhabited both land and water.) In any case, the contemporaneity of *Rodhocetus* (shorter femur and deeper water) and *Indocetus* (longer femur with life in shallower water) illustrates the diversity that already existed in cetacean evolution. Evolution, as I always say, no doubt to the point of reader's boredom, is a copiously branching bush, not a ladder.

FUNCTION. *Rodhocetus* lacks tail vertebrae, so we can't tell for sure whether or not this whale had yet evolved a tail fluke. But evidence of the beautifully preserved spinal column—particularly the unfused sacral vertebrae, "making," in the words of Gingerich et al., "the lumbocaudal [back to tail] column seamlessly flexible"—indicates strong dorsoventral (back to belly) flexion at the rear end of the body—the prerequisites for swimming in the style of modern whales (with propulsion provided by a horizontal tail fluke, driven up and down by bending the vertebral column). I was particularly pleased by this result, since I closed my essay with a mini-disquisition on multiple adaptive peaks and the importance of historical legacies, as illustrated by vertical tail fins in fishes vs. horizontal flukes in whales—both solutions working equally well, but with whales limited to this less familiar alternative because they evolved from terrestrial ancestors with backs that flexed dorsoventrally in running. Gingerich and colleagues conclude: "This indicates that the characteristic cetacean mode of swimming by dorsoventral oscillation of a heavily muscled tail evolved within the first three million years or so of the appearance of the archaeocetes."

A tangential comment in closing: The sociology of science includes much that I do not like, but let us praise what we do well. Science at its best is happily and vigorously international (see essay 20)—and I can only take great pleasure in the following list of authors for research done in an American lab based on fieldwork in Asia, supported by the Geological Survey of Pakistan: Philip D. Gingerich, S. Mahmood Raza, Muhammad Arif, Mohammad Anwar, and Xiaoyuan Zhou. Bravo to you all. I also couldn't help noting the paper's first sentence: "The early evolution of whales is illustrated by partial skulls and skeletons of five archaeocetes of Ypresian (Early Eocene) . . . age." The geological time scale is just as international, for our fossil record is a global scheme for correlating the ages of rocks. So a layer of sediments in Pakistan may be identified as representing a time named for a place that later became the bloodiest European battle site of World War I—the dreaded Ypres (or "Wipers" as British soldiers named and pronounced their hecatomb).

But so much for lugubrious and sentimental thoughts. Let's just end in the main essay's format for our new case of *Rodhocetus:*

Case Five. Open and shut.

Verdict: sustained in spades, wine and roses.

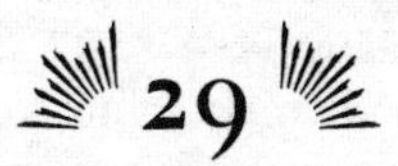

A Special Fondness for Beetles

JUST AS THE LORD holds the whole world in his hands, how we long to enfold an entire subject into a witty epigram. The quotable one-liner is a mainstay of culture, not an innovation in our modern era of sound bites. How could we grasp the eternal truths of nature and humanity if we couldn't ask Sam to play it again, or didn't know that nice guys finish last?

The most widely quoted one-liner in evolutionary biology brilliantly captures the central fact about life's exuberant variety and composition. According to an older tradition that Darwin overturned, we should be able to infer both God's existence and his benevolence by studying the organisms that he created. This idea of "natural theology" dominated British zoology, at least from Robert Boyle in the late seventeenth century to William Paley in the generation just before Darwin (see essay 9 in *Eight Little Piggies*). The natural theologians sought God's handiwork not merely in the good design of organisms, but especially in the supposed arrangement of nature to reflect human superiority and domination.

As a powerful corrective to this arrogant tradition of natural theology, evolutionists argued, early and often, that nature's undoubted order is neither benevolent in our terms (but "red in tooth and claw"—see essay 6), nor established with us in mind or at the helm. The kind of

God implied by nature's actual composition might not be a deity worthy of our worship.

At this point in the argument, almost any evolutionist will turn to our canonical one-liner for epigrammatic emphasis and support. J. B. S. Haldane (1892–1964), author of the phrase, was a founder of modern Darwinism (see his 1932 book, *The Causes of Evolution*) and a distinguished man of letters as well. I cite the famous words from the standard source—not Haldane himself, but a footnote on the first page of the most widely read paper in modern evolutionary biology: "Homage to Santa Rosalia, or why are there so many kinds of animals" (*American Naturalist*, 1959), by G. Evelyn Hutchinson, the world's greatest ecologist and the only twentieth-century British biologist who could match Haldane in brilliance and wit. Hutchinson wrote:

> There is a story, perhaps apocryphal, of the distinguished British biologist, J. B. S. Haldane, who found himself in the company of a group of theologians. On being asked what one could conclude as to the nature of the Creator from a study of his creation, Haldane is said to have answered, "An inordinate fondness for beetles."

Lovely line, but did Haldane utter the words—and if so, when, where, and how? The standard source illustrates the problem—not hard copy with a byline, but a secondary report, frankly labeled as "perhaps apocryphal." Haldane was a brilliant and copious writer, but he was an even more fluent barroom wit—and great comments in this venue end up either scratched into soggy napkins, or dimly remembered in the midst of a subsequent (and consequent) hangover.

Haldane's line—an inordinate fondness for beetles—is now so famous and standard that we really do yearn to pin down the source. Yet nothing is so elusive as a canonical well-turned phrase, for the vast majority of such quips are either misstated or misattributed (see *Nice Guys Finish Seventh: False Phrases, Spurious Sayings and Familiar Misquotations*, by my old college buddy Ralph Keyes). I tried to slip two of them by you in my first paragraph. Leo Durocher did not confine angels to the cellar; and Humphrey Bogart (as Rick in *Casablanca*) never told Sam to play it again (though Woody Allen purposely used this standard error as the title for a later film).

Thanks to a charming, if somewhat cranky, English tradition—lengthy and passionate exchanges in letters-to-the-editor on the minute details of smallish subjects—we finally have both as good a resolution as we can get, and a catalog of the usual mistakes that make canonical quotations so hard to trace. The hubbub started in the October 5, 1989, issue of *Nature*, when my friend (and Oxford professor) Bob May reviewed a meeting on interactions between ants and plants under a title that parodied Haldane's quip—"an inordinate fondness for ants." May began his article: "Haldane's best-remembered remark, that God has 'an inordinate fondness for beetles,' was elicited by Jowett's question, at high table at Balliol, as to what his studies had revealed about the deity." This claim elicited a firestorm—for reasons that will soon be obvious—and the letters columns of both *Nature*, Britain's leading professional journal in general science, and *The Linnean* (newsletter of the Linnean Society of London) erupted in magmatic frenzy.

I do not trace the history of this canonical line in the interests of antiquarian pedantry, but because the enterprise can yield such rich rewards in teaching us about crucially important, and often unrecognized, biases in our modes of thought and styles of storytelling. The pervasive errors made in citing and attributing canonical quotations are not random, but follow a clear and sensible pattern. Basically, most errors are aggrandizements in three categories: misattributions to more-famous people; recasting to render a quote more pithy or pungent; and alteration of circumstances to make the relatively mundane either funnier or more heroic. Haldane's quip about beetles has wallowed in all three categories of error.

WHO SAID IT? Haldane was sufficiently famous to win exemption from the "magnet effect"—the directed migration of good quotes to more-celebrated mouths. No biologist of Haldane's generation could have made the quote more notable by assuming false parentage. But any evolutionary one-liner in English must eventually wander toward the greatest of all prose stylists in our profession, Thomas Henry Huxley. I have four misattributions of the beetle line to Huxley in my files (this essay has been gestating for more than a decade), and the same error may have prompted the howler that unleashed the recent round of discussion in *Nature* and *The Linnean*—Bob May's statement that Haldane made the quip to Benjamin Jowett at Balliol.

Jowett was the greatest English classical scholar of his generation. His erudition and arrogance, as master of Balliol College, prompted a notable couplet in the annals of humorous verse:

I am the master of this college
What I know not is not knowledge.

The pious and conservative Jowett would have been a perfect foil for Haldane's remark—but for one small problem. Jowett died in 1893, before Haldane reached his first birthday. (My freshman philosophy course in college still used Jowett's translation of Plato's *Republic*). Jowett must have entered the story through misattribution of the beetle quote to Huxley (a contemporary who died in 1895), or perhaps to Haldane's father, a famous physiologist in his own right.

Bob May is an Australian, not an upper-class English Oxbridgian. But his response to learning about his mixup of generations, published in *Nature* on October 26, 1989, was both gorgeously arch and blessedly brief: "Mundane constraints of time and space do not apply to stories about Oxford."

In what circumstances? As May noted, good stories require the transmutation of the mundane into the charming, the amusing, or the dramatic. Haldane, as we shall see, made the quip several times, but always among friends. Yet the story improves immensely if the line can be recast as a spontaneous riposte to a specific taunt, or to a query from a worthy adversary. Most versions therefore add this common element of myth. May used the shade of Jowett as an impossible foil. Hutchinson reported that Haldane had uttered the line in response to a specific query from theologians. A. J. Cain, another distinguished British biologist who knew Haldane well, wrote in 1987, "It was Haldane, not Huxley, and he told the story to me himself . . . Some solemn ass asked him what could be inferred of the work of the Creator from a study of the works of Creation . . . and got the crushing reply 'An inordinate fondness for beetles.' "

With what words? The flurry of letters in *Nature* and *The Linnean* have validated Haldane as the source of our finest one-liner; at least no earlier version has been uncovered. Moreover, we can be fairly confident that Haldane did not utter the line in the "best story" situation as a riposte or retort; at least no victim or witness has come forward. But as we resolve these questions of personality and circumstance, the basic issue of Haldane's actual words remains elusive. Haldane was a great writer and the author of several popular volumes of scientific essays (most based on his columns in the communist *Daily Worker*—another intriguing aspect of his iconoclastic career, but best saved for another essay). He apparently loved his quip about beetles and used it often in casual conversation and public addresses. But no one has found any evidence that he ever wrote the line down—and we therefore do not know exactly what he said (or if he varied the words).

The closest approach to an "official" version emerged as a result of all the recent correspondence. On April 7, 1951, Haldane filled in for an indisposed colleague, the great physicist J. D. Bernal, to deliver an address to the British Interplanetary Society. He did not publish his remarks, but a report of his speech, written by A. E. Slater, the society's secretary, did appear in volume 10 of the *Journal of the British Interplanetary Society*—a publication that will not be found in your local library, not to mention your corner drugstore (and another reason for delayed documentation of Haldane's remark). I present the full citation:

> Coming to the question of life being found on other planets, Professor Haldane apologized for discoursing, as a mere biologist, on a subject on which we had been expecting a lecture by a physicist. He mentioned three hypotheses:
>
> (a) that life had a supernatural origin
> (b) that it originated from inorganic materials, and
> (c) that life is a constituent of the Universe and can only arise from preexisting life.
>
> The first hypothesis, he said, should be taken seriously, and he would proceed to do so. From the fact that there are 400,000 species of beetles on this planet, but only 8,000 species of mammals, he concluded that the Creator, if he exists, has a special preference for beetles.

Fine. But have we now lost the delicious words in our usual citation? Did Haldane really say "special preference," and not the much more pungent and ironic "inordinate fondness"? Is our usual version just another example of a falsely "promoted" quotation? Or did the secretary of the Interplanetary Society either misremember or downgrade in the ancient tradition of British understatement? Or did Haldane say different things at different times? We shall never know, but a letter in *The Linnean* (August 1992) from Haldane's friend Kenneth Kermack restores our hope for accuracy of the usual version:

> I have checked my memory with Doris [Kermack's wife], who also knew Haldane well, and what he actually said was: "God has an inordinate fondness for beetles." J.B.S.H. himself had an inordinate fondness for the statement: he repeated it frequently. More often than not it had the addition: "God has an inordinate fondness for stars and beetles." . . . Haldane was making a theological point: God is most likely to take trouble over reproducing his own image, and his 400,000 attempts at the per-

> fect beetle contrast with his slipshod creation of man. When we meet the Almighty face to face he will resemble a beetle (or a star) and not Dr. Carey [the Archbishop of Canterbury].

So pay your money and take your choice. You can either select the duller "special preference" for beetles alone, or you must share the wittier "inordinate fondness" with all the celestial multitudes. I have made a hybrid compromise in the title of this essay.

But what about the facts underlying the phrase? How inordinate is God's fondness, how special his preference, for the Coleoptera? How many species of beetles do inhabit our planet—and what, pray tell, should we make of the number?

In a recent summary of data, British Museum entomologist Nigel E. Stork reports that the total number of formally named species of animals and plants (excluding the diverse kingdoms of fungi, bacteria, and other unicellular creatures) now stands at approximately 1.82 million. Of this totality, more than half are insects (57 percent)—and nearly half of all named insect species are beetles. Thus, beetles represent about 25 percent of all named species in the plant and animal kingdoms—good candidature, I trust we would all agree, for a creator's inordinate fondness.

But this compendium of available names only provides a beginning, a tip of the proverbial iceberg. All taxonomists agree that the vast majority of earth's species remain undiscovered and unnamed. In his recent book, *The Diversity of Life*, my colleague Ed Wilson writes:

> How many species of organisms are there on earth? We don't know, not even to the nearest order of magnitude. The numbers could be close to 10 million or as high as 100 million. Large numbers of new species continue to turn up every year. And of those already discovered, over 99 percent are known only by a scientific name, a handful of specimens in a museum, and a few scraps of description in scientific journals. It is a myth that scientists break out champagne when a new species is discovered. Our museums are glutted with new species. We don't have time to describe more than a small fraction of those pouring in each year.

So how inordinate is nature's fondness for beetles when we try to estimate actual numbers, rather than relying on the paltriness of published information? The brief for beetles now grows mightily in strength—not only for the obvious absolute gain in number of species, but primarily for increasing domination in relative frequency, or percentage of species. For beetles, by their size and favored habitats, rank with the most undercounted groups of organisms.

A complete census of species will not add membership equally and across the board—for in some groups we nearly have them all; while in others we have barely begun to count. The worldwide company of birdwatchers, for example, has been so assiduous—and objects of their study generally so conspicuous—that we expect no great increase in the nine thousand or so named species of birds. The influx of new bird species has already dwindled to the merest trickle, with only a species or two added each year. Similarly, the ledger of four thousand or so mammalian species, while not so diligently cataloged as birds, will experience no massive gain in numbers.

But beetles are small and mostly inconspicuous—and the prominence of some species as agricultural pests does not lead to discovery for the majority, especially since most beetle species have limited geographic ranges in restricted habitats of the world's most lush and understudied environments: the tropical rain forests. (When we realize how many species remain unknown in this abode, and when we recognize how many are being lost daily as human rapacity uproots these rich environments, usually for the short-term profits of a few, we can appreciate the appropriate focus of the environmental movement on tropical rain forests, even though these habitats may seem so distant from most of our immediate concerns and locales.)

We may get some handle on the probable number of beetle species by considering the basis for lowest and highest estimates of the world's fauna and flora. Ed Wilson cited 10 to 100 million as his large ballpark. I have read estimates ranging from 1.87 to 80 million for insects alone—leading to some 3½ million to more than 150 million animal and plant species altogether, if insects form about half the total.

The basis for this small biological industry of estimation lies in the remarkable work of the American entomologist Terry Erwin, published in the early 1980s. (Erwin first gave us a reasonable quantitative estimate for the incredible, and uncataloged, diversity of tropical rain forests—a vital contribution both to biological knowledge and to strategies of the environmental movement.) In 1982, Erwin presented a surprising number that, although initially hard to believe, has since become a standard figure, cited in countless textbooks and newspaper articles. Erwin concluded that 30 million species of arthropods alone dwell in tropical rain forests—and he based his estimate on beetles.

Erwin's number comes from hard work in the field, not from a pocket calculator consulted in an armchair. He began by recognizing that insects cataloged from rain-forest trees are a pitiful fraction of the enormously diverse community actually living in these tropical heights. But how can a scientist census all (or even most) of the species in a tall tropical tree, since so many of the species are rare, inconspicuous, and downright secretive in their habits? Erwin therefore used a drastic approach: he fogged entire trees with volleys of strong insecticide, and collected everything that fell out. (I don't mean to sound peremptory or facetious; such work is rigorous and difficult, both conceptually and muscularly. How do you climb trees to fog? How do you collect the resulting bounty? How do you know that you have recovered most of the species, for some die deep in the bark and do not fall out? Above all, how do you identify the plethora of previously unknown forms, especially since no one can be an expert on all groups?)

Erwin reached his figure of 30 million by extrapolating from the beetles on a single species of tropical tree, *Luehea seemannii*. Consider his argument in eight steps, and you will begin to appreciate both the difficulty of the enterprise and the reasons for such a wide range of estimates.

1. Working during all three seasons recognized in the Panamanian rain forest, Erwin fogged nineteen trees of *Luehea seemannii*—thus getting a handle on variation among trees and seasons.

2. He counted the total number of beetle species recovered at some 1,200.

3. Erwin then assigned each species to one of four "guilds," or ecological roles in habitats: herbivores (plant eaters), fungivores (feeders on fungi), predators, and scavengers.

4. As a key problem in moving from beetles on a tree to insects in a forest, one must know how many of those beetles live exclusively on one kind of tree, and how many are more cosmopolitan. (If, for example, all 1,200 beetles lived only on *Luehea*, then the total number in the forest may be as high as 1,200 times the number of tree species. But if all 1,200 beetles live on all forest tree species, then the total number of beetles may be just 1,200, period.) Erwin made his division into guilds in order to estimate this degree of "endemicity" (defined here as confinement to a single tree species). He arrived at estimates of 20, 10, 5, and 5 percent for his four guilds of herbivores, fungivores, predators, and scavengers.

5. Applying these indices of endemicity to the 1,200 beetle species

collected on *Luehea*, Erwin estimated that 163 species might be confined to life on *Luehea*.

6. Worldwide diversity of tropical trees probably stands at some 50,000 species. If 163 is a reasonable average for endemic beetles per tree species, then tropical trees house 50,000 × 163, or 8,150,000 species of beetles.

7. Since beetles represent some 40 percent of total arthropod diversity, tropical trees may house some 20 million species of arthropods.

8. This estimate of tropical diversity only counts species in tree canopies. Erwin then argued that canopy species might outnumber ground-dwelling species by about two to one—adding another 10 million arthropods for the forest floor, and raising the final estimate to 30 million.

The lower and higher figures of 1.87 and 80 million for rain forest arthropods arise from other data or from different estimates upon figures used by Erwin in extrapolating, not from challenges to his empirical counts in fogging *Luehea* trees. For example, Nigel Stork's highest estimate of 80 million arises from two modifications of Erwin's figures, both substantially raising the number of estimated species. He believes that beetles represent far less than 40 percent of canopy arthropod species. Stork's preferred figure of 20 percent immediately doubles the total estimate for rain-forest arthropods. Stork also argues that Erwin overestimated the percentage of species in the canopy vs. the forest floor—thus yielding the still larger total of 80 million species when using a higher percentage for ground dwellers in the total estimate.

The lowest figure of 1.87 million species can be found in a 1991 article by I. D. Hodkinson and D. Casson that used, for its title, a parody on Haldane's quip: "A lesser predilection for bugs: Hemiptera (Insecta) diversity in tropical rain forests." "Bug" may be a vernacular term for any creepy-crawlie (not to mention errors in computer programs and illicit listening devices). But, to a zoologist, "bug" is a technical term for insects of the order Hemiptera (often called "true bugs" to gain proper distance from ordinary usage). Hodkinson and Casson used bugs rather than beetles to make their worldwide estimate, and their sly title is a double entendre, for bugs are not nearly so speciose as beetles in nature, and the authors' estimate of total arthropod diversity is also far smaller than most others offered of late).

As Erwin extrapolated from beetles on nineteen trees, Hodkinson and Casson worked upward "from an intensive study of the bug (Hemiptera) fauna of a moderately large and topographically diverse area of tropical rain forest in Sulawesi Utara, Indonesia." In broadest outline, Hodkinson and Casson followed the same logic that Erwin had employed. They collected 1,690 species of bugs from Sulawesi and determined that 62.5 percent had been previously unknown. If the five hundred described species of Sulawesi trees yielded 1,056 new species (the total of 1,690 species times 62.5 percent for the proportion of newly discovered forms), then the worldwide figure of approximately fifty thousand species on tropical trees might yield one hundred times as many new species of bugs, or 105,600. Add these new species to the 81,700 species already described, and we arrive at a total estimate of 187,300 bug species worldwide. Since bugs include about 10 percent of all insect species, the worldwide fauna of insects might stand at some 1.87 million.

How can two estimates based on the same style of argument arrive at such different figures? The logic of inference, as all people engaged in this work know only too well, is "iffy" in the extreme, for conclusions are only sound if the premises be true—and why should endemic beetles on one kind of tree in Panama, or true bugs in one small region of Indonesia, provide a model or average for estimating an entire world's fauna? Erwin's estimates may be way too high because his tree species houses more insects than most others (and a figure derived from his species alone will vastly overestimate worldwide diversity), or because he greatly overstated the number of beetles unique to each species of tree (the most common and cogent criticism of Erwin's estimate). Hodkinson and Casson's estimates may be way too low because their area of Indonesia is relatively poor in species, or because they did not use so comprehensive a collecting method as fogging trees, and may therefore have missed a substantial amount of diversity.

In any case, we certainly learn that nature's fondness for beetles must be vastly more inordinate than a simple count of 400,000 formally named species (Haldane's stated basis for his quip) would indicate. We also understand, from our great difficulty in estimating the true number of species on earth, and from the substantial differences among figures offered by our best experts, how precious little we know about the natural history of our planet. The next time someone tells you that taxonomy is a dull subject because we only need to fill in a few details for an earth already well known—laugh in his face.

In the midst of this ignorance, we should take comfort in two conjoined features of nature: first, that our world is incredibly strange and therefore supremely fascinating (the key point, I think, behind Haldane's quip that ultimate meaning must reside in the unparalleled diversity of a group that rarely rivets our attention); second, that however bizarre and arcane our world might be, nature remains potentially comprehensible to the human mind.

I should end by cementing these two cardinal precepts with their canonical one-liners. Einstein spoke for the possibility of grasping natural complexity when he wrote, in a theological metaphor second only to Haldane's on beetles: *Raffiniert ist der Herr Gott, aber Boshaft ist er nicht* (The Lord God is subtle, but he is not malicious). As for the joy of nature's strangeness, we cannot do better than a famous line by a chap named J. B. S. Haldane—and this time we know what he said because he wrote it down! (in *Possible Worlds*, 1927). "My suspicion is that the universe is not only queerer than we suppose, but queerer than we *can* suppose."

If Kings Can Be Hermits, Then We Are All Monkeys' Uncles

WE LEARN FROM OUR ERRORS, perhaps most of all from our shameful mistakes. I therefore begin with a story at my own expense. Many years ago, one of my students told me about her father's brother, a severely retarded man of childlike disposition. When she described him as "my uncle," I did a mental double take (and fortunately said nothing, so the shame of my error remained internal until now). I said to myself, "Uncles are wise people who render free advice (not always worthwhile) and take you to baseball games; how can a person with such limits be an uncle?" I then kicked myself (also metaphorically) and continued the soliloquy: "He is her father's brother; he is therefore her uncle pure and simple; *uncle* is a genealogical term of relationship, not a functional concept of action; he is as good and as true an uncle as any man who ever lived."

Evolutionary relationships are also genealogical, not primarily functional. We all understand that whales are mammals by history of common descent, not fishes because they swim in the ocean. In genealogical terms, closeness is defined by position in a sequence of branchings—what Darwin called "propinquity," or relative nearness. I may look and act more like my cousin Bob than my brother Bill, but Bill is still closer to me by genealogy. Function and appearance need not correlate strongly with genealogical propinquity. To cite a classic example: all evolutionists agree that genealogical relationships among trout, lungfishes, and cows are correctly described in the accompanying diagram. Terrestrial vertebrates branched off the line of early fishes at a point near the ancestry of modern lungfishes; trout evolved much later from a persisting earlier line of fishes. Therefore, if we choose to classify purely by genealogy, lungfishes and cows must be placed together in a group separate from trout. Many of us rebel against such an idea because our conventional classifications mix functional and strictly genealogical relationship. We may say, "A lungfish looks like a fish, swims like a fish, acts like a fish, and (presumably, for I have never had the pleasure) tastes like a fish. Therefore, it is a fish." Perhaps so; but, by propinquity, lungfishes are closer to cows.

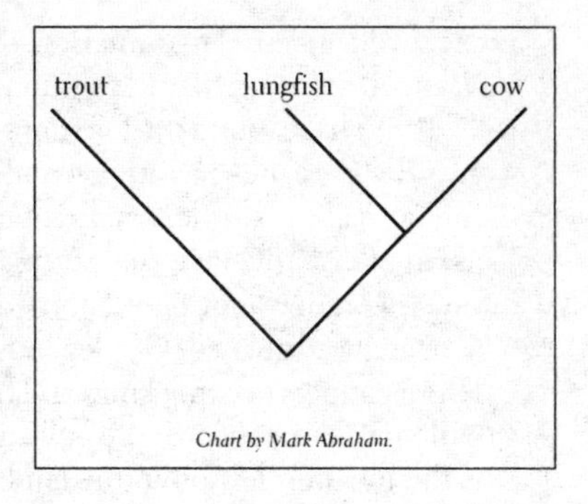

Chart by Mark Abraham.

I don't wish to pursue this theoretical point in classification any further in this essay—although aficionados should note that this issue now pervades the science of systematics as the great debate about "cladism." Cladists advocate classification by pure genealogy (branching order), with no attention whatsoever to traditional concepts of similarity in function or biological role. For this essay, we need only carry away the lesson that genealogical and functional similarity are different concepts, and that we can be terribly fooled when we make a mistaken equation—particularly when we assume a closeness in branching (propinquity) from evidence of common appearance or behavior.

(Sorry, but I must add one more didactic paragraph: if we call a whale a fish, we make a simple error by misunderstanding the evolutionary phenomenon of "convergence." The fishlike characters of whales evolved separately and independently in a line derived from fully terrestrial vertebrates. But the fishy similarities of trout and lungfishes are genuine evolutionary marks of common ancestry. These similarities don't forge a closer genealogical bond between lungfish and trout than between lungfish and cow because such shared features are common characters of *all* early vertebrates; propinquity is marked by shared characters of *later* derivation. I would not, for example, use the character "five fingers" to unite humans and dogs, while placing seals in another group, for dogs and seals are genealogically close as members of the order Carnivora. Possession of five fingers is a shared character of all ancestral mammals; such traits cannot help us make divisions *within* later mammalian evolution.)

If you have found the foregoing lesson abstract and dull, let me now reward your patience with a wonderful story that becomes even better when you absorb the lesson. In functional terms, we would acknowledge maximal disparity between a king in his castle and a hermit in a hovel. But, as I argued above, functional and genealogical similarity need not be strongly correlated. Our legends are replete with scenarios of rags to riches: paupers become kings and frogs turn into princes. All the world's opulence does not debar the possibility that a king's closest cousin might be the meanest hermit in the land.

Let us now contrast the kings and hermits, explicitly so called, of the world of crabs. We could scarcely find two more apparently different creatures in this admittedly limited domain. The king crab (*Paralithodes camtschatica*), a paragon of size within the brotherhood, lives in Arctic and north temperate waters from the northern tip of Vancouver Island, all around Alaska, over to Siberia, and down the eastern Pacific margin as far as Japan. To borrow my earlier formulation for lungfishes, it looks like a crab, moves like a crab, acts like a crab, and certainly tastes like a crab—the basis for a prosperous Alaskan "fishery" (the correct English word, if taxonomically inaccurate in this case), which, at its height in the early 1960s, yielded 180 million pounds per year and a revenue equal to 40 percent of what the lucrative salmon trade provided. (Severe recent declines may be due to parasitism, disease, or overfishing—there we go again, with our chauvinistic vertebrate-centered etymologies.) The largest individual ever captured had a leg spread of just under five feet, and weighed 24½ pounds. Ten pounders with a three foot leg span are common in the trade.

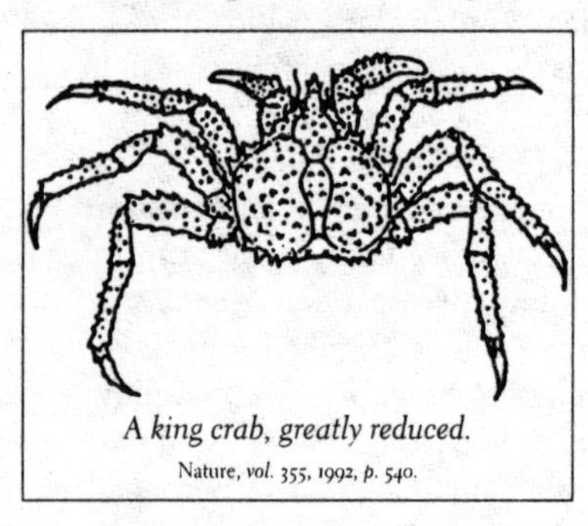
A king crab, greatly reduced.
Nature, *vol.* 355, 1992, *p.* 540.

By contrast, consider the humble hermit crab—actually a large group of related forms, including some eight hundred species in more than eighty genera. Most are an inch or two in length and live curled up inside an empty snail shell (which they eventually outgrow and "trade in" for a larger model). However great the difference in size and habits, the disparity in form between king crabs and typical hermit crabs is even more pronounced. A king crab looks like an ordinary crab: its carapace (outer shell) is flattened and widened, and it bears a pair of claws up front and three pairs of long sturdy legs behind (most crabs have four pairs behind the claws).

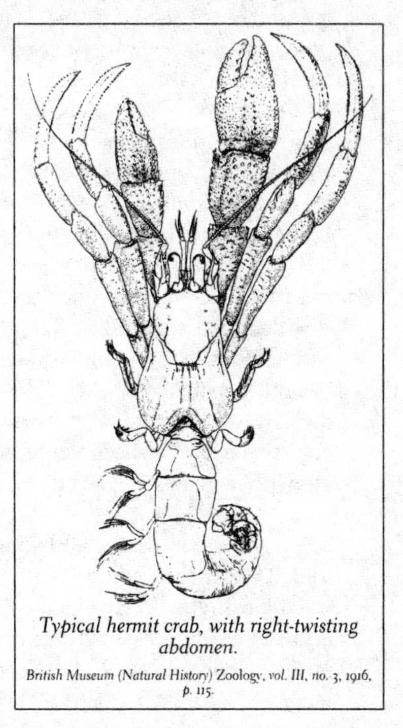

Typical hermit crab, with right-twisting abdomen.

British Museum (Natural History) Zoology, vol. III, no. 3, 1916, p. 115.

By contrast, I don't know why anyone ever decided to designate the hermits as crabs in the first place. Crabs form one of three major divisions—with lobsters and shrimps as the others—within the large group of marine crustaceans called Decapoda. (The Arthropoda, biggest of all phyla, contains three great groups: the Crustacea; the Uniramia, including insects, millepedes and centipedes; and the Chelicerata, including spiders, scorpions, and horseshoe crabs.) True crabs belong to the order Brachyura, meaning "short tail." As a defining feature, their abdomens (rear ends) are shortened and narrowed, tucked around the back of the body and firmly pressed against the underside. The flat and wide crab shell corresponds only with the front part of the body in a lobster or shrimp. The abdomen, source of all good eating, extends out and back in shrimps and lobsters, but disappears from sight (as a remnant tucked underneath the body) in crabs. Take the front end of a lobster, flatten it out, pull from both sides until the shell becomes wider than long, reduce the tail, tuck it under the body—and, *voilà*, you have a crab. (The relationship among the three major decapod groups becomes clearer in the light of this thought experiment.)

So why call a hermit a crab at all? Hermit crabs are not members of the Brachyura by genealogy, but form a separate group, called Anomura, in a never-never land between conventional shrimps and other decapods. Their bodies are elongate, as in shrimps. They have only two strong pairs of legs behind the frontal pair of claws (the two highly reduced pairs farther behind function to hold the animal within its borrowed shell). Most important, the abdomen is not reduced or folded up under the body, but rather curved and extended, strongly altered and well adapted to fit into snail shells. Hermit crab abdomens are soft and decalcified, all the better to slip neatly into the shell. Moreover, the abdomen is coiled to one side, mimicking the shell that will serve as its abode. In fact, the primary taxonomic division among hermit crabs recognizes a twofold split in the direction of abdominal coiling—to the left (when seen from above) for the less common Coenobitoidea, and to the right for the standard, garden-variety (I suppose I should say seashore-variety) Paguroidea. So why are these creatures called hermit *crabs?* Doesn't hermit *shrimp* fit appearances better?

And yet, experts have suspected for a long time that king crabs are not true brachyurans either, and that these Alaskan giants (and other members of their family Lithodidae) are closest cousins of hermit crabs. But how could such propinquity in genealogy permit the development of such maximal disparity in form and function? And how, given this disparity, did anyone ever suspect propinquity in the first place? Three arguments have been presented, and they make a strong, if not entirely convincing, case.

1. The abdomen of adult king crabs, though diminished in size and folded under the body as in true crabs, is asymmetrical in form, recalling the rear end of hermit crabs. A few other features of adult anatomy also suggest affinity with hermit crabs. For example, the crustacean group of crabs, lobsters, and shrimps is called Decapoda, meaning "ten-legged." In true crabs, a frontal pair of claws and four posterior pairs of legs make up the complement of ten. In hermit crabs, as stated above, only two pairs of strong legs follow the claws, with the final two pairs reduced to small protuberances that grasp the borrowed snail shell. In king crabs, the first pair of reduced legs is secondarily enlarged to form a third pair of strong legs behind the claws; but the second pair remains small and inconspicuously situated under the body.

2. These sporadic similarities in adult form would never have built a strong case by themselves. Confidence in this odd linkage of kings and hermits rose, however, with the discovery of profound and pervasive likenesses in the larval forms of the two groups. Adult animals are often so strongly specialized and differentiated that most signs of ancestry become hidden or obliterated. But early larvae or embryos often retain the ancestral mode of development, in part because the complex assembly of adult from egg leaves little flexibility for substantial modification, and in part because larval environments often remain stable while adult habitats change. I like to call this common (but by no means invariable) phe-

nomenon of larval conservation the "*Sacculina* principle" to honor a famous parasite of crabs (as it happens). As an adult, *Sacculina* is little more than a blob of formless reproductive tissue in the host's body, but free-living larval stages retain clear features of barnacle ancestry.

The accompanying illustration, from a crucial 1957 article by MacDonald and colleagues, shows glaucothoe (late larval) stages of *Pagurus bernhardus*, a standard hermit crab, and *Lithodes maia*, a smaller relative of king crabs. Some of the adult differences are already established (though the abdomens have yet to assume their asymmetry): the king crab cousin, for example, has developed its characteristic spines and elongated the third pair of legs behind the claws. But the striking similarities of form overwhelm the differences at this early stage of development.

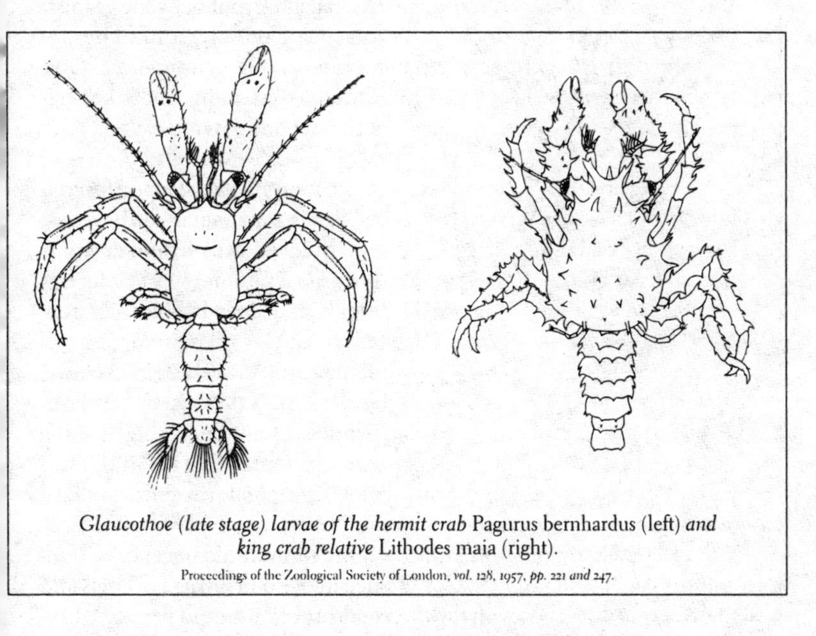

Glaucothoe (late stage) larvae of the hermit crab Pagurus bernhardus (left) *and king crab relative* Lithodes maia (right).

Proceedings of the Zoological Society of London, *vol.* 128, 1957, *pp.* 221 *and* 247.

3. In the most fascinating and general point of all, convergent evolution to crablike form is an oft-repeated trend in decapod crustaceans. I will not speculate on either the advantages or ease of such a transformation, but merely record the multiple occurrences. Flatten the carapace and pull it out to the sides; suppress the abdomen and curl it under the body—and a crablike creature results. This trend is common enough to have a special name, conferred in 1916 by the celebrated British zo-

ologist L. A. Borradaile—carcinization ("crabification" in the less dignified vernacular. Remember that we also call a cancer-producing substance a carcinogen, and that *cancer* itself comes from the Latin word for crab—a reference to the central mass and clawlike extensions of many tumors).

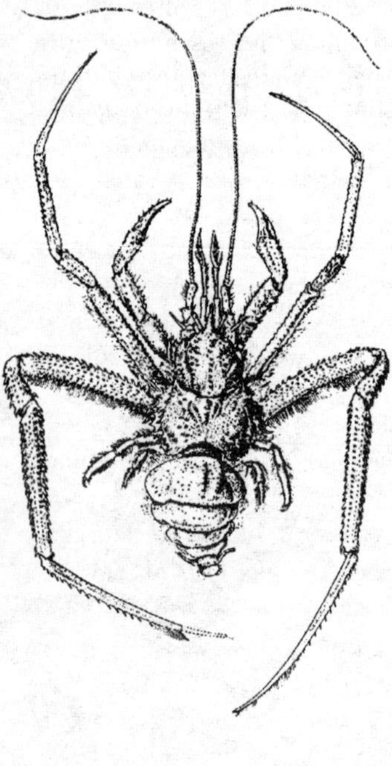

Partly carcinized deep-sea hermit crab Probeebei mirabilis, *with less twisted and calcified abdomen and strong legs for free walking.*

Galathea Report, *vol. 4, 1961, p. 13.*

Many evolutionary lines of hermit crabs have undergone carcinization. In some, the result is only partial—but halfway points give us valuable insight into the full process. Consider the accompanying picture of *Probeebei mirabilis,* a partly carcinized hermit crab properly classified in 1961 by the Danish zoologist Torben Wolff. The abdomen is still asymmetrical and twisted to the right, but it has become secondarily calcified. The two pairs of legs behind the claws are now strongly developed for free walking and extended to the sides, rather than protruded from the front (the proper position for sticking out from a snail shell). The basic reasons for such a change are clear in this case. *Probeebei* lives in deep waters off the coast of Costa Rica (Wolff's specimens came from a depth of more than ten thousand feet). Snail shells (and other potential homes) are rarely available at such depths, and this crab has reverted to a free-living ancestral mode of life.

Two other cases hold special interest for understanding the frequent occurrence of carcinization. *Porcellanopagurus,* the subject of Borrodaile's original study, develops a shortened and fairly symmetrical abdomen. But this creature uses a clam shell, rather than a snail, for a cover—and you don't have to twist to fit under a basically flat plate. *Birgus latro,* the large and well-known "robber" or "coconut" crab of Pacific islands, displays much of the process in its own growth. The adult is fully terrestrial and crablike in appearance, but juveniles still have twisted abdomens and inhabit snail shells at the shoreline.

We may move from these partially carcinized lines of hermit crabs to four cases of virtually complete carcinization in decapod crustaceans in general. Most successful, of course, are the true crabs themselves (brachyurans), with thousands of species and worldwide distribution. But three other lines of fully carcinized crustaceans then arose from a more restricted hermit crab ancestry—two groups little known to non-specialists (the families Lomisidae and Porcellanidae, the porcelain crabs), and the family Lithodidae, including the king crab (and fifty-two other species in sixteen genera—mostly much smaller animals and denizens of cold waters).

If doubts of close propinquity between hermit crabs and king crabs persisted, they were recently dispelled, and convincing new proof provided, in an elegant study published in 1992 ("Evolution of king crabs from hermit crab ancestors," by C. W. Cunningham, N. W. Blackstone, and L. W. Buss).

This study, done in the laboratory of my friend and colleague Leo Buss of Yale University, takes advantage of the revolution in taxonomy now underway thanks to recent technological advances that allow us to sequence DNA cheaply and rapidly (see the next essay). Conventional taxonomy struggles with fewer morphological, physiological, and behavioral traits often frustratingly subject to convergence. Sequencing of DNA and RNA provides hundreds or thousands of newly available characters (the ordering of strings of nucleotides, often highly conserved in evolution). These molecular data are, of course, equally subject to convergence and other forms of confabulation, but what a bounty of novel evidence!

Buss and colleagues sequenced part of an important gene that codes for ribosomal RNA, and they found 108 "phylogenetically informative" positions—an enormous increase in the number of useful characters for classification. They developed a matrix of similarities for all pairs of comparisons among twelve species of hermit and king crabs and a thirteenth more distant relative, chosen to anchor the tree (and called an "outgroup" in our jargon; they used the brine shrimp *Artemia salina*). They then applied a variety of standard tree-building techniques to this matrix of relative similarities. They achieved the same basic result with either of the two most common methods for tree-building: distance

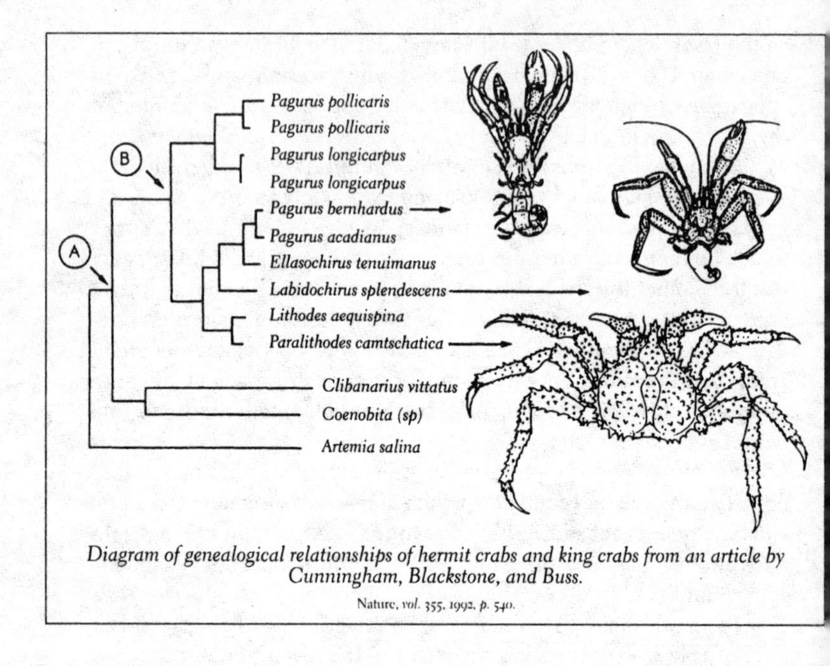

Diagram of genealogical relationships of hermit crabs and king crabs from an article by Cunningham, Blackstone, and Buss.

Nature, vol. 355, 1992, p. 540.

analysis, which works only with measured degrees of overall similarity; and parsimony, which constructs trees with a minimal number of evolutionary steps. The invariant result from different procedures greatly increases our confidence in their findings.

Their remarkable and enormously satisfying result appears in the accompanying diagram, reproduced directly from their article. *Artemia* is, as expected, separated from the twelve other species at the oldest division. The next dichotomy (A on their diagram) follows traditional classification by dividing hermit crabs with left-twisting abdomens (the two lower species) from their right-twisting cousins. The large upper clump of ten species represents ordinary, traditional hermit crabs of the family Paguridae. The next division (B on their diagram) makes a basically geographic separation within the Paguridae.

We now come to the truly remarkable point: Notice the two lower species in the upper clump—the king crab *(Paralithodes camtschatica)* and its close relative *Lithodes aequispina*. Now consider the species in the two major subclumps of this larger group formed at the separation of right and left-twisting lines at division A. The upper subclump represents species of the genus *Pagurus*, the standard hermit crab of any textbook or local seashore. But now look at the lower subclump—and note that it includes two further species of the genus *Pagurus* along with the two species from the king crab line. In other words, king crabs are so close to hermit crabs by the proper criterion of propinquity that they actually branch off *from within* a narrowly restricted genealogical grouping so conventional in form and behavior that all species have been included in the canonical genus *Pagurus!*

The form of the tree also allows us to make a reasonable inference for the time of splitting between the king crab and the conventional hermit crab lineages. Point A, the splitting of left and right-twisting hermit crabs, can be estimated at some 73–78 million years from independent evidence of fossils. The geographic division of regions within right-twisting hermit crabs (point B) occurred some 35–40 million years ago, again by independent geological and paleontological evidence. By extrapolation downward, king crabs split from the genus *Pagurus* some 13–25 million years ago—a good stretch of time, but not a great deal (geologically speaking) for the evolutionary work accomplished.

One tangential point before leaving this elegant study. Creationist critics often charge that evolution cannot be tested, and therefore cannot be viewed as a properly scientific subject at all (see the next essay for a fuller discussion of this important issue). This claim is rhetorical nonsense. How could one ask for a better test, based on a very risky prediction, than this? The counterintuitive link between king and hermit crabs was postulated on the basis of classical evidence from morphology (the arguments detailed previously in this essay as points 1–3). This prediction was then tested by the completely independent data set of DNA sequence comparisons—and confirmed in spades, with even closer propinquity than suspected between king crab and hermit crab lines.

I regard this story of king and hermit crabs as one of the most elegant I have learned of late in evolutionary biology—a lovely combination of a fascinating and counterintuitive tale; a multifaceted, rigorous and convincing pile of supporting data; and a lesson of intriguing generality (the difference between genealogical propinquity and any functional meaning of similarity—and the overriding importance of propinquity). But can I bring you along with me in the face of an obvious demurral that many readers will offer? Yes, I do grasp the story, but I'm not moved very much by crabs. They just don't intersect my life very often, so why should I care? Let me try to overcome this reticence by giving another example of precisely the same evolutionary phenomenon—one to which you simply cannot be indifferent.

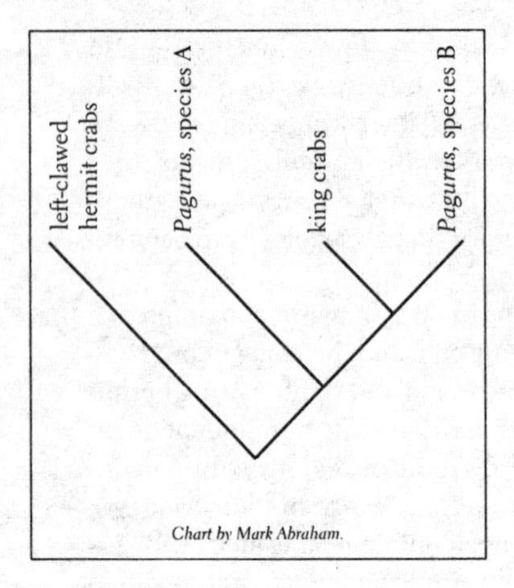

Chart by Mark Abraham.

The accompanying genealogical diagram epitomizes our surprising crab story. King crabs branch off from the hermit line within the space of the genus *Pagurus*, the most ordinary and conventional of all hermit crabs. Who would have thought that such a difference could be achieved in so little genealogical room? Who would have imagined that kings and hermits could be so close by the most important of all evolutionary criteria—propinquity, or genealogical distance?

I now reproduce the exact same diagram. I have made no changes whatever in the positions and orderings of branches. I have, however, substituted different names, for I now wish to depict our best knowledge of propinquity in the so-called "higher primates." The genealogical story of humans and our closest primate relatives corresponds exactly with the tale of propinquity between king crabs and hermit crabs!

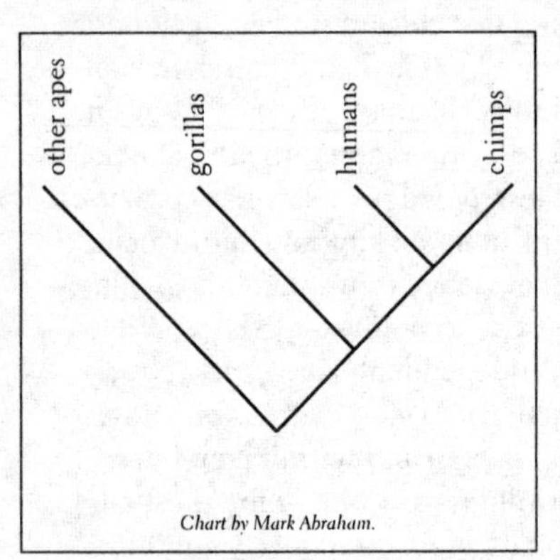

Chart by Mark Abraham.

Darwin correctly surmised, and little doubt has been entertained by scientists ever since, that chimps and gorillas are our closest relatives. But Darwin, and nearly everyone else until recently, assumed that chimps and gorillas formed the closest genealogical pair among the three species—only reasonable, after all, given the evident similarity between the two apes, and our obviously exalted separateness. (But remember that functional and genealogical similarity need not correspond!) Although evidence remains imperfect, and still subject to wide debate, most of the latest information suggests that we have been wrong—and that chimps and humans form the closest genealogical pair, with gorillas branching off a bit earlier.

Chimps and gorillas are conventionally classified in the family Pongidae, with humans in the separate family Hominidae. But if my diagram is correct, then humans arise *within* the space of the Pongidae, and cannot therefore represent a separate family, lest we commit the genealogical absurdity of uniting two more-distant forms (chimps and gorillas) in the same family and excluding a third creature (humans) more closely related to one of the two united species. I surely cannot claim to be more closely related to my uncle than to my brother, but we make exactly such a statement when we argue that chimps are closer to gorillas than to humans—see the third diagram of identical topology.

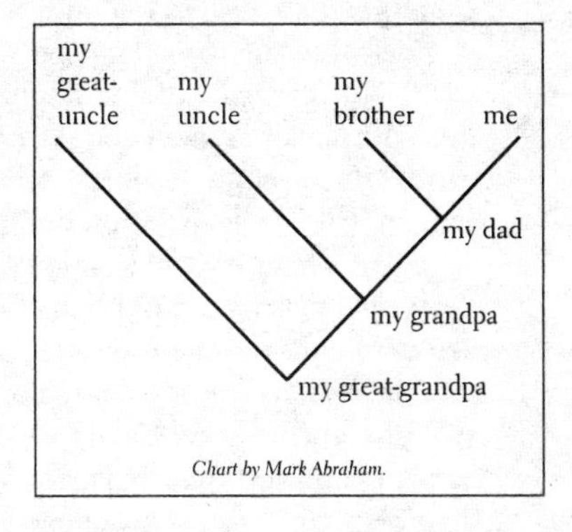

Chart by Mark Abraham.

Hermit crabs and king crabs tell precisely the same story. Our instincts inform us that the two groups should be separated in classification to recognize their profound differences in form and function. But king crabs arise genealogically within the domain of hermit crabs—indeed within the very restricted space of the canonical genus *Pagurus!* So how can we place king crabs in one evolutionary group, and hermit crabs in another? Similarly, how can we continue to rank humans in glorious separateness, while we unite chimps and gorillas in the family Pongidae?

We may legitimately inquire how such apparent difference could arise in such a restricted genealogical space. But we can be fooled by appearances, and the underlying disparity may not be so great (or perhaps the differences are truly profound, and evolutionary rates have been enormously accelerated in the king crab and human lines). Small underlying changes can yield large accumulated effects if introduced early in growth, with cascading consequences thereafter. Maybe carcinization is not so big a change after all—flatten and widen the carapace, reduce and tuck in the abdomen. Maybe all these changes can occur as consequences of a single coordinated transformation in growth. After all, *Birgus latro,* as discussed above, undergoes such a transformation during its own growth—living with a twisted abdomen in a snail shell during its youth, but carcinizing and becoming free-living as an adult. Moreover, the large size of king crabs, however impressive, need not represent a major evolutionary change. Abandon the need to find a shell for a house, and limits to size will be abrogated. Any carcinized free-living hermit crab may possess the capacity for greatly expanded bulk.

Similarly for humans: Are we really so different from chimps as we so confidently and arrogantly assert? In appearance, of course (reduced hair and erect posture have a strong visual impact). In brainpower, undoubtedly (chimps are as smart as could be, but they will never ponder the genealogical position of king crabs). But the underlying biological differences need not be so great. Strengthen and straighten the legs, enlarge the brain. The consequences have been enormous, and unprecedented in all the history of life; but I am not so sure that the topological and genetic transformations have been so profound. Consequences are effects, and effects are not equatable with generating forces and morphological results. Small changes can have cataclysmic effects.

People of goodwill and intelligence readily acknowledge our kinship with apes and monkeys. We know this affinity in our heads and can parrot the appropriate phrases. But we have never incorporated this vital knowledge deep within our guts, into visceral understanding—in large part because we have mistakenly assumed that functional and genealogical distance must be strongly correlated. If humans look so unlike chimpanzees, then we must really be very different, whatever our kinship. But if we ever grasp the primacy of genealogical distance as an evolutionary measure, and if we ever understand the potentially illusory nature of outward appearance as an indicator of fundamental difference, then we might reassess to our enormous (if humbling) advantage. Kings can be hermits, and humans can be closest brethren to chimps.

The motto of the functional view proclaims: *Der Mann ist was er isst.* (You are what you eat.) But an evolutionist must add the defining voice of history: You are what you have been and what your closest genealogical nexus shares. Think kinship. We will be a bit freer, a bit more enlightened, a bit readier to work for planetary preservation with the rest of kindred life, when we truly know why each and every last one of us is a monkey's uncle.

31

Magnolias from Moscow

I HAD TO GO ALL THE WAY to Moscow to complete my roster of fifty states—Moscow, Idaho, that is. I spent my entire childhood in the northeastern United States. I wandered more in my twenties, hit the outposts of Hawaii and Alaska in my thirties, and filled in nearly all the blanks in my forties—Montana as forty-eighth, Mississippi as forty-ninth. But I had always missed Idaho, often by just a few miles (during visits to Yellowstone National Park, for example). So an invitation to complete fifty at age fifty could not be refused, and I happily presented a talk at the University of Idaho in Moscow. While ruminating upon the symbolism of double fifty, I did fear that something terrible might happen once I crossed the state line after landing at the nearest airport in Spokane, Washington. Perhaps God would ask me to intone Simeon's prayer—"Oh, Lord, now lettest thou thy servant depart in peace"—and then strike me dead. But I guess he doesn't care much about artificial boundaries on his bounteous real estate (or maybe he just isn't spending every moment watching over me). Frankly, westernmost Idaho looks just like immediately adjacent easternmost Washington.

Moscow certainly sounds vestigially radical. They must have had some trouble during the McCarthyite hysteria of the 1950s. The local pizza joint is named, with good humor, Karl Marks's. (The town's label records only a distant Russian link. A local postmaster in the late nineteenth century had naming rights, and he so designated the town because local topography reminded him of the terrain back home in Pennsylvania—around a village also named Moscow, and presumably for a Russian connection.)

In fact, Moscow seems the very antithesis of dangerous and discordant foreign values. Could we possibly find anything more stereotypically apple-pie American than a town that advertises itself as the "dry pea and lentil capital" of America, and has a "skyline" dominated by grain elevators at the railroad tracks. Summer noncredit courses at the university include "Bake a Pie for the Fair," "Pickle Making for the Fair," "Growing Big Pumpkins," and, for the truly risqué and adventurous, "Wine Tasting in Spokane," the nearest city of any size.

From a loyal and lifelong Eastern urbanite like myself, a New Yorker no less, such observations might elicit the most contemptible form of parochialism—the silly name-calling that elevates one's own insecurity to supposed superiority, and precludes any understanding of different styles and terrains. I'll take a good bowl of lentil soup and a nice slice of pie at the county fair any day over much of the tedious, incomprehensible, and self-congratulatory debate that passes for profundity in urban intellectual centers. Moreover, we Eastern urbanites from traditional elite universities have the actual situation backwards when we imagine ourselves immersed in more stimulating discourse than our colleagues at isolated state colleges. In fact, few of us at Harvard talk much to our colleagues, especially those in fields even marginally distant from our own immediate work. We are all insanely busy with our own overcommitments, and we don't bother to set up any continuing network of intellectual discussion with local colleagues, if only because we sit upon a crossroads, and compatriots from all over the world eventually come to us if we wait long enough.

But the faculty of Moscow, Idaho, must make everything happen locally. Visitors are not so frequent, and intellectual stimulation has to be indigenous. I have seen the same phenomenon in many places off main routes. Lubbock, in western Texas, for example, is perhaps the most isolated population center in America (the real Texas is as big as the legend, and Lubbock is actually closer to Denver than to Houston). I greatly admire my colleagues at Texas Tech University for the beehive of intellectual activity—seminars, discussion groups, reading circles, local performances—that they have established entirely by and with their own effort. I now extend this praise all the way to Moscow.

As another advantage offered by relatively isolated places, visitors receive truly warm and welcome treatment. I once gave a prestigious lecture series at Yale; no one was mean to me (and the talks were well attended), but no one could figure out what to do with me, either. Everyone was busy, and I visit New Haven for other reasons frequently enough anyway. So I gave the lectures, went right down to the railroad station, and Amtrak'd my way back to Boston. In Moscow, I just knew that folks were really happy to see me, and nothing makes a man feel better.

As one sign of hospitality, my host Valerie Chamberlain and other geologists took me on a field trip to their most precious local site (where "local" means a couple of hours' driving)—the Miocene (17 to 22 million year old) lake beds exposed near the town of Clarkia. Western fieldwork conjures up images of struggle on horseback (or the minimal mount known as shank's mare)—toughing it out on one canteen a day as you labor up and down mountains. The value of a site is supposedly correlated with the difficulty of getting there. This, of course, is romantic drivel. Ease of access is no measure of importance. The famous La Brea tar pits are right in downtown Los Angeles. To reach the Clarkia lake beds, you turn off the main road at Buzzard's Roost Trophy Company and drive the remaining fifty yards right up to the site. Francis and Vickie Kienbaum, who own and run the company, gave their property a street address—85th and Plum, because, as they explained, it is about eighty-five miles from Spokane "and plum out in the middle of nowhere."

Francis Kienbaum discovered the lake beds in 1971 when he was bulldozing his land to make a skimobile racetrack (now named the Fossil Bowl). The family business of trophy making began as a cost-cutting measure for supplying symbols of victory to winners at the track—and then expanded to the usual roster of bowling clubs, Little League baseball, and so forth. The Kienbaums still purchase the simulated metal (actually plastic) tops—the homunculi who bowl, bat, and race—but they manufacture the wooden bases on their wonderful collection of restored antique machining and carpentry tools. (I am always fascinated to meet people involved in businesses that most of us never explicitly consider, though they occupy an occasional corner of our lives. After all, I represent a profession—paleontology—that most people absolutely never think about at all. I once had a remarkable airplane conversation with a woman who sells department-store mannequins—a larger and more serious business than you might imagine, complete with trade journals, juicy gossip, and a range of issues: high-priced lines for yuppie boutiques vs. lower lines for Kmart; colors and sizes geared to local differences.)

If the Kienbaums now make their living from the abundant wood in surrounding forests, their great discovery of 1971 extended our knowledge of local botany 20 million years back. Kienbaum, noting fossil leaves that peeled off the rock layers as black films and blew off in the wind, and recognizing the leaves as unrelated to any trees now living in the area, called the geology department in Moscow. Fortunately, Charles J. Smiley ("Jack" to his friends and colleagues), an expert paleobotanist, was on hand and delighted to take up the work. Jack Smiley has focused his research on the Clarkia site ever since, and has enlisted an impressive and international array of experts to collaborate on the bounty.

Fossil leaves from ancient lake beds are not rare (the site also contains less abundant insects and fish, and a large array of microscopic forms). The fame and uniqueness of Clarkia rest upon the extraordinary preservation of the fossils. The shales are waterlogged and very finely bedded. Mr. Kienbaum, who greeted us on our visit, broke out large blocks with his bulldozer. Jack Smiley handed me several chunks and told me to split the layers and look for leaves. I expected to grab a large chisel and geological hammer, and proceed to bash away in the usual manner. Instead, I was handed an ordinary kitchen knife and told to slice parallel to the bedding as if the block were a giant seven-layer cake. I laughed with incredulity, for I don't believe in using ridiculously flimsy tools for hefty work (a legacy of a teenaged month in France spent at a "work camp," euphemistically so called, trying to demolish some hypersturdy barracks, built to last a thousand years, with mallets).

Well, the slab looked like a big rock, but it sliced liked butterscotch pudding. The rock split cleanly at bedding planes, revealing abundant leaves at each fracture. I now saw first-hand what has been amazing my colleagues for twenty years: the leaves look as fresh as the day they fell into the lake. They retain their original color, usually green, but often an autumnal red or brown. After just a few minutes of exposure and drying out, they oxidize before your eyes to a black film. This quick change in modern air teaches us that the leaves have never been exposed to oxygen since their burial. They must have fallen into a stagnant lake, been rapidly covered in bottom sediments devoid of oxygen, and then remained buried (and waterlogged) for all 20 million years since.

The Clarkia lake formed when a lava flow blocked a stream valley, forming a narrow body of water some twenty miles long and surrounded by forests. The lake gradually filled during its one thousand years or so of existence, leaving a sedimentary sequence some ten feet thick. The composition of the flora clearly indicates much warmer climates then than now, for many of the trees are closest relatives of forms now living in southern Appalachian swamps and uplands (bald cypress, tupelo, and magnolia, for example). Interestingly, several of the Clarkia genera now live only in eastern Asia, indicating a previously wider range for these warm temperate forests, followed by later constriction, as the earth cooled (and eventually went into ice-age cycles), with some genera surviving only to the west in Asia, and others remaining in eastern areas of North America. Of Asiatic components in the Clarkia beds, the genus *Metasequoia*, the dawn redwood, has excited most interest—for this form was first discovered as a fossil, and only later found living in a few remote valleys of central China.

Preservation with this degree of fidelity requires an unusual set of circumstances. (I can hardly describe the thrill and eerie feeling of splitting a 20 million year old rock and finding a leaf clothed in unaltered autumnal colors!) In this case, three happy circumstances conspired to preserve fossil leaves that look as fresh as the day they fell off their trees. First, many of the leaves blew directly into the lake; they were not altered and mashed up in long-distance transport. Second, the stagnant lake bottom contained no oxygen, hence no organisms to promote decay of the leaves. They were quickly buried and sealed in fine-grained sediments. Third, the rocks have remained unoxidized and water-saturated ever since. When we split the rocks today, we expose the leaves to oxidation for the first time—and just a minute or two changes a glistening autumnal red into a dry black film, thus accomplishing in a moment what 20 million years didn't even begin.

Of course, Smiley and his colleagues noted the gorgeous and extraordinary preservation right away. They published several articles on details of cellular structure (including such organelles as nuclei and chloroplasts) retained by these ancient leaves, fruits, seeds, and stems. Other scientists then began to realize that if original form could be preserved to such fine detail, perhaps the leaves might preserve some aspects of original chemistry as well. Karl Niklas, my brilliant paleobotanical colleague from Cornell University, then published several papers (with a number of co-workers) during the late 1970s and mid-1980s, reporting many remarkable details of "fossil chemistry," as unchanged by time and burial as the fine features of form.

These studies are the preludes to the latest chapter of Clarkia's fame, the first discovery that brought these wonderful fossils into an international and public spotlight (we professional paleontologists have delighted in these ancient plants from the first)—the extraction and sequencing of DNA from chloroplasts of leaves in two genera, *Magnolia* and *Taxodium* (the bald cypress). (I have read all the major press reports of this discovery. They are generally accurate in describing the results, but often misleading in not reporting the antecedents. One might think that the leaves were discovered yesterday, popped into a fancy machine in a modern molecular biology lab, and turned into a string of DNA base pairs. Science thrives on continuity; science requires endless patience; scientists, following Edison's famous quip, perspire much more than they inspire. We never would have gotten to the DNA without Francis Kienbaum's bulldozer, Jack Smiley's unparalleled expertise in traditional systematics, and Karl Niklas's skills in chemical analysis. Moreover, the DNA sequence would mean very little without these equally important supporting data from so many other disciplines, most not so flashy and trendy as molecular biology.)

We have extracted DNA from ancient creatures before, but nothing that would inspire more than a "harumph" from a real paleontologist. Egyptian mummies, quaggas (zebra relatives that became extinct about a century ago)—even frozen mammoths—have yielded DNA. Before Clarkia, the record belonged to a thirteen thousand year old sloth. By moving from thirteen thousand to more than 13 million, we have extended the range of preservation for DNA more than a thousandfold. Such an enormous gain must inspire doubt amid wonderment. Many biologists had previously regarded DNA as incapable of survival without degradation for more than a few million years at most; consequently, some colleagues originally doubted the Clarkia results (though all, I think, are now satisfied). Only the extraordinary preservation of the Clarkia leaves—the main and repeated theme of this essay—has allowed this impossible dream to come true.

The first report, from Mike Clegg's lab in the University of California at Riverside, was published in the April 12, 1990, issue of *Nature* ("Chloroplast DNA sequence from a Miocene *Magnolia* species," by E. M. Golenberg and six other authors). Golenberg and colleagues sequenced an 820 base pair fragment of DNA from a chloroplast gene called *rbc*L in the Clarkia species *Magnolia latahensis.* (Most DNA resides on chromosomes in the nucleus. But the mitochondria, or "energy factories," and chloroplasts, or photosynthetic organelles, also contain small DNA programs.)

This discovery could not have been made ten years ago. For our current ability to extract and sequence DNA rapidly, we owe thanks to a revolutionary technique called PCR (polymerase chain reaction), which can isolate and amplify tiny amounts of DNA. Even with PCR, we might not have found any Clarkia DNA except for two happy features of chloroplasts. (No nuclear or mitochondrial DNA has yet been extracted from these leaves.) First, chloroplasts are present in many copies per cell. Second, for reasons not well understood, chloroplasts are preserved far better than any other cell organelle at Clarkia. (In a 1985 paper, K. J. Niklas, R. M. Brown Jr., and R. Santos randomly sampled 2,300 cells from leaves in the Clarkia beds. They found that 90.1 percent contained chloroplasts, 26.0 percent mitochondria, and only 4.3 percent nuclei.) Thus, if any DNA has been preserved at Clarkia, chloroplast genes provide our best prospects.

In comparing this 820 base pair sequence with the same region in a closely related living species, *Magnolia macrophylla*, Golenberg and colleagues found differences at only seventeen positions. The nature of these differences confirms a central fact about the evolution of DNA (not surprising, but always gratifying to have direct evidence from ancient DNA). Each position of a DNA string may contain any of four base pairs (called A,G,C, and T for adenine, guanine, cytosine, and thymine). Each sequence of three base pairs codes for an amino acid (and a chain of amino acids makes a protein, the building blocks of organic matter). The DNA code is "redundant" in the third position—that is, a change in base pair in the last position of a triplet code does not alter the amino acid built from the triplet, whereas most changes in the first or second position do lead to a different amino acid. Base-pair changes that do not alter the amino acid are called "silent substitutions" because they don't change the chemical structure of the organism—and natural selection, working on organisms, will not notice such silent alterations of DNA. Since natural selection regulates rates of evolution, and since the stability of well-adapted design is so vastly more common than change (see essay 10), selection works to maintain existing arrangements almost all the time. Therefore, silent substitutions (that natural selection can't detect) should be far more common than first and second position changes (called nonsynomymous) that yield a different amino acid, thereby altering the resulting protein. The *Magnolia* data provide a striking confirmation for this expectation. Of the seventeen substitutions between the fossil Clarkia magnolia DNA and the homologous sequence in a living species, thirteen are silent and only four change the resulting amino acid.

The second study, published in 1992, offers more data and even more direct insight into evolution ("An *rbc*L sequence from a Miocene *Taxodium* (bald cypress)," by P. S. Soltis, D. E. Soltis, and C. J. Smiley). Soltis and her colleagues resolved a sequence of 1,320 base pairs from the *rbc*L gene in chloroplasts of Clarkia bald cypress. (The entire gene contains 1,431 base pairs, so the fossil sequence is most of a totality, not a mere fragment.) They found only eleven base-pair changes between the fossil and the homologous sequence in modern *Taxodium distichum*—and all substitutions were silent third-position changes. Thus, the amino-acid sequence for *rbc*L in bald cypress has been unchanged for 20 million years! Moreover, they compared this 1,320 base-pair sequence with the same gene in three other genera with increasing taxonomic and morphological distance from bald cypress, and they found a perfect correlation between total amount of DNA change and inferred evolutionary separation. For example, the dawn redwood *Metasequoia*, next in degree of resemblance, differs from fossil *Taxodium* by thirty-eight base-pair substitutions, twenty-nine silent and eight nonsynonymous.

The difference in amount of change—less than a percent of base pairs in *Taxodium*, more than 2 percent in *Magnolia*—is intriguing. Perhaps *Magnolia* simply evolves faster, but the two cases are not really comparable, and *Taxodium* may be more revealing. In *Magnolia*, the fossil and modern species differ, and we are not certain that the modern species is the direct descendant, removed by only one event of speciation from *Magnolia latahensis* at Clarkia. Perhaps several episodes of branching intervened; perhaps the two species are not parent and direct daughter on the evolutionary tree of *Magnolia*. Thus, the 2-percent difference records an uncertain evolutionary distance. But modern *Taxodium* may belong to the same species as the Clarkia leaves. In other words, for *Taxodium* we may be looking at an unbroken and unbranched evolutionary lineage—a true continuity over 20 million years—and the smaller percentage of changes, with no alterations at all in amino acids, may record the actual architecture of evolutionary stability.

Extraction of fossil DNA will not make the traditional paleontology of overt fossils obsolete. First of all, to make the obvious (but not adequately appreciated) philosophical point, DNA code and organism represent disparate biological objects. The gene is not more "basic" than the organism, or closer to the "essence of life," whatever that means. Organisms have DNA codes, and they maintain external forms and behaviors. Both are equal and fundamental components of being. DNA does not even build an organism directly, but must work through complex internal environments of embryological development, and external environments of surrounding conditions. We will not know the core and essence of humanity when we complete the human genome project.

Second, to mention the practical point, extraction of DNA from Clarkia leaves is a precious rarity, not the harbinger of a general revolution in practice. DNA quickly degrades in almost all geological formations that preserve fossils. Only the gorgeous and exquisite preservation of the Clarkia leaves permits the preservation of a molecule usually incapable of such long survival. Even at Clarkia, successful extraction of DNA is exceptional and difficult. Scientific papers traditionally do not report failures (see essay 10); you have to ask around to get the full picture. I made the requisite inquiries, and learned that extraction of DNA is no picnic or panacea. The great majority of fossil leaf scrapings yield nothing, and the procedure is therefore long, expensive, and frustrating. Moreover, only chloroplast DNA has been recovered so far at Clarkia, and chloroplasts preserve better than any other organelle at this locality. Unfortunately, nuclei are least stable, and most of the DNA resides therein. We are not about to sequence the entire genetic program of a fossil magnolia.

Is the tale of Clarkia DNA nothing but an isolated story in natural history, an episode with neither extension nor generality? Not at all, for this discovery of ancient DNA raises several deep issues in evolutionary theory, at the most basic level that professionals rarely consider in their day-to-day work, but must form the heart of a series of essays for general readers. Above all, the sequencing of fossil DNA provides a striking illustration of the best evidence we can produce for the factuality of evolution itself.

Our creationist detractors charge that evolution is an unproved and unprovable charade—a secular religion masquerading as science. They claim, above all, that evolution generates no predictions, never exposes itself to test, and therefore stands as dogma rather than disprovable science. This claim is nonsense. We make and test risky predictions all the time; our success is not dogma, but a highly probable indication of evolution's basic truth. As in any historical science, most predictions refer to an unknown past (technically called "postdictions" in the jargon). For example, every time I collect fossils in Paleozoic rocks (550 to 225 million years old), I predict that I will not find fossil mammals—for mammals evolved in the subsequent Triassic period (while young-earth creationists, claiming that God made life in six days of twenty-four hours, should expect to encounter mammals in all strata). If I find fossil mammals, particularly such late-evolving creatures as cows, cats, elephants, and humans, in Paleozoic strata, our evolutionary goose is cooked.

As a paramount prediction from the different subdiscipline of taxonomy, we hold that well-established evolutionary schemes of classification and phylogeny should be affirmed in broad outline by any new and independent criterion of evidence. We have a reasonable classification of forest trees, based on tried and true Linnaean criteria of external morphology in complex shapes and structures of leaves and flowers. With respect to this established taxonomy, biochemical data of DNA sequences provide an entirely separate domain of new evidence.

We predict that an evolutionary tree built from DNA resemblances should be congruent with the traditional classification based on morphology. Creationists, on the other hand, should expect no such similarity, for God may do as He pleases. Why should entirely different criteria record the same supposedly nonexistent pattern of evolutionary branching? (Of course, one might say that God chose to create with such congruence, but then God becomes meaninglessly omnipotent—that is, so flexible, so available for invocation to achieve any conceivable result that He becomes truly impotent for failure to illustrate any distinctive act that might test the nature of His unique ways and power.) I am thrilled that we can extract 20 million year old DNA and learn that the sequences fit so well with evolutionary relationships determined by the anatomy of leaves and flowers. What better evidence can we offer, what prediction more triumphantly affirmed, than a fossil record of DNA changes fully congruent with the independent evidence of external anatomy?

The Clarkia work also helps us to understand the proper relationship of reinforcement and mutual respect among subdisciplines of evolutionary science. Molecular biology is trendy and expensive; fossil collecting, and curation in museums, strike many people (including, unfortunately, the press) as stuffy and antediluvian. But DNA is no Holy Grail, and the specific sequence of base pairs in a chloroplast gene of a magnolia is no more fundamental to the species than the shape and cellular architecture of its leaf.

The saddest aspect about second-class citizenship lies in self-hate and deprecation among the afflicted. Despised groups often tell jokes about themselves. Rural people often cede to urban opinion and do not assert their own legitimate concerns. This culture of poverty tends to afflict my own colleagues who work with whole organisms in museum collections. We too often bow before the icons of expensive molecular research, and start to agree that the big boys really should get all the new buildings and funds; we almost apologize for our existence, ask for nothing, and seem content so long as existing powers don't fold our jobs to fund the expansion of a parking lot, or a new administrative building.

We need to be more assertive. Our work is just as modern, and just as important, as anything in evolutionary science. We may use molecular data on differences between humans and chimps to infer that we last shared a common ancestor 6 to 8 million years ago; but no evidence from modern genes will reveal the appearance of that ancestral creature. We will need fossils (which we don't yet have) for this crucial information. We can extract DNA from Clarkia leaves, but the gorgeous red autumnal colors of many leaves record the same phenomenon at a different scale—the extraordinary geological circumstances that preserved both color and chemistry. Praise Mike Clegg with his PCR and his banks of lab equipment. And praise Jack Smiley with his kitchen knife at the outcrop.

Two threads of this essay therefore unite at the end. I heard too many jokes about backwardness and insignificance in Idaho. I encounter too much resignation and self-hate among my colleagues in museums. Enough apology. Museums are our intellectual (and visceral) glory (see the essays in Part Five)—and Idaho was fiftieth on my list because any savvy person always saves the best for the last.

PART EIGHT

LINNAEUS AND DARWIN'S GRANDFATHER

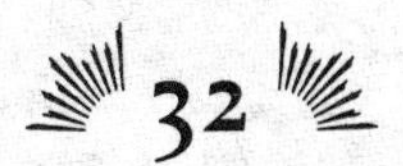

The First Unmasking of Nature

I CAN'T IMAGINE two people in more intimate contact—just the tiniest sliver of a millimeter's separation—than the figures on the front and back of a banknote. We all know, of course, that the fall of any flower's petal reverberates through the universe to disturb the most distant galaxy; therefore, these closer juxtapositions cannot be without deep meaning.

During a recent visit to Sweden, I was delighted to encounter the common European practice of placing scientists, rather than an uninterrupted parade of statesmen, on banknotes—for Linnaeus graces two Swedish bills, just as Italy features Galileo, and the old British pound note had Newton backing up the Queen. (America does so honor both Jefferson and Franklin, but not, I think, primarily for paleontology and electricity.) I was, however, puzzled by a juxtaposition on the fifty-kronor note: Why (as shown in the accompanying illustration) do Linnaeus and King Gustav III stand in such minimal proximity (they even look directly at each other if you hold the bill up to the light and scan both figures at once).

On one immediate level, the union poses no problem: both men were eminent and contemporary Swedes; Linnaeus lived from 1707 to 1778, and Gustav III, born in 1746, reigned from 1771 to 1792. Their personal ties were not close, but they certainly liked and appreciated each other. The older scientist flourished in the atmosphere of the Swedish

1989

9031078835

Enlightenment, so strongly promoted by this artistically inclined king (who collaborated on an opera while not overly engaged in affairs of state); while the young king basked in the prestige, and endless string of honors, won by Sweden's most famous naturalist and scholar. In 1774, after Linnaeus had suffered a stroke and lost his legendary zeal for work, Gustav sent him a collection of plants from Surinam packed "in hogsheads of spirit." Linnaeus, according to legend, immediately left his bed and went back to work, describing the two hundred species sent by His Majesty. When Linnaeus died, four years later, Gustav eulogized him before the Swedish legislature: "I have lost a man who did honor to his fatherland as a worthy citizen, being celebrated all over the world."

Yet the principle of petals and galaxies tells us that a deeper connection, something lovely and downright hermeneutical, must also link the two men—and I have found it!

We must begin by asking, Where do the majority of decently educated folks (like me and thee) encounter Gustav III, when our knowledge of things Scandinavian tends to be largely (and lamentably) blank between Thor and Ingrid Bergman? The answer, of course, is Giuseppi Verdi—for Gustav III is the subject of *A Masked Ball* (*Un Ballo in Maschera*). On March 16, 1792, King Gustav III was shot and mortally wounded while attending a midnight masquerade at the Stockholm Opera House. His assailant, Jakob Johan Anckarström, representing an aristocratic conspiracy opposed to Gustav's reforms, was arrested, tried, convicted, flogged, and then beheaded after the offending hand that held the pistol had been hacked off.

This wonderful opera is now usually performed in its proper Swedish time and setting. But not at the debut of 1859. An attempt had just been made on the life of Napoleon III, and the ever-watchful censors of Naples decreed that Verdi must change the locale and demote the king to some lesser station, lest any potential assassin be wrongfully inspired by a night at the opera. Verdi was no stranger to such official intrusions. Seven years earlier, in Venice, the censors had issued a similar edict for his opera about a jester named Triboulet in the court of Francis I of France. (We now know the opera for the renamed jester Rigoletto, while the French king was demoted to the Italian duke who sings "La Donna è Mobile".)

This time Verdi did an even better job of masking, for he switched the locale to, of all places, my present home of Boston—where his opera became our anachronistic regional response to Puccini's later *Girl of the Golden West* (*La Fanciulla del West*). Gustav III was transmogrified, teleported, and demoted to the mythical Riccardo, "Governor of Boston" at some unspecified colonial time. (Boston had no governor, though Massachusetts did; while the thought of a glittering masked ball at the opera house of this puritanical city continues to provoke endless amusement.) The conspirators were recast as a pair of miscreants named Sam and Tom (Samuele and Tommaso), traditionally played, we must note with sadness, either as Indians or blacks.

We therefore usually meet Gustav III in double masquerade—first, literally at the ball (where Anckarström needs half an act to find him); second, by Verdi, in transferring him a hemisphere away and demoting him to governor. And now I know the deep connection to Linnaeus—for nature was doubly masked when Linnaeus met her, and though he only succeeded in removing the first disguise (we needed Darwin for the second), uncovering an entire world must rank as a greater challenge than discovering a king in a crowd of revelers.

Linnaeus is certainly acknowledged and honored among biologists. In the broadest realm, he developed the system of binomial nomenclature still used (with no substantial change since his formulation) to designate and classify all organisms. In the most parochial region, he gave us our own name, *Homo sapiens*. Yet I believe that we systematically underestimate Linnaeus by measuring him against a false view of how scientific knowledge grows. We see him as a great organizer of information, but, in an important sense, as a codifier of error—for he believed that he had classified God's created order, not (as we now know) the products of a genealogical system built by evolutionary change. Some commentators have even viewed his role as retrogressive, for Linnaeus's creationist convictions canceled an older folk tradition of mutability, sometimes mistaken as a natural antecedent to evolution, if only the Linnaeans had not intervened. (This older folk tradition spoke more of occasional monstrosities and weird hybrids between distantly related species than of ordered systems changing by natural laws. Fables about mutating beasts, and travelers' tales about fabulous creatures in distant lands, do not qualify as prototypical evolutionary theories.)

In this oversimplified view of scientific progress, we advance along a pathway of accumulating knowledge, guided by a timeless method of accurate observation and relentless logic. The classic expression of this view can be found in the preface to a book that must be on everyone's short list of contenders for the greatest work of popular science ever written: T. H. Huxley's *The Crayfish*, his marvelous monograph on how to teach the most abstruse principles of science by developing the details of a single example, in this case the anatomy and physiology of a common animal.

Huxley begins by telling us that "science is simply common sense at its best; that is, rigidly accurate in observation, and merciless to fallacy in logic." He then argues that the study of organisms has progressed through the same three stages followed by all sciences in their development: an initial phase of gathering information without theoretical guidance (Huxley calls this first step *Natural History*, defined as "accurate, but necessarily incomplete and unmethodized knowledge"); a second stage of systemizing and organization, though still without guiding theory (called *Natural Philosophy*); and, finally, the third rung and synthetic climax of *Physical Science*, "this final stage of knowledge, [where] the phenomena of nature are regarded as one continuous series of causes and effects."

In this system of three steps from unorganized description to causal interpretation, Linnaeus occupies the middle rung. We are better off in Linnaeus's world than in the first stage, because our previously unmethodized knowledge has been ordered into a coherent scheme, but we are not at the apogee of stage three because we have no decent theory for the causes of order. Huxley, in fact, argued that this third stage had only begun in his own century, following the death of Linnaeus and his creationist approach: "The conscious attempt to construct a complete science of Biology hardly dates further back than . . . the beginning of this century, while it has received its strongest impulse, in our own day, from Darwin."

I would agree with most modern historians of science in branding this view as misleading, and as unfair to our predecessors. I do not deny that science progresses in the crucial sense of achieving more accurate and comprehensive explanations of empirical reality, but two aspects of the older positivist view (so well exemplified by Huxley) lack validity and impede understanding: the notion of a timeless scientific method based on rigorously objective observation and logic, and the idea that earlier systems were either theory-free or theory-poor because explanation can only follow accurate description.

Theory-free science makes about as much sense as value-free politics. Both terms are oxymoronic. All thinking about the natural world must be informed by theory, whether or not we articulate our preferred structure of explanation to ourselves. The old fabulists of Huxley's first phase had a theory—if only the folk idea that starlight or serious fright could influence the form of a fetus in the womb. The taxonomists of Huxley's second phase had a theory—that God had created a timeless order for human ingenuity to find. These theories may have been wrong, but they were as pervasive (and restrictive) in the structuring of knowledge as any more accurate and later system. Linnaeus's *Homo sapiens* is a thinking machine (or a thinking reed, if you prefer the botanical metaphor); we cannot collect information without a theory to organize our searches and observations.

Moreover, theory is always, and must be, colored by social and psychological biases of surrounding culture; we have no access to utterly objective observation or universally unambiguous logic. With this perspective, we may return to the subject of unmasking and the comparison of Linnaeus and Gustav III—for if scientific progress depends more upon replacing theories than adding observations (and waiting until they coalesce into a proper explanation), and if all theories are bolstered by cultural biases, then any process of replacement requires an unmasking of previous structures (protective clothing, whatever their virtues).

We must remove two disguises to reach Gustav III. We may also epitomize the history of our knowledge about organic order as a double unmasking. Linnaeus is the proper focus and symbol of the first great unmasking, Darwin of the second. This perspective challenges our antiquated and disrespectful view of Linnaeus as an old worthy who ultimately failed despite an admirable passion for order. We may then become free to see the great Swede as a scientist with a brilliant and coherent system that fruitfully replaced a severely constraining theory.

Darwin tore away nature's second mask by establishing the explanatory basis for natural order in a theory of evolutionary transformation. But you cannot have a proper theory of order before you know the order that must be explained—and Linnaeus established this foundation with his method and practice of taxonomy. If Linnaeus had merely gathered and codified all the disorganized information long accumulating in Huxley's first stage, then we might say, "So what? Someone had to do it eventually. Linnaeus lived at the right time, and was lucky enough to possess the right combination of zeal and tidiness." But Linnaeus didn't just codify; he unmasked. His system didn't just gather together; it replaced a different principle of organization that had hidden nature's order from our sight. We cannot conceive of Darwin without Linnaeus, of evolution without prior knowledge of taxonomic structure.

The first mask covered nature with our parochial penchant for viewing the universe as constructed either for us or in our terms. Artificial orderings by some system of human judgment, or some principle of our mental or linguistic usages, can only mask nature's truly genealogical arrangement. Consider the monographs of the most famous systematizers at the inception of modern natural history—the great sixteenth-century scholars Ulysse Aldrovandi in Bologna, and Conrad Gesner in Zurich. Aldrovandi used an eclectic system based on multiple, and sometimes contradictory, criteria organized only by some notion of importance to humans (or just noticeability by humans). He began his *History of Quadrupeds* with the horse, *quod praecipuam nobis utilitatem praebeat* (which exhibits particular use to us). His volume on birds mixes a range of criteria in sequencing by human interest—from the noble (eagles and hawks considered first), to the wise (owls), the similar (bats, falsely placed), the big (ostriches), the awesome (gryphons), and on to parrots, crows, and, finally, all the small things that go tweet.

Gesner makes even less appeal to any notion of natural order by simply proceeding through the Latin alphabet in his 1551 volume on mammals—from *De Alce* (on elks) to *De Vulpe* (on foxes). Gesner does follow the equally anthropocentric, but at least not quite so artificial, principle of the "chain of being" in sequencing his subsequent volumes—two on birds, three on coldblooded terrestrial vertebrates. But volume 4 then amalgamates all water-dwelling creatures, with a primary emphasis on fishes (next step down the vertebrate ladder), though also including mermaids, medusae, and octopuses.

I do not say that no one questioned these anthropocentric systems before Linnaeus, or that his work marks a sudden break from this older tradition. Rather, Linnaeus represents the culmination and codification of more than a century's work by natural historians throughout Europe, laboring on collections gathered throughout the world. But his legendary zeal and energy for work, his formidable memory and capacity for synthesis, led to a series of books—looking more like the products of an industry than the output of one man—that established both the practice and structure of modern taxonomy.

Linnaeus produced his unmasking of nature at two levels—first, by designating species as basic units and establishing principles for their uniform definition and naming; and, second, by arranging species into a wider taxonomic system based on a search for natural order rather than human preference or convenience. Linnaeus's binomial method has been used, ever since his *Systema Naturae* (first edition published in 1735, definitive edition for animal taxonomy in 1758), as the official basis for naming organisms. Linnaeus gave each species a two-word (or binomial) name, the first (with a capital letter) representing its genus (and potentially shared with other closely related species), and the second (called the trivial name and beginning with a lowercase letter) as the unique and distinctive marker of a species. (Dogs and wolves both reside in the genus *Canis*, but each must have a separate trivial name to designate the species—*Canis familiaris* and *Canis lupus* respectively, in this case.)

Linnaeus did not invent this system from whole cloth while seated in his armchair. He evolved his nomenclature from the usual convention of representing species by a string of Latin words epitomizing their distinctive characters. In this system, the first word (since it began the phrase) was capitalized, while the others remained lowercase. Linnaeus first experimented with regularizing the form and number of the word string (one of his earlier systems allowed a maximum of twelve words per species).

He then decided that a summary of the summary, restricted to two words, might work best as a device for tabulation and standardization. At first he regretted giving up the key idea that a word string should accurately describe the species's key features—for two words are not enough, and may not even turn out to be appropriate. (Many thinkers have noted that Linnaeus may have erred spectacularly in his most famous decision, to name us *Homo sapiens* in honor of our supposed wisdom.) But Linnaeus later realized that he had accomplished something enormously useful and clever without realizing why at first. The Linnaean species name is not a description but a place-holder—a legal device to keep track and to confer a distinctive name upon each natural entity. Any comprehensive system based on millions of unique items must use such a mechanism, and Linnaeus finally understood that he had brought a necessary and fundamental principle of naming through the back door of a search for modes of epitomized description.

But Linnaeus's definition of species—not his mechanism for naming them—codified the change that unmasked nature. For Linnaeus's definition fractured the conceit of a human-centered system with basic units defined in terms of our needs and uses. Linnaeus proclaimed that species are the natural entities that God placed on earth at the creation. They are His, not ours—and they exist as they are, independent of our whims. (We may have trouble recognizing and defining species, but our difficulties do not alter God's actions.) In a famous maxim (number 157) of the *Fundamenta botanica* (1736), Linnaeus proclaimed: *Species tot sunt diversae quot diversas formas ab initio creavit infinitum Ens* (there are as many species as the Infinite Being created diverse forms in the beginning).

Since creationist-bashing is a noble and necessary pursuit these days, readers may wonder why I am praising such an invocation of God's power to create immutable entities all at once—especially since

Linnaeus substituted this idea for earlier notions of looser definition and mutability. But, as I argued above, the history of science progresses in such a manner—from theory to theory along a complex surface with a slant toward greater empirical adequacy, not along a straight and narrow path, pushed by a gathering snowball of factual accumulation. The conceptual change was surely enormous, but Darwin's invocation of natural selection in steps as a replacement for God all at once did not require any major overhaul in practice. Species are real whether created by God or constructed by natural selection—and Darwin's conceptual shift, the second unmasking, required little revision in Linnaean methods.

(Linnaeus later moved away from this rigid insistence that God had created all species at the beginning of time. In his early work he proclaimed over and over again, *nullae species novae* [no new species]. In later books, Linnaeus argued that new species could form by hybridization between pairs from the original creation. He even toyed with the idea that God may have created only a common source for each genus, or even for each order, and then allowed subsequent species to form as hybrids. Some commentators, unfairly imagining that we may only honor those we deem right by today's standards, have grasped at these later views to recast Linnaeus as at least a closet evolutionist. But such a strategy must be rejected on two grounds: First, Linnaeus clearly remained a creationist. What can be more different than a fully genealogical system, based upon constant change and common ancestry for all life, and a claim that God made fewer than all his forms at first, and then let the others slide into slots of a preordained system by combination? Second, Linnaeus was a great scholar in his own terms, a man who pulled the first mask from nature as surely as Darwin snatched the second. We don't need to clothe him at Armani's in order to respect his accomplishments.)

At the second level of organizing species into a larger taxonomic system, Linnaeus also fractured previous schemes of anthropocentric arrangement by insisting that relationships among species be ties of natural order, not human convenience. God had created according to a rational scheme; species are the items that build God's framework. Taxonomists have been given the sublime job of discovering God's plan in the interrelationships among his species. (The ties are ideological in Linnaeus's system, not genealogical as in Darwin's—but chains of implication among ideas are no less firm than links of matter in physical continuity.) This key point was beautifully expressed by a surprising source: Dag Hammarskjöld in an address on the two hundred fiftieth anniversary of Linnaeus's birth, for no Swede, whatever his profession, could neglect such a national hero: "Here, man is no longer the center of the world, only a witness, but a witness who is also a partner in the silent life of nature, bound by secret affinities to the trees."

The history of science is studded with monumental egos, but none holds a candle to Linnaeus. I doubt that anyone before Muhammad Ali surpassed Linnaeus's own third-person assessment from one of his several autobiographical documents:

> God has suffered him to peep into his secret cabinet.
>
> God has permitted him to see more of His created work than any mortal before him.
>
> God has bestowed upon him the greatest insight into nature-study, greater than anyone has gained . . .
>
> None before him has so totally reformed a whole science and made a new epoch.
>
> None before him has arranged all the products of nature with such lucidity.

Arrogant, yes. But note that the cabinet is God's, and the products thereof. If someone with an ego so unbounded still stated that he had only discovered God's order, not constructed his own from the transcendent brilliance of his unique psyche, then I must suppose he truly embraced the idea of natural order independent of the human mind (even at its Linnaean apogee). I would be suspicious of such protestations from a more modest man.

In practice, Linnaeus classified his plants by the form, number, and arrangement of their organs of fructification—the so-called "sexual system" (see the next essay for a full exposition). Basically, he divided plants into classes by the number and position of stamens, and he then divided classes into orders by the number of pistils. Ironically, he knew that such a system must be artificial—an almost numerological imposition of human logic upon nature's greater complexity. All his life, Linnaeus searched for a *methodus naturalis*, or "natural method," to capture God's objective arrangement in his hierarchy of names—and he never succeeded. A solution to this problem awaited the second unmasker, for no rational order of divine intelligence unites species. The natural ties are genealogical connections along contingent pathways of history—and such ties, although recoverable once you know how the system works, do not fall into patterns of beautiful symmetry or complex geometry that creationists like Linnaeus anticipated, and therefore could never find. The masks discussed in this essay are not camouflage placed by nature over her products, but impediments that we construct with false theories.

Since this essay began with a curious conjunction, I must end with one of the most striking of all meaningful coincidences. Where did the second unmasking begin? Where did the world first hear the sounds of Darwin's revolution? In London, yes, for Darwin became a homebody after his *Beagle* voyage, and never crossed the English Channel. But where in London?

When Linnaeus's son and successor died in 1783, his mother and sisters decided to sell the great collection—specimens, cabinets, books, letters, and manuscripts—to the highest bidder. The remains of Linnaeus's intellectual life were purchased by James Edward Smith, a young English naturalist, and son of a rich manufacturer in Norwich, for the incredible bargain price (even by eighteenth-century standards) of just over a thousand pounds.

We must now consider one last Gustavian connection. The king was in Italy and France during hurried negotiations that led to the sale. Many historians speculate that, had he been in Stockholm and known of plans to deport such a national treasure, he would probably have intervened (especially since a Swedish buyer had been found, but only after the sale had been legally concluded). The most famous nineteenth-century Swedish biography of Linnaeus states, "Had he [Gustav] been informed in proper time it is certain that he would have strongly exerted himself and rescued these precious collections for the fatherland, especially when one considers his care for Sweden's honor, and the great admiration he entertained for Linnaeus." According to one apocryphal tale—actually no more than a persistent rumor—Gustav III dispatched a warship to intercept the brig that had already sailed with Linnaeus's collections on board, but the brig had too great a head start and reached London safely.

In any case, Smith treasured Linnaeus's artifacts all his life, and kept them in good order. When he died, in 1828, the collections and books were bought by a fledgling organization called the Linnean Society of London. They still form the shrine and centerpiece of this organization, Britain's foremost society for the study of natural history. The collections reside in Burlington House, right on Piccadilly in the center of London. I once visited this shrine and workplace for a consummately practical reason. Linnaeus himself named the type species of the genus of snails that forms the subject of my technical research. This species, *Cerion uva*, comes from Curaçao, and I wanted to make sure that Linnaeus's specimen truly represents this distinctive species (see essay 27). I was ushered into the vault of the holy of holies and shown the specimen. Linnaeus was right again; it is *Cerion uva,* from Curaçao.

When Charles Darwin received Wallace's manuscript from Ternate, and realized that his younger colleague was about to scoop twenty years of work on natural selection, he appealed to his friends to find some honorable way that might recognize both his priority and Wallace's discovery. His friends proposed a joint presentation of Wallace's paper with some of Darwin's earlier, unpublished writings. Darwin did not attend this meeting in 1858, and remained at home mourning the death of his young son. But the meeting was held in London—at the rooms of the Linnean Society, in the building that housed Linnaeus's own specimens. The joint papers were published in the 1858 volume of the *Proceedings of the Linnean Society of London.*

Thus did the second unmasking begin in the transposed home and living presence of the first—and the ghost of Linnaeus smiled (being more generous and ecumenical than the great Swede in life) as Darwin's words revealed a causal basis for the natural order that Linnaeus had codified. And the psalmist intoned his ancient song: "Behold, how good and how pleasant it is for brethren to dwell together in unity!"

33

Ordering Nature by Budding and Full-Breasted Sexuality

William Hayley (1745–1820), poet, biographer, and patron of the arts, owes his little corner of immortality almost entirely to the heroic couplet written by William Blake at the termination of their relationship:

> Thy friendship oft has made my heart to ache:
> Do be my Enemy for Friendship's sake.

The wealthy Hayley had engaged Blake to engrave figures for his books, and had housed the great poet and illustrator in a cottage on his estate near Chichester. But Hayley never understood Blake's idiosyncratic brilliance, and threatened to throttle his artistic genius with philanthropic kindness attached to expectations for tamer poems and figures. Calling Hayley "an enemy of my spiritual life while (pretending) to be the friend of my corporeal," Blake penned his famous couplet about artistic integrity and moved back to London.

At about the same time, another acquaintance of Hayley was also writing heroic couplets. Consider these lines from 1789 about a lovers' meeting on a mountaintop. The woman reaches the summit first, and her love follows:

The steepy path her plighted swain pursues,
And tracks her light step o'er the imprinted dews;
Delighted Hymen gives his torch to blaze,
Winds round the crags, and lights the mazy ways;
Sheds o'er their secret vows his influence chaste,
And decks with roses the admiring waste.

Heroic couplets, with their stilted images and forced rhymes in iambic pentameter, tend to attract ridicule these days (though I, as a great fan of Alexander Pope, do not ally myself with this consensus). The lines of Hayley's friend, quoted above, do seem hard to defend, especially when you realize that he is not talking about human lovers stealing a kiss in solitude before a magnificent view, but about lichens—yes, those patches on rocks representing the complex symbiosis of an alga with a fungus.

Nonetheless, Mr. Hayley thought highly of his botanical colleague, for he wrote a poem to introduce his friend's versified descriptions of plant sexuality:

Thus Nature and thus Science spake
In Flora's friendly bower;
While Darwin's glory seemed to wake
New life in every flower.

Now scarcely an essay goes by in this series without some mention of Charles Darwin. Moreover, since my hero wrote many botanical books (mostly late in his life), readers might assume that Hayley praises the author of natural selection in these lines. But unless Mr. Hayley had unusual insight into future events, he cannot be speaking of Charles, who was born on Lincoln's birthday in 1809, twenty years after a different Darwin wrote lyrics for lichens. Who, then, is this earlier Darwin?

To achieve best prospects for a ripe old age, one should choose long-lived parents. Similarly, brilliant forebears provide the best predictors for intellectual success (rich doesn't hurt either). Charles Darwin was certainly fortunate in his choice of ancestors. His father, Robert, was a highly respected physician and local financial wizard. Even better, his grandfather Erasmus (1731–1802), source of Hayley's praise, was one of England's leading intellectuals—physician, scientist, philosopher, and prominent member in a movement of progressive industrialists and scholars, centered in Birmingham. These men of the "lunar society," named to honor their monthly meetings at the full moon, advocated a variety of liberal reforms in politics and economics, but eventually ran afoul of public opinion when the French Revolution, which they had vigorously supported in its early days, started to suppress religion and execute kings. (As a parochial footnote for Americans, Darwin's good friend and fellow society member, the clergyman-chemist Joseph Priestley, watched his house, library, and laboratory go up in smoke when crowds rioted in Birmingham on July 14, 1791, the second anniversary of the fall of the Bastille. Priestley, supported by his friends John Adams and Thomas Jefferson, eventually settled in Pennsylvania. But Antoine Lavoisier, the other chief figure in the discovery of oxygen, fared far worse by feeding the guillotine during the height of the Reign of Terror [see essay 24 in *Bully for Brontosaurus*].)

William Hayley wrote his puff to introduce a later edition of a book that Erasmus Darwin first published in happier times, as the French Revolution began auspiciously in 1789—*The Loves of the Plants.* Darwin stated that he wrote this extended poem on the sexuality of plants for the most pervasive and conventional of authors' reasons: " 'The Loves of the Plants' pays me well, and . . . I write for pay, not for fame." Darwin was not a dabbling amateur, but a well-known and widely read English poet, however much his chosen style and substance provoked a later erasure from the history of art. The young Wordsworth wrote of the "dazzling manner of Darwin," though Coleridge stated in 1796 that "I absolutely nauseate Darwin's poem," while allowing that Darwin had at least "accumulated and applied all the sonorous and handsome-looking words in our language." Interestingly, Darwin published the first edition of *The Loves of the Plants* anonymously, fearing that the risqué subject might injure his lucrative medical practice.

Why would anyone write four cantos and 238 pages of heroic couplets about plants, and why would anyone buy such a work in the late eighteenth century? Darwin may have written for money, but he never tired of defending his own broadly based philosophy of natural history. We must view *The Loves of the Plants* primarily as a work of advocacy—for its point of view, not some magnificently antiquated poetry, has given this book a substantial place in the history of science. Darwin's footnote to his couplets on lichens provides insight into his larger purposes (Darwin accompanies each verse with a prosaic explanation of his grand metaphors in small type):

> Lichen . . . Clandestine marriage. This plant is the first that vegetates on naked rocks, covering them with a kind of tapestry, and draws its nourishment perhaps chiefly from the air; after it per-

> ishes, earth enough is left for other mosses to root themselves; and after some ages a soil is produced sufficient for the growth of more succulent and large vegetables. In this manner perhaps the whole earth has been gradually covered with vegetation, after it was raised out of the primeval ocean by subterraneous fires.

As scholars say so often (though the message never really sinks in, because people seem to crave clean points of origin and heroic initiators), Charles Darwin did not invent the concept of evolution. Instead, he took this most common of nineteenth-century biological heterodoxies, gathered more copious and persuasive evidence than ever before, and discovered a plausible mechanism of change in natural selection. Charles's own grandfather Erasmus ranks among the most prominent (and vocal) of his precursors. (Charles never knew Erasmus personally, for his grandfather died in 1802, but Erasmus remained a strong presence in the Darwin household, and Charles read and admired his writings.)

Erasmus's thoroughly historical and evolutionary view of nature—not rare during the Enlightenment, but by no means orthodox either—shines forth in his commentary on lichens, which presents, in its remarkable last sentence, an epitome of the earth's progressive history: from bare land raised from the ocean by internal fires, to formation of soil through the erosive action of lichens, to growth of larger and more complex plants upon the substrate made by these lowly organisms. We need both the footnote and some knowledge of Erasmus's evolutionary view to interpret his last line of verse— "And decks with roses the admiring waste"—a reference to this later growth of higher plants upon the soil produced by lichens.

But advocacy of evolution provides only a subsidiary theme in *The Loves of the Plants.* Darwin wrote his long poem for a different theoretical purpose—and the key lies in the first two words of his footnote on lichens: "clandestine marriage." Darwin composed *The Loves of the Plants* as a popular presentation of Linnaeus's so-called "sexual system" for botanical taxonomy, using the age-old literary devices of pleasant form (rhymed couplets) and accessible imagery (the metaphor of personification). In the first paragraph of his "advertisement," placed at the very beginning of his volume, Darwin explained his purposes:

> The general design of the following sheets is to enlist Imagination under the banner of Science; and to lead her votaries from

> the looser analogies, which dress out the imagery of poetry, to the stricter ones, which form the ratiocination of philosophy. While their particular design is to induce the ingenious to cultivate the knowledge of Botany, by introducing them to the vestibule of that delightful science, and recommending to their attention the immortal works of the celebrated Swedish Naturalist, Linnaeus.

Linnaeus's sexual system for botanical classification, first proposed in the 1730s, became a focal point for popular education in the natural sciences—an important cultural theme in the late eighteenth century, as the Enlightenment swept Western Europe. Not everyone cared for Linnaeus's decision to base the taxonomy of plants on numbers and arrangements of male and female organs in flowers. Predictably, some conservatives detested an explicit sexual basis for anything, fearing a collapse of public morals—a fragile thing indeed, if subject to such easy undermining by a mere counting of stamens and pistils! Professor Johann Siegesbeck of St. Petersburg decreed that God would never have based his natural arrangements on such a "shameful whoredom"—and Linnaeus responded by naming a small and ugly weed *Siegesbeckia* in his honor. Linnaeus himself developed severe doubts about the sexual system because he recognized the artificial basis of his own categories—for application of the rules often gathered unrelated plants into the same group. But Linnaeus never found success in his search for a natural system, and he continued to use his sexual classification *faute de mieux*.

Linnaeus's sexual system found its greatest advantage in utility—easy to learn and easy to practice. Proponents of popular education therefore became strong advocates. As a leading liberal thinker, committed to increasing Britain's commercial power through the spread of scientific knowledge among the public, Erasmus Darwin looked upon Linnaeus's sexual system with particular favor. Darwin established the Botanical Society of Lichfield (his home town in Shropshire) especially to sponsor the translation into English of two Linnaean books on botanical taxonomy—the *Species plantarum* and the *Genera plantarum*. (These translations were duly accomplished and prominently advertised in *The Loves of the Plants*.)

The sexual system of Linnaeus is not based on anything so salacious as the mechanics of observed fertilization, but rather upon the dry anatomy of numbers and arrangements of stamens and pistils, the male (pollen-bearing) and female (ovary and stigma-bearing) parts of flowers.

(Very little was known about plant sexuality in Linnaeus's time; even the existence of separate sexual organs and fertilization remained controversial.)

Linnaeus's original hierarchical system for all of nature utilized four categories—classes, orders, genera, and species in descending levels. (Modern taxonomy maintains the same structure, but introduces more categories—kingdom and phylum on top, and family between order and genus, for example.) The key to Linnaeus's sexual system lies in his primary division of plants into twenty-four classes, based almost entirely on numbers and arrangements of male organs.

The first thirteen classes include hermaphrodite flowers (containing both male and female parts), and base divisions on the number of stamens (most flowers have one pistil, but several stamens of roughly equal length). The first ten—named *Monandria, Diandria, Triandria,* etc., or One Male, Two Males, Three Males, etc.—simply refer to the number of stamens. Class 10, *Decandria,* thus contains flowers with ten stamens. Class 11, *Dodecandria,* has twelve; Class 12, *Icosandria,* has twenty; while Class 13, *Polyandria,* or Many Males, includes all higher numbers from twenty to one hundred. The next two classes, called "powers" by Linnaeus, recognize flowers with stamens of two distinct sizes. Class 14, *Didynamia,* or Two Powers, has four stamens, two short and two tall; while class 15, *Tetradynamia,* or Four Powers, has six stamens, two short and four tall. The next five classes, 16 through 20, honor the mode and degree of adhesion among stamens, not their number. Classes 16 through 18 refer to different arrangements of joining by filaments, or "stems" of the stamens: *Monadelphia* (One Brotherhood) for one group, *Diadelphia* (Two Brotherhoods) for two groups, and *Polyadelphia* for three or more groups. Class 19, *Syngenesia,* or Confederate Males, includes flowers with stamens joined by their anthers, or top parts. Class 20, with the loaded name of *Gynandria,* or Feminine Males, contains flowers with stamens attached to the pistil.

Three of the last four classes accommodate plants with separate male and female flowers. For classes 21 and 22, Linnaeus coined two words still prominently in use by botanists—*Monoecia* (or One House) for separate male and female flowers on the same plant, and *Dioecia* (or Two Houses) for male and female flowers on separate plants. Order 23, *Polygamia,* includes plants with both hermaphrodite and separate-sex flowers. Finally, Linnaeus named his last class *Cryptogamia,* or Clandestine Marriage, for plants with sexual parts that he could not observe. Botanists still use "cryptogam" as an informal term for plants that reproduce by spores or gametes rather than seeds—a heterogenous and nongenealogical group including ferns, mosses, and algae. We can now finally understand Darwin's initial description of lichens as "clandestine marriage," and his poetic image of males sneaking up on females. He is referring to the status of lichens at the base of Linnaeus's list, in class number 24.

Metaphor is a dangerous, if ineluctable, device (see the next and last essay). We use images and analogies to foster understanding of complex and unfamiliar subjects, but we run the risk of falsely infusing nature with the baggage of our parochial prejudices or idiosyncratic social arrangements. The situation can become truly insidious, even perversely so (in the literal rather than pejorative sense), when we impose a human institution upon nature by false metaphor—and then try to justify the social phenomenon as an inevitable reflection of nature's dictates!

Arguments of this sort have long been prominent in the sensitive areas of sexual and racial politics, where dominating groups tend to seek biological rationales for their truly transient and unjustifiable social status. To choose a perhaps silly but not really trivial example, all engineers on a ship that once served as my home for nine months referred to electrical plugs as "he" and receptacles as "she"—thus reinforcing a notion that males penetrate and females accept.

Or, to choose a more recent and concrete case, some biologists have identified (though, to be fair, certainly not defended) human rape as natural because brutally forced copulation in some birds, particularly mallard ducks, has long been called "rape" in the technical literature. But use of the human term for birds is pure metaphor, not a statement backed by evidence for common causality. We have no reason for linking two such different phenomena—one (in humans) primarily about social power, the other (in birds) clearly about reproduction—by the same name merely because they share a superficial resemblance in who does what to whom. Yet by falsely describing an inherited behavior of birds with an old name for a deviant human action, we subtly suggest that true rape—our own kind—might be a natural behavior with Darwinian advantages to certain people as well.

Since sex is so powerful and pervasive in our lives, the error of reading our own socially conditioned ways into nature becomes most apparent when we use metaphors based on human sexuality—especially when we borrow from the charged (and usually sexist) realm of power based on gender. In one sense, Linnaeus's sexual system might be deemed relatively free of such human interference, since he mostly just counts stamens and pistils and records their lengths.

But a closer look reveals that Linnaeus's botanical macrocosm becomes, through his sexual system of classification, a biased reflection of the human social microcosm. Most notably, he follows the oldest convention of sexism by regarding male differences as primary and female variation as secondary—for Linnaeus based his highest division into twenty-four classes on male characters, and his subsequent finer division into many more orders on female traits! (These points have been forcefully made in two excellent articles that helped me greatly in preparing this essay: Janet Browne's "Botany for Gentlemen; Erasmus Darwin and *The Loves of the Plants*," and Londa Schiebinger's "The Private Life of Plants: Sexual Politics in Carl Linnaeus and Erasmus Darwin.")

Moreover, as my earlier epitome of his sexual system illustrates, Linnaeus infused his classification with human terms and concepts. Conjoined stamens are "brotherhoods," and single-sex flowers live in either one or two "houses." Such metaphorical terminology pervades Linnaeus's writing. He refers to fertilization as an act of marriage, and speaks of stamens and pistils as husbands and wives. Flower petals are bridal beds, and infertile stamens become eunuchs, presumably guarding the wife (pistil) for other fertile stamens (husbands). Schiebinger cites a fascinating passage from an essay that Linnaeus wrote in 1729:

> The flowers' leaves . . . serve as bridal beds which the Creator has so gloriously arranged, adorned with such noble bed curtains, and perfumed with so many soft scents that the bridegroom with his bride might there celebrate their nuptials with so much the greater solemnity.

But Linnaeus, at least in his public persona, was a social conservative and something of a prude on sexual issues. He did not allow his four daughters to study French, for fear that they might thus learn the liberal standards of that land. When his wife sent one daughter to school, Linnaeus promptly withdrew her, declaring such education for women "nonsensical." He also refused the Queen of Sweden's offer to receive one of his daughters at court, for fear that such an environment might promote moral laxity. (He may well have been right.) We should therefore not be surprised that Linnaeus's sexual metaphors for plants tended to be drab (however lyrically expressed), and rather strictly within the married state.

In the maximally different persona of Erasmus Darwin, however, sexual imagery for plants reached an apogee. Darwin was crusadingly liberal in politics, probably a genuine atheist in religion, and remarkably broad-minded in his views on sexual behavior. He had twelve children in two happy marriages, and lived openly with a third woman in between, fathering two additional daughters who then lived in harmony with him and his second wife. He favored schooling for women, and wrote a work entitled *Plan for the Conduct of Female Education* when his two illegitimate daughters set up a seminary for girls. He was no egalitarian—but then, who advocated such a position among wealthy eighteenth-century British males? He wanted to educate intelligent women so that they could become better companions for their husbands, not to prepare them for professions.

In short, Erasmus Darwin viewed sex as a natural part of a healthy and rational world order—and these attitudes provided a much broader base for metaphors beyond legitimate marriage when he chose images to describe the sexuality of plants. Moreover, in deciding upon the consistent and fully explicit device of treating stamens as male persons and pistils as female persons, Erasmus Darwin developed a degree of human imagery for plants that neither Linnaeus nor any other prominent botanist writing for the public ever essayed—and all in flowery heroic couplets to boot!

Darwin's liberality served this metaphorical tradition admirably, for Linnaeus's images didn't work well in a restricted world of legal marriages. After all, most flowers have several stamens for a single pistil—meaning many husbands for one wife, if the metaphor be carried to its logical conclusion. We are therefore stuck with the rarely acknowledged phenomenon of polyandry, and not with the more conventional malfeasance immortalized in William James's amazing epitome of human relationships:

> Higgamus hoggamus, woman's monogamous
> Hoggamus higgamus, men are polygamous

A contemporary critic rightly identified this weakness, in rejecting Linnaeus's images "because the great concourse of husbands to one wife is so unsuitable to the laws and manners of our people." But Erasmus Darwin, not so restricted, could talk of numerous suitors, beaux, and swains for one young female beauty—so his more varied personifications didn't violate both law and custom.

The Loves of the Plants is no libertine debauchery, but rather a tale of pleasantly courting rustics and gentlefolk in Arcady. Still, Darwin does describe some remarkable social situations in metaphor—and his imposition of human affairs upon plants becomes an amusing example of a common practice that serves as the focus of this essay, and can become decidedly unfunny, even pernicious, in a variety of historical examples from the annals of racism and sexism (including the case of mallard "rape" described earlier).

Darwin, for example, does not always avoid the nuptial terminology of Linnaeus's earlier uses, and he does acknowledge the consequent, and unconventional, polyandry. He writes of *Iris*, in the class of Triandria, or Three Males:

> The freckled *Iris* owns a fiercer flame,
> And three unjealous husbands wed the dame.

He laments the fate of plants in the class Monoecia, for males and females dwell in different beds (flowers) within the same house (plant), and must therefore be estranged in some way:

> *Cupressus* dark disdains his dusky bride,
> One dome contains them, but two beds divide.

But truly separated lovers in the class Dioecia (male and female flowers on different plants) win his admiration for their tenacity in achieving a treacherous meeting across great distances. The naked pistil grows up and calls to her beaux:

> Each wanton beauty, tricked in all her grace,
> Shakes the bright dew-drops from her blushing face;
> In gay undress displays her rival charms,
> and calls her wondering lovers to her arms.

As one might expect, stamens in the class Gynandria, or Feminine Males (stamens attached to pistils) do not fare well, and are branded as both effete and subjugated:

> Gigantic Nymph! the fair *Kleinhovia* reigns,
> The grace and terror of Orixa's plains . . .
> With playful violence displays her charms,
> And bears her trembling lovers in her arms.

Inevitably, I suppose, Darwin depicts males of the classes Didynamia and Tetradynamia (Two and Four Powers, with one group of stamens shorter than the other) as belonging to disparate social classes:

Two knights before thy fragrant altar bend,
Adored *Melissa*! and two squires attend.

And, for Tetradynamia, with four bosses and two underlings:

Pleased round the Fair four rival Lords ascend
The shaggy steeps, two menial youths attend.

But for all his toleration of variety and multiplicity, Darwin does draw the line at one point. Hybrid matings between genera, yielding no fertile seed, do violate all moral propriety:

Caryo's sweet smile *Dianthus* proud admires,
And gazing burns with unallowed desires;
With sighs and sorrows her compassion moves,
And wins the damsel to illicit loves.

As a final example, Janet Browne notes that Darwin, perhaps unconsciously, grants different status to pistils (females) depending upon how many stamens grow in the same flower (or, in the chosen metaphorical image, how many males fall under her control). Flowers of the class Monandria (one stamen and one pistil) are united in monogamous and connubial bliss; he writes of *Canna*, or Indian Reed, uncomfortably transported from the tropics to cold Britain, but able to survive the chill by the warmth of committed love:

First the tall *Canna* lifts his curled brow
Erect to heaven, and plights his nuptial vow;
The virtuous pair, in milder regions born,
Dread the rude blast of Autumn's icy morn;
Round the chill fair he folds his crimson vest,
And clasps the timorous beauty to his breast.

Darwin also treats "lowly" plants of the last class, Cryptogamia, as similarly virtuous, for at least they hide whatever they do. He even shows proper compassion in acknowledging that the short and the ugly can also enjoy sexual pleasure (more than Disney would later grant to his Seven Dwarfs). He writes of underground truffles:

Deep, in wide caverns and their shadowy aisles
Daughter of Earth, the chaste *Truffelia* smiles;

On silvery beds, of soft asbestos wove,
Meets her gnome-husband, and avows her love.

At an intermediary number of four to six stamens, Darwin invokes a variety of images about pursuit, lust, and coyness. Males are depicted as vigorously courting swains and beaux; females may be flirtatious or tough, but they are not easily won:

Meadia's soft chains five suppliant beaux confess,
And hand in hand the laughing belle address;
Alike to all, she bows with wanton air,
Rolls her dark eye, and waves her golden hair.

A tougher sister exerts more control:

With charms despotic fair *Chondrilla* reigns
O'er the soft hearts of five fraternal swains;
If sighs the changeful nymph, alike they mourn;
And, if she smiles, with rival raptures burn.

But females accompanied by large numbers of males (ten or more stamens to one pistil) are invariably depicted either as truly dismissive, or as fully regal:

Sweet blooms *Genista* in the myrtle shade,
And ten fond brothers woo the haughty maid.

With twenty males at her beck and call, the pistil must be a powerful goddess:

While twenty Priests the gorgeous shrine surround
Cinctured with ephods, and with garlands crowned,
Contending hosts and trembling nations wait
The firm immutable behests of Fate.

We may laugh at these particular Darwinian reveries, but the underlying theme of our tendency to impose a human social scheme upon supposedly objective nature, especially in the charged domain of sex and gender, teaches us something important about the cultural construction of knowledge. I therefore close with another Linnaean example from the other end of our conventional taxonomic ladder. I knew that Linnaeus had invented the term Mammalia for our vertebrate class in his *Systema naturae* of 1758, but I thought that he had simply promoted an old vernacular word to a new technical meaning. However, in an important article, written in 1993, Londa Schiebinger shows that Linnaeus truly invented the word—and that no language had ever before referred to the group of warm-blooded, hair-sporting, live-bearing vertebrates as mammals.

All previous systems had treated and named our relatives differently. Aristotle established a vertebrate group called Quadrupedia, with a primary subdivision into Oviparia (scaly and egg-laying, including reptiles and some amphibians, and Viviparia (hairy and live-bearing, thus including most mammals, but also excluding such creatures as bats, whales, and, most importantly, humans). By Linnaeus's time, our group had a better definition, but no recognized name. John Ray, for example, the greatest of Linnaeus's predecessors, had suggested Pilosa, meaning "hairy," as a way of annexing obviously related animals that did not exhibit Aristotle's defining feature of four legs.

So why did Linnaeus choose a new name, and why, particularly, did he select such a peculiar term as Mammalia, referring, obviously, to the female breast? We must first grasp the extreme unconventionality of Linnaeus's decision. Most generally, and for the usual sexist reasons, we tend to personify active phenomena as male—and organisms judged most complex should certainly fall under this convention. (In contemporary English, we still generally refer to an unsexed animal as "he"—as in "isn't he cute," or "look at 'im go.") If Linnaeus had been an explicit egalitarian, out to sink a bad habit by example, he might have chosen Mammalia for this overt political reason. But Linnaeus, as we have seen, was a social conservative and a conventional sexist. More particularly, zoologists have long translated this cultural convention into technical practice by treating male organisms as the canonical icon for any species.

Why, then, did Linnaeus choose a female trait to define the highest group, apparently adding insult to male injury by selecting a feature that males also possess, but in a rudimentary and useless state. Schiebinger argues cogently that Linnaeus made his decision for an ideological reason—one very distant from any notion of sexual equality. Linnaeus had been deeply engaged in a different and equally important battle, this time (or so most of us would judge today) on the right side: his campaign to classify humans *into* nature with other animals, at a time when many naturalists still insisted on a separate human kingdom for beings with a soul and created in God's image.

Propagandists have always recognized that an adroit choice of name can convey great power of persuasion. Nature has almost invariably been personified as female, in a cultural and linguistic tradition dating at least to Chaucer. (See the *Oxford English Dictionary*'s entry on Nature; Chaucer wrote in 1374 that "Nature had grete joy her to behelde.") If one wishes to gain rhetorical advantage in a struggle to place humans within nature, then choose a female feature to define our larger group, thereby emphasizing our closeness to Mother Earth and her other animate productions. Interestingly, in the same work that defined us within nature as mammals, Linnaeus also sought to distinguish us as a species for our mental prowess—and here he chose a male designation, as *Homo sapiens* (although the Latin *homo* may be read more generally in the old sense of "man" as "humankind," while *vir* is, more specifically, a male person—from which we obtain, and for sexist reasons, the notions of virtue and virility).

We are therefore mammals, rather than hairy beings, as yet another legacy of pervasive sexual politics. (I am not doing justice to the complexity and subtlety of Schiebinger's full argument. She presents many other sources of evidence, including Linnaeus's active involvement in a movement against wet-nursing—accomplished largely by trying to persuade wealthy women that they should not view departure from nature as a virtue, but should get in touch with our common heritage by suckling their own babies.)

If we are surrounded by such a miasma of social prejudice at the heart of what we regard as objective knowledge, what hope do we have for understanding a natural world truly external to us in so many respects? I have no simple answer, but great thinkers offer us tidbits to integrate into a solution. Since William Blake chose to preserve his integrity against the siren song of Hayley's welcome patronage, he provides several insights in his idiosyncratic and often oracular verses. Consider two thoughts from his *Songs of Innocence*, published in the same year as Darwin's *Loves of the Plants*. Nature and social life are not entirely commingled, and much injustice lies on our personal doorstep:

> When my mother died I was very young,
> And my father sold me while yet my tongue
> Could scarcely cry "'weep! 'weep! 'weep! 'weep!"
> So your chimneys I sweep, and in soot I sleep.

And lest we continue our old attempts to blame our worst traits upon biological heritage, we should also remember that what we call human nature embodies much potential for good amid the tendencies we might choose to brand as evil:

> For Mercy has a human heart,
> Pity, a human face,
> And love, the human form divine,
> And Peace, the human dress.

34

Four Metaphors in Three Generations

MY FIRST TRIP to Greece became my most exhilarating celebration of the ordinary—for I discovered that many of our most abstract and highfalutin words, often invented by pedants to showcase their knowledge of ancient languages and intended to debar the general public from such rarefied information, refer to plain and ordinary things in the land of their etymological birth. For a blessed moment of joyful hubris, I thought that all of Greece had decided to celebrate our theory of punctuated equilibrium, with its key claim for what Niles Eldredge and I call "stasis," or lack of substantial change during the geological history of most species. (Yeah, we might have called the phenomenon "stability," but "stasis" had that nice, slightly jargony ring, and stood close enough to the vernacular for comprehension.) But then I saw all these large rectangular vehicles pulling up to the "stasis" signs, and realized that local governments were only marking their bus stops!

I also learned a lot about derivations and change in meaning. I wondered why the *domestic* airlines terminal bore the name "esoteriki"—for esoteric things are strange and obscure and should therefore mark, or so I supposed, the entrance point for foreigners at the international terminal. But I learned later, from my *Oxford English Dictionary*, that *esoteric* came to mean "obscure" because the word designates information vouchsafed only to initiated members of an inner circle, and there-

The author learns the meaning of stasis *in Greece.*

fore mumbo-jumbo to others. "Eso" mean "within" in Greek, and therefore becomes appropriate for domestic flights (they ought to know, after all). I also couldn't figure out why the area of a church set aside for candles and votive offerings bore the apparently contradictory sign "anathemata." But *anathema,* in Greek, is an offering (literally "a thing set up"). Greek churches still use the original sense of "an object devoted or consecrated to divine use"; only later did the word acquire a restricted and usually opposite meaning of "a thing dedicated to evil" (in ecclesiastical Latin, an *anathema* is an excommunicated person).

One day, as I sat at an alfresco lunch spot enjoying a view of the Acropolis, a small truck pulled up to the curb and blocked the Parthenon. I was annoyed at first, but later wonderfully amused as I watched the moving men deliver some furniture to the neighboring house. Their van said "metaphora." "Of course," I realized: *phor* is the verb for carrying, and *meta-* is a prefix meaning "change of place, order, condition, or nature." A moving truck helps you change the order of something by carrying it from one spot to another—and is surely a metaphor. I then discovered that all sorts of carriers are metaphors, including the wheels for your airplane luggage, shown in the accompanying snapshot.

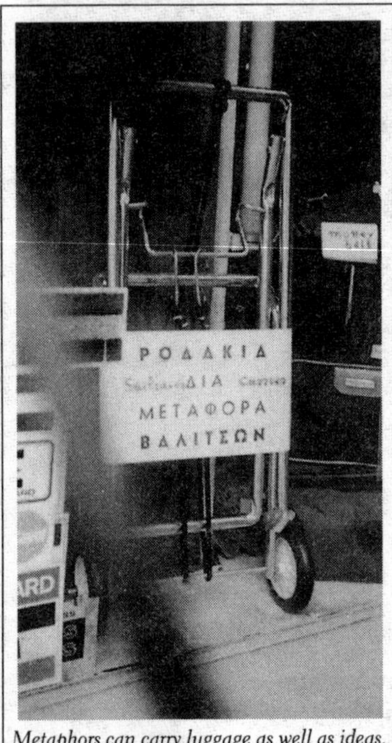

Metaphors can carry luggage as well as ideas.

How lovely. In school, I was frustrated almost to tears in struggling to grasp the differences among various kinds of images: metaphors, similes, eponyms, metonyms, synecdoches—and all this after mastering the distinction between iambs and dactyls. (I learned these words, by the way, at P.S. 26, Queens, not some fancy private school. The New York City public schools actually taught this stuff not so long ago.) If only I had known about that little, concrete moving van! A metaphor carries you from one object (that may be difficult to understand) to another (more accessible, and therefore helpful, by analogy, in grasping the original concern).

I love metaphors; I use them all the time in these essays, and have even written a column or two specifically on the subject of metaphors and their uses. The human mind is a wondrously obtuse and circuitous instrument. We think of ourselves as logical, as able to move in deductive order through a set of arguments from beginning to ineluctable conclusion. But we almost never work in such an idealized way. We always encounter obstacles and chasms that require a creative leap. And we need carriers, or metaphors, to make these imaginative jumps. Moreover, a scholar's choice of metaphor usually provides our best insight into the preferred modes of thought and surrounding social circumstances that so influence all human reasoning, even the scientific modes often viewed as fully objective in our mythology.

But metaphors also present twin dangers, literary and intellectual. On the structural side, some metaphors are so silly or obtuse that we end up laughing (or scratching our heads) rather than gaining enlightenment. Among famous examples, I would submit the opening verse of Psalm 42: "As the hart panteth after the water brooks, so panteth my soul after thee, O God." (I had particular reservations about this image because I first learned the psalm through oral recitation, pictured "heart" instead of "deer," and then couldn't imagine what one of my inner organs would do with a river. I also had trouble with souls panting—dogs, yes; ethereal versions of me, never.) On the conceptual side, many images are flat wrong—from the ladder of life, to the king of beers, to the evil empire.

The distinction of metaphors, between useful and ineffective, brilliantly insightful and dismally misleading, therefore becomes an important consideration in making or evaluating any complex intellectual argument. But how shall we set up a catalog of criteria? I will not address this vital subject in the abstract, but will devote this essay instead to providing matched examples—in the hope that we might extract some general guides. I will discuss two celebrated and related authors, separated by generations—a grandfather and his more famous grandson. I will present four metaphors central to the primary argument of each writer. I will claim that grandfather Erasmus Darwin gained little but verbiage from his efforts, whereas grandson Charles deftly used metaphors not only to illuminate, but also to advance, his epochal theory of evolution by natural selection.

Erasmus Darwin (1731–1802)—prominent liberal thinker of Britain's Enlightenment, successful physician and scientist, respected poet in his era's favorite style of heroic couplets (written in iambic pentameter, not in dactyls; you see, Mrs. Ponti, I do remember the difference)—published his most remarkable work in 1789, *The Loves of the Plants,* highlighted in the previous essay. In four cantos and 238 pages of verse, he described and illustrated Linnaeus's sexual system for classifying plants by personifying stamens as male humans, pistils as female humans, and then drawing out the consequences in heroic couplets. If this device were not metaphor enough, Erasmus Darwin then added another layer by comparing the personified flower to some item of classical mythology or current events. I will illustrate his technique by discussing four of these secondary metaphors, admittedly selected for their more forceful imagery (but by no means atypical of Darwin's poetic strategy).

1. Sequential fertilization in *Gloriosa*, and the suicide of Ninon's son. *Gloriosa* belongs to Linnaeus's class Hexandria, or six sta-

mens for a single pistil in each flower. In Darwin's first metaphor, he must therefore describe six lovers with one woman. Since fertilization is sequential in *Gloriosa*—one group of three stamens act first, followed later by the second group of three—Darwin invokes the risqué image of Mrs. Robinson, an older seductress, using wiles to captivate the second group of three after her physical beauties have faded:

> When the young Hours amid her tangled hair
> Wove the fresh rose-bud, and the lily fair,
> Proud *Gloriosa* led three chosen swains,
> The blushing captives of her virgin chains.—
> When Time's rude hand a bark of wrinkles spread
> Round her weak limbs, and silver'd o'er her head,
> Three other youths her riper years engage,
> The flatter'd victims of her wily age.

Darwin then cements this metaphor with another image from recent history. Ninon de Lenclos (1620–1705), lover of more famous Frenchmen than this essay has space to list, ended life as a respectable socialite (Voltaire's father managed her business dealings, and she left money for the young Voltaire in her will), but accumulated quite a string of legends from her earlier years, including this tale of sexually enticing her own son (with dire consequences), recounted by Erasmus Darwin as a metaphorical comparison with *Gloriosa*—quite inappropriate, since *Gloriosa*'s stamens lie within one flower and represent a single generation, whatever the spacing of fertilization:

> So, in her wane of beauty, Ninon won
> With fatal smiles her gay unconscious son.—
> Clasp'd in his arms she own'd a mother's name,—
> "Desist, rash youth! restrain your impious flame,
> "First on that bed your infant form was press'd,
> "Born by my throes, and nurtured at my breast."
> Back as from death he sprung, with wild amaze
> Fierce on the fair he fix'd his ardent gaze;
> Dropp'd on one knee, his frantic arms outspread,
> And stole a guilty glance toward the bed;
> Then breath'd from quivering lips a whisper'd vow,
> And bent on heaven his pale repentant brow;
> "Thus, thus!" he cried, and plung'd the furious dart,
> And life and love gush'd mingled from his heart.

2. SPREADING SEED IN *IMPATIENS*, AND THE SLAYING OF MEDEA'S CHILDREN. Darwin writes exactly four lines on *Impatiens*, the touch-me-not—describing how this plant, if touched when the seeds are ripe, "suddenly folds itself into a spiral form . . . and disperses the seeds to a great distance." Darwin follows with three pages on Euripides' tale of Medea, who killed her own children by Jason (of Golden Fleece fame), when Jason abandoned her for the daughter of King Creon. The comparison with *Impatiens* seems awfully forced, for *Impatiens*'s seeds are fertile, and what does growing a plant hold in common with killing children—except that Medea threw her slain offspring to the ground, just as *Impatiens* tosses seeds violently to earth. Moreover, with four lines on *Impatiens*, and three pages on Medea, one must suspect that Darwin is using botany as an excuse for retelling the gory details of Euripedes' tragic figure, rather than invoking the human story to illuminate the plant, as advertised:

Thrice with parch'd lips her guiltless babes she press'd,
And thrice she clasp'd them to her tortur'd breast;
Awhile with white uplifted eyes she stood,
Then plung'd her trembling poniards in their blood.
"Go, kiss your fire! go, share the bridal mirth!"
She cry'd, and hurl'd their quivering limbs on earth.

3. PLANTS THAT DISPERSE THEIR SEEDS BY WIND, AND MONTGOLFIER'S BALLOON. At the beginning of his second canto, Darwin describes several plants with windblown seeds. Just as we knew the excitement of humanity's first flights into space, and our grandparents felt the thrill of aviation's earliest days, Erasmus Darwin's generation experienced the elation of our first elevation above the earth's surface—the discovery of ballooning. The brothers Montgolfier in France, the great pioneers of this art, became cultural heroes beyond the Wright brothers, beyond Lindbergh, beyond Neil Armstrong's giant leap for mankind. Darwin really wanted to write about his great contemporaries, and one feels again that the plants serve as an excuse to describe the human achievement, not vice versa as his metaphorical method supposedly advocates. Darwin waxes particularly lyrical in describing the flight of Montgolfier's balloon above the cosmos—so much so that we will need some footnotes (to be given after the citation) to sort out the flowery images:

So on the shoreless air the intrepid Gaul
Launch'd the vast concave of his buoyant ball.—

Journeying on high, the silken castle glides
Bright as a meteor through the azure tides; . . .
Rise great Montgolfier! urge thy venturous flight
High o'er the Moon's pale ice-reflected light;
High o'er the pearly Star, whose beamy horn
Hangs in the east, gay harbinger of morn;
Leave the red eye of Mars on rapid wing,
Jove's silver guards, and Saturn's crystal ring;
Leave the fair beams, which, issuing from afar,
Play with new lusters round the Georgian star; . . .
For thee Cassiope her chair withdraws,
For thee the Bear retracts his shaggy paws.

To make sense of all this, you must realize that Montgolfier's balloon is receding farther and farther from earth. The spacecraft first passes Venus, the morning star ("gay harbinger of morn") which, like the moon, has phases ("beamy horn"). It then moves through the planets in order—Mars, Jupiter, and Saturn—and out to the stars, where the constellations of Cassiopeia (a lady sitting in a chair) and Ursa Major (the Great Bear, including the Big Dipper in its tail) move aside in homage to let the balloon pass. But now we must track back to the hard item (two bits to any reader who knew the answer before reading this paragraph).[1] What is the "Georgian star"? Answer: the planet Uranus, next in line after Saturn. When the great astronomer William Herschel discovered Uranus (the first planet found by telescope) in 1781, just eight years before Darwin's book, he named the new object Georgium Sidus (or Georgian Planet) to honor his patron, King George III. Thankfully for Americans, who have never been particularly keen on mad King George, especially since Jefferson penned such a powerful list of his wrongdoings in the Declaration of Independence, this exercise in chauvinistic designation didn't stick. All the other planets bore classical names that honored no modern nation or monarch, so astronomers eventually accepted the suggestion of J. E. Bode, made in the same year as Herschel's discovery, that the new planet be named for Uranus, the father of Saturn (just as Saturn was the father of Jupiter, thus preserving a proper correlation of distance with generation).

1. Unfair! Unfair! My colleague Owen Gingerich, distinguished historian of astronomy, called up and claimed the prize. I tried to explain that I had advanced the challenge to lay readers, not to professionals—of course *he* would know! But he correctly replied that I had made no such explicit specification, that I would have to pay, and that he wanted a check to show his friends, not a coin with another George on the "heads" side. Well, I am a man of honor—so I coughed up. But at least I have the satisfaction of fair assurance that he ain't gonna cash the thing (even if he embarrasses me by hanging it on the wall)!

4. MIGRATING PLANTS FROM THE BIRTH OF MOSES TO THE ABOLITION OF SLAVERY. The end of the third canto features Darwin's most extended and farfetched metaphor. He speaks first of American plants whose seeds can cross the Atlantic and still germinate in Europe:

Where vast Ontario rolls his brineless tides,
And feeds the trackless forests on his sides,
Fair *Cassia* trembling hears the howling woods,
And trusts her tawny children to the floods . . .
Soft breathes the gale, the current gently moves,
And bears to Norway's coasts her infant loves.

The comparison to Moses seems obvious enough, for the great prophet was placed into the Nile in a basket and floated to safety:

With paper-flags a floating cradle weaves,
And hides the smiling boy in Lotus-leaves;
Gives her white bosom to his eager lips,
The salt-tears mingling with the milk he sips;
Waits on the reed-crown'd brink with pious guile,
And trusts the scaly monsters of the Nile.

The transition to slavery is a bit forced. Moses lifted the yoke of his nation ("let my people go" and all that), while slaves, just like seeds, are transported across oceans.

E'en now, e'en now, on yonder Western shores
Weeps pale Despair, and writhing Anguish roars:
E'en now in Afric's groves with hideous yell
Fierce Slavery stalks, and slips the dogs of hell.

But England's leaders have power to end this scourge:

Ye bands of Senators! whose suffrage sways
Britannia's realms, whom either Ind obeys;
Who right the injured, and reward the brave,
Stretch your strong arm, for ye have power to save! . . .
Hear him, ye Senates! hear this truth sublime,
"He, who allows oppression, shares the crime."

Admirable sentiments, forcefully (if a bit floridly) expressed—but quite a stretch from floating seeds.

Charles Darwin, so unlike his grandfather, made no attempt at lyrical writing. He composed the *Origin of Species* (1859) as a general work for a literate public, not as a technical treatise for scientists. Still, Darwin's prose, while clear, tends to be dry and measured. Much of the *Origin* is a list of supporting facts for evolution.

Nonetheless, Charles Darwin understood that difficult and controversial concepts could not be grasped by factual accumulation alone. He knew that he had to make a convincing verbal case as well. (In a famous passage, Darwin calls his book "one long argument.") Darwin also sensed that imagery and metaphor must be used as indispensable tools in the art of persuasion. At several crucial moments in the book, he therefore introduces key metaphors to illuminate essential points. I have long admired these metaphors and have often discussed them in these essays. I will, therefore, not enumerate the details here, but will describe the decisive role played by these devices.

Charles's strategy is so different from his grandfather's. Erasmus tried to drown us in a gushing flood of florid language, creating a soup so thick that an occasional tidbit of exquisite taste gets lost in the glop. Charles knew the value of numerous metaphors about needles in haystacks or roses amid the thorns. He grasped the virtue of rarity, and therefore strikes us between the eyes every few dozen pages with a metaphor so appropriate, and so expansive, that natural selection becomes a familiar friend rather than an arcane and distant concept. Consider my four favorites, including the keystone of the entire book.

1. SUPERFICIAL APPEARANCES AND DEEPER REALITIES OF NATURE. Darwin faced the substantial problem of reorienting his public's fundamental concept of nature's order. Most people viewed nature as basically benevolent, created in perfection by a loving God and largely for human benefit. Darwin's theory required a stunning reversal—to a world devoid of intrinsic purpose, developing limited order only as a side consequence of nature's true causality: the struggle among organisms for personal reproductive success. To facilitate this flip in perspective, Darwin developed a brilliant metaphor to contrast external appearances with internal reality:

> We behold the face of nature bright with gladness, we often see superabundance of food; we do not see, or we forget, that the birds which are idly singing round us mostly live on insects or seeds, and are thus constantly destroying life; or we forget how

> largely these songsters, or their eggs, or their nestlings, are destroyed by birds and beasts of prey.

2. THE DRIVING FORCE OF COMPETITION IN A CROWDED WORLD. For natural selection to work relentlessly, the world must be full of competitors; otherwise, the ill-adapted might live forever by moving to some previously unoccupied real estate. At a crucial point in his argument, Darwin first states the idea of perennial competition directly, and then reinforces his key point with a metaphor about natural crowding: new forms can only enter a full terrain by driving others out (new wedges insinuate themselves at their thin edges—to cite another common image—and eventually force others out).

> In looking at Nature, it is most necessary . . . never to forget that every single organic being around us may be said to be striving to the utmost to increase in numbers; that each lives by a struggle at some period of its life; that heavy destruction inevitably falls either on the young or old, during each generation . . . The face of Nature may be compared to a yielding surface, with ten thousand sharp wedges packed close together and driven inwards by incessant blows, sometimes one wedge being struck, and then another with greater force.

3. THE TREE OF LIFE. Darwin presents the theory of natural selection in the first four chapters of the *Origin*. (He devotes the rest of the book to refuting objections and marshaling evidence for evolution.) As an epitome for his entire argument, Darwin chooses to end chapter 4 with an elaborate metaphor based on a comparison of life's history to the growth of a tree. Darwin, of course, did not invent the tree of life, an old biblical image (Proverbs 3:18, for example) that had also been employed to express taxonomic relationships among organisms. But Darwin added a dynamic element in comparing living and dead branches with successful and failed lineages subject to natural selection (with extinct groups of the fossil record depicted as broken limbs at the tree's base)—a brilliant and complex image combining taxonomic structure with causal production.

> The affinities of all the beings of the same class have sometimes been represented as a great tree. I believe this simile largely speaks the truth. The green and budding twigs may represent

> existing species; and those produced during each former year may represent the long succession of extinct species. At each period of growth all the growing twigs have tried to branch out on all sides, and to overtop and kill the surrounding twigs and branches, in the same manner as species and groups of species have tried to overmaster other species in the great battle for life. The limbs divided into great branches, and these into lesser and lesser branches, were themselves once, when the tree was small, budding twigs; and this connection of the former and present buds by ramifying branches may well represent the classification of all extinct and living species in groups subordinate to groups.

Darwin then ends the chapter with a rare flourish of truly lyrical prose:

> As buds give rise by growth to fresh buds, and these, if vigorous, branch out and overtop on all sides many a feebler branch, so by generation I believe it has been with the great Tree of Life, which fills with its dead and broken branches the crust of the earth, and covers the surface with its ever branching and beautiful ramifications.

4. DARWIN'S CENTRAL ARGUMENT IS BASED ON ANALOGY, NOT DIRECT EVIDENCE. The preceding three metaphors are helpful images, designed to clarify unfamiliar—even socially objectionable—concepts crucial to the acceptance of natural selection. But Darwin employs metaphor in a much deeper way by rooting his entire argument in a fundamental analogy. Darwin begins his book by discussing domestic pigeons, and then presents voluminous information—well known, fully documented, and not subject to doubt—about substantial historical change achieved by human intervention in domesticating animals and improving food crops. But Darwin knew that he could not prevail by recounting this incontrovertible information; he had to convince readers that much greater changes through vastly longer times could also occur by transmutation. But geological time is not available for direct experimentation; one might document the results of large-scale evolution, but how can you demonstrate causality by natural selection? Darwin therefore proceeded by a grand analogy: If we know that humans can make small changes in limited time by a process that Darwin called artificial selection, then, surely, the much more powerful forces of nature can accomplish any amount of evolution in the ample time allowed—by a

process that Darwin named, by analogy, natural selection. Our little but palpable world of domestication and agriculture therefore becomes a metaphorical microcosm for nature's unobservable grandness:

> As man can produce and certainly has produced a great result by his methodical and unconscious means of selection, what may not nature effect? Man can act only on external and visible characters: nature cares nothing for appearances . . . She can act on every internal organ, on every shade of constitutional difference, on the whole machinery of life . . . How fleeting are the wishes and efforts of man! how short his time! and consequently how poor will his products be, compared with those accumulated by nature during whole geological periods.

If we ask why Erasmus's metaphors seem so unhelpful, even obtrusive, and why Charles's are so apt and useful, two major differences stand out: First, Erasmus's comparisons seem either forced (expulsion of *Impatiens* seeds to Medea's murdered children, seeds floating to Norway to Moses sailing down the Nile), or pedestrian (aerial seeds to Montgolfier's balloon); and the linkage therefore doesn't help us to understand the biological example—whereas Charles's imagery illuminates his science (the tree of life as both dynamic and genealogical, artificial selection as an extended analogy to what nature might do to greater effect). Second, Erasmus's balances seem out of whack if the metaphors are meant to clarify the science—for his metaphors often extend for pages after a few telegraphic lines about plants, and one suspects that botany has become an excuse for florid versification about classical tragedies and modern triumphs (though we thought that the verses were meant to illuminate the flowers). Charles's metaphors, on the other hand, are short and pungent, often only a phrase in length ("the face of nature bright with gladness"), and always subservient to a scientific point thus clarified.

Perhaps I am being unfair to Erasmus in this comparison. After all, his chosen genre is so different from his grandson's. He composed a long poem on purpose, so of course his imagery must be more extensive, whereas Charles wrote a more conventional prose work of science. But Erasmus also presents an explicit theory of metaphor that virtually guarantees the limited scientific utility of his images—and on this basis we may make the comparison justly. Erasmus develops these views in sections of *The Loves of the Plants* that have won practically no attention from scholars and commentators (who love to quote the heroic couplets): the three prose interludes, placed between the four cantos of the books, and written in the form of dialogues between a poet (Darwin himself) and a bookseller (who, he hopes, will flog the volume successfully).

Erasmus Darwin introduces his book by defending the search for apt comparisons, arguing that the complexity of natural objects produces a plethora of relationships among them, all worthy of exploration: "Since natural objects are allied to each other by many affinities, every kind of theoretic distribution of them adds to our knowledge by developing some of their analogies." But, in the first Interlude, the poet then drives a wedge between poetry and science by banishing the loose imagery of metaphor from science, where logic must hold sway, while defending such verbal pictures for poetry, whose purpose Darwin conceives as primarily visual—"bringing objects before the eye, or expressing sentiments in the language of vision": "Science," Erasmus writes, "is best delivered in Prose, as its mode of reasoning is from stricter analogies than metaphors or similes."

Erasmus then tells us that his poetic images are not even intended to clarify his botany, but are invented for two other purposes. First, Erasmus argues, both poets and painters succeed by idealizing nature, not by accurate depiction: "The farther the artist recedes from nature, the greater novelty he is likely to produce; if he rises above nature, he produces the sublime; and beauty is probably a selection and new combination of her most agreeable parts" (see essay 15 for Burke's distinction of the sublime and the beautiful, followed here by Darwin). Second, metaphors and similes, as poetic visions, should be digressions from nature, not illuminations of her actuality. Using a lovely archaic word—to "quadrate," or literally to "square with"—Erasmus defends the first great poet:

> But the similes of Homer have another agreeable characteristic; they do not quadrate, or go upon all fours (as it is called) . . . Any one resembling feature seems to be with him a sufficient excuse for the introduction of this kind of digression; he then proceeds to deliver some agreeable poetry on this new subject, and thus converts every simile into a kind of short episode.

Finally, the poet lays it on the line when the bookseller asks, "Then a simile should not very accurately resemble the subject?"

> No; it would then become a philosophical analogy, it would be ratiocination instead of poetry: it need only so far resemble the

> subject, as poetry itself ought to resemble nature. It should have so much sublimity, beauty, or novelty, as to interest the reader; and should be expressed in picturesque language, so as to bring the scenery before his eye; and should lastly bear so much verisimilitude as not to awaken him by the violence of improbability or incongruity.

With such a wedge driven between the reasoning of science and the visual imagery of poetry, we should not be surprised that Erasmus Darwin's metaphors (explicitly situated in his visual rather than his philosophical world) do not enlighten his botanical examples. Charles, unencumbered by such an aesthetic theory, used all the literary devices at his command, including metaphor, to advance his singular purpose—his "one long argument" about evolution.

One other contrast between Erasmus and Charles should help us to understand why Erasmus's heroic couplets have been forgotten, while Charles's book revolutionized human thought. Charles Darwin also knew about Linnaeus's observation of American seeds floating to Norway and germinating there. He wrote, in an 1855 note to the *Gardeners' Chronicle and Agricultural Gazette*: "The seeds which are often washed by the gulf Stream to the shores of Norway, with which Linnaeus was well acquainted, float, as I have lately tried." From Erasmus, this observation only inspired a flood of forced words about Moses in the reeds and the evils of slavery. But Charles recognized that the possibility of long-distance transport could remove a major obstacle to the acceptance of evolution—and he therefore undertook a laborious series of remarkably simple but highly ingenious experiments to learn how long seeds could stand a saltwater bath.

Evolution clearly required a single region of origin for each species, whereas creationists argued that God might simultaneously situate new forms in many places. Thus, the common occurrence of identical species in widely distant places seemed to speak against evolution. Darwin countered by trying to find plausible mechanisms for long-distance transport, so that a species evolved in one region might make the required journey. Since plants provided many challenging examples, dispersal of floating seeds across oceans seemed promising—and Darwin decided upon an experimental approach. In an article titled "Does Sea-Water Kill Seeds?" published in 1855 as the first of seven notes and papers on the subject, Darwin wrote cautiously (still not willing to expose the evolutionary beliefs that he would continue to hide for another three years):

> As such experiments might naturally appear childish to many, I may be permitted to premise that they have a direct bearing on a very interesting problem, . . . namely, whether the same organic being has been created at one point or on several on the face of our globe.

Darwin constructed a series of experimental containers: "The sea-water has been made artificially with salt procured from Mr. Bolton . . . which has been tested by better chemists than men, namely, by numerous sea animals and algae having lived in it for more than a year." He immersed eighty-seven different kinds of seeds and found that nearly three fourths of the species would still germinate after a twenty-eight-day bath. Pepper seeds survived best, for thirty of fifty-six seeds "germinated well after 137 days immersion." Darwin then consulted a book of oceanic currents and determined that seeds could float the requisite thousands of miles and grow on distant continents.

But still Darwin worried. He could keep the seeds alive long enough in water, but they sank too soon, and therefore could not float unaided to their putative destinations. So he began a second series of experiments and observations, in his usually thorough, almost obsessive way. What natural rafts, floats, or airborne devices might carry a seed the required distance? Darwin wrote to a sailor who had been shipwrecked on Kerguelen to find out if he remembered any seeds or plants growing from driftwood on the beach. He asked an inhabitant of Hudson Bay if seeds might be carried on ice floes. He studied the contents of ducks' stomachs; he was delighted to receive in the mail a pair of partridge's feet caked with mud; he rooted through bird droppings to see if seeds could germinate after a run through the digestive tract. He even followed a suggestion of his eight year old son that they float a dead and well-fed bird. Darwin wrote in a letter that "a pigeon has floated for 30 days in salt water with seeds in crop and they have grown splendidly." In the end, Darwin found more than enough mechanisms to move his viable seeds.

Erasmus's words and Charles's actions! By their fruits ye shall know them. I am led, in conclusion, to remember an event of my youth, as I graduated from Junior High School 74 in 1955 (middle schools also have numbers in New York City). A local bus driver had been particularly kind to us, a rarity among municipal employees in this harried city; he treated students with respect, deigned to speak with us, and even, wonder of wonders, waited when he saw us running, rather than shutting the door in our faces. I didn't even know his name, but one morning I asked him to sign my graduation autograph book, and he kept a busload

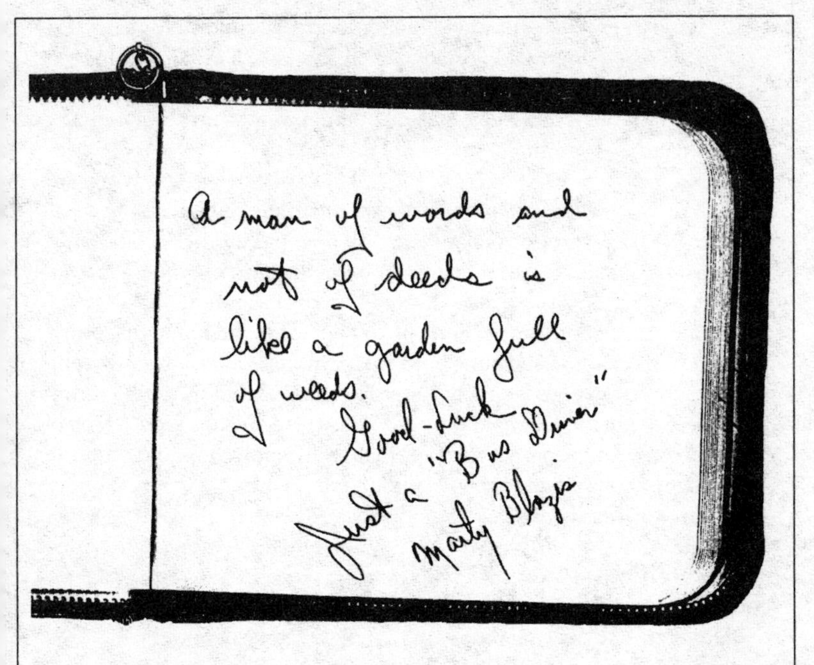

The doggerel verse written by a Queens bus driver in the author's autograph book thirty-eight years ago.

of adult commuters waiting. (Most of them seemed charmed by the incident and, for once, didn't begrudge the slight delay.) He signed his page, "Good Luck. Just a Bus Driver. Marty Blazis." Mr. Blazis, if you are still around,[2] may I offer this belated thanks from all of us for your kindness; it mattered a great deal. May I also report his words, one of those old and classical doggerel verses that captures so much wisdom in its familiarity, for I also note a comparison of Charles Darwin's triumph with Erasmus's lesser impact in these lines (and I do appreciate a last botanical metaphor for this essay). Mr. Blazis wrote:

> A man of words and not of deeds
> Is like a garden full of weeds.

2. Marty Blazis died a few years ago, but I received a lovely note from his son thanking me for the mention, and reinforcing my memories of his father's lifelong and pervasive kindness.

Bibliography

Abele, L. G., W. Kim, and B. E. Felgenhauer 1989. Molecular evidence for inclusion of the phylum Pentastomida in the Crustacea. *Molecular Biology and Evolution* 6: 685–91.

Aldrovandi, Ulyse 1648. *Aldrovandi Museum Metallicum.* Bologna.

Allen, D. E. 1980. The woman members of the Botanical Society of London, 1836–1856. *British Journal for the History of Science* 13: 240–54.

Alvarez, Luis, Walter Alvarez, Frank Asaro, and Helen V. Michel 1980. Extraterrestrial cause for the Cretaceous-Tertiary extinction. *Science* 208: 1095–1106.

Bacon, Francis 1605. *Advancement of Learning.* London: Herie Tomes.

——— 1620. *Novum Organum.* London: Ionannem Billium.

Baker, H. B. 1924. Land and freshwater molluscs of the Dutch Leeward Islands. *Michigan University Museum of Zoology Occasional Papers*, no. 152.

Barber, Lynn 1980. *The Heyday of Natural History 1820–1870.* New York: Doubleday.

Begg, Colin B., and Jesse A. Berlin 1988. Publication bias: a problem in interpreting medical data. *Journal of the Royal Statistical Society* 151: 419–63.

Bickerton, Derek 1981. *Roots of Language.* Ann Arbor, Michigan: Karoma.

——— 1990. *Language and Species*. Chicago: University of Chicago Press.

Borradaile, L. A. 1916. Crustacea. Part II. *Porcellanopagurus:* An instance of carcinization, *British Antarctic. "Terra Nova." Expedition Zoology* 3: 111–26.

Bova, Ben 1976. *Millennium: A Novel About People and Politics in the Year 1999*. New York: Ballantine Books.

Bowring, S. A., J. P. Grotzinger, C. E. Isachsen, A. H. Knoll, S. M. Pelechaty, and P. Kolosov 1993. Calibrating rates of early Cambrian evolution. *Science* 261: 1293–98.

Box, Joan Fisher 1978. *R. A. Fisher, The Life of a Scientist*. New York: Wiley.

Brasier, Clive 1992. Champion thallus. *Nature* 356: 382–83.

Brown, Thomas 1836. *The Conchologist's Text-book*. Glasgow: Archibald Fullerton and Co.

Browne, Janet 1989. Botany for gentlemen: Erasmus Darwin and *The Loves of the Plants*. *Isis* 80: 593–621.

Brusca, Richard C., and Gary J. Brusca 1990. *Invertebrates*. Sunderland, Mass: Sinauer Associates.

Budd, Ann F., and Anthony G. Coates 1992. Non-progressive evolution in a clade of Cretaceous *Montastraea*-like corals. *Paleobiology* 18: 425–46.

Buffon, Georges 1778. *Les époques de la Nature*. Paris: L'imprimerie Royale.

Burbank, Luther 1907. *The Training of the Human Plant*. New York: Century.

Burke, Edward 1757. *Philosophical Inquiry into the Origin of Our Ideas of the Sublime and Beautiful*. London: Printed for R. and J. Dodsley.

Büttner, J. W. E. 1956. *Leben und Wirken des Naturforschers Johann Gotthelf Fischer von Waldheim*. Berlin: Akademie Verlag.

Carro, R. J., H. N. Poinar, N. J. Pieniazek, A. Acra, and G. O. Poinar 1993. Amplification and sequencing of DNA from a 120–135 million year old weevil. *Nature* 363: 536–38.

Colbert, Edwin Harris 1991. *Evolution of the Vertebrates: A History of the Backboned Animals Through Time*. New York: Wiley-Liss.

Crichton, Michael 1990. *Jurassic Park*. New York: Knopf.

Cunningham, C. W., N. W. Blackstone, and L. W. Buss 1992. Evolution of king crabs from hermit crab ancestors. *Nature* 355: 539–42.

Darwin, Charles R. 1855. Does sea-water kill seeds? In Paul H. Barrett (ed.), *The Collected Papers of Charles Darwin*. Chicago: University of Chicago Press, 55.

——— 1859. *On the Origin of Species.* London: John Murray.

——— 1862. *On the Various Contrivances by which British and Foreign Orchids Are Fertilized by Insects.* London: John Murray.

Darwin, Erasmus 1789. *The Botanic Garden. Part II. The Loves of the Plants.* Dublin: J. Moore.

Davis, P., D. H. Kenyon, and C. B. Thaxton 1989. *Of Pandas and People.* Dallas: Haughton Publishing.

de Beus, J. G. 1985. *Shall We Make the Year 2000: The Decisive Challenge to Western Civilization.* London: Sidgwick and Jackson.

De Salle, R., J. Gatesy, W. Wheeler, and D. Grimaldi 1992. DNA sequences from a fossil termite in Oligo-Miocene amber and their phylogenetic implications. *Science* 257: 1933–36.

de Vries, W. 1974. Caribbean land molluscs: notes on Cerionidae. *Studies Fauna Curaçao and other Caribbean Islands* 45: 81–117.

Draper, John William 1874. *History of the Conflict Between Religion and Science.* New York: Appleton.

Eberhard, William G. 1985. *Sexual Selection and Animal Genitalia.* Cambridge, Mass.: Harvard University Press.

Eldredge, N., and S. J. Gould 1972. Punctuated equilibria: an alternative to phyletic gradualism. In T. J. M. Schopf (ed.), *Models in Paleobiology.* San Francisco: Freeman, Cooper and Co., 82–115.

Erdoes, Richard 1988. *AD 1000.* New York: Harper & Row.

Erwin, T. L. 1982. Tropical forests: their richness in Coleoptera and other arthropod species. *Coleopterists Bulletin* 36: 74–75.

Erzinclioglu, Zakaria 1992. Ancient error. *Nature* 355: 195.

Estling, Ralph 1992. Old argument. *Nature* 355: 667.

Fausto-Sterling, Anne 1985. *Myths of Gender: Biological Theories About Women and Men.* New York: Basic Books.

Fischer von Waldheim, Gotthelf 1813. *Zoognosia tabulis synopticis illustrata, in usum praelectionum Academiae imperialis medico-chirugicae mosquensis.* Moscow: Nicolai S. Vsevolozsky.

Fisher, Ronald Aylmer 1930. *The Genetical Theory of Natural Selection.* Oxford: The Clarendon Press.

——— 1957. Dangers of cigarette-smoking. *British Medical Journal*, 43.

——— 1957. Dangers of cigarette-smoking. *British Medical Journal*, 297.

——— 1958. Lung cancer and cigarettes? *Nature* 182: 108.

Flynn, James T. 1988. *The University Reform of Tsar Alexander I, 1802–1835*. Washington, D.C.: Catholic University of America Press.

Foçillon, Henri 1969. *The Year 1000*. New York: Frederick Ungar.

Gamow, George 1940. *Mr. Tompkins in Wonderland: Or, Stories of C, G, and H*. New York: Macmillan.

Gilbert, M. 1986. *The Holocaust*. London: Collins.

Gillispie, C. C. 1978. Laplace. *Dictionary of Scientific Biography*, vol. xv, Supplement I. New York: Charles Scribner's Sons.

Gingerich, P. D., N. A. Wells, D. E. Russell, and S. M. Ibrahim Shah 1983. Origin of whales in epicontinental remnant seas. *Science* 220: 403–6.

Gingerich, P. D., B. H. Smith, and E. L. Simons 1990. Hind limbs of Eocene *Basilosaurus:* evidence of feet in whales. *Science* 249: 154–57.

Gingerich, P. D., S. M. Raza, M. Arif, M. Anwar, and X. Zhou 1993. Partial skeleton of *Indocetus ramani* (Mammalia, Cetacea) from the Lower Middle Eocene Domanda Shale in the Sulaiman Range of Punjab (Pakistan). *Contributions from the Museum of Paleontology of the University of Michigan* 28: 393–416.

——— 1994. New whale from the Eocene of Pakistan and the origin of cetacean swimming. *Nature* 368: 844–47.

Gish, Duane 1985. *Evolution: The Challenge of the Fossil Record*. San Diego: Creation Life Publications.

Glen, William 1982. *The Road to Jaramillo*. Stanford, Calif.: Stanford University Press.

——— 1994. *The Mass Extinction Debates: How Science Works in a Crisis*. Stanford, Calif.: Stanford University Press.

Goilo, E. R. 1962. *Papiamentu Textbook*. Aruba: De Wit.

Golenberg, E. M., D. E. Giannasi, M. T. Clegg, C. J. Smiley, M. Durbin, D. Henderson, and G. Zurawski 1990. Chloroplast DNA sequence from a Miocene *Magnolia* species. *Nature* 344: 656–58.

Gould, Stephen Jay 1973. The misnamed, mistreated, and misunderstood Irish elk. *Natural History Magazine* 73 (March): 10–19.

——— 1974. The origin and function of "bizarre" structures: antler size and skull size in the "Irish elk," *Megaloceros giganteus*. *Evolution* 28: 191–220.

——— 1977. *Ever Since Darwin*. New York: W. W. Norton.

——— 1981. *The Mismeasure of Man.* New York: W. W. Norton.

——— 1984. Covariance sets and ordered geographic variation in *Cerion* from Aruba, Bonaire and Curaçao: a way of studying nonadaptation. *Systematic Zoology* 33: 217–37.

——— 1985. *The Flamingo's Smile.* New York: W. W. Norton.

——— 1989. *Wonderful Life: The Burgess Shale and the Nature of History.* New York: W. W. Norton.

——— 1991. *Bully for Brontosaurous.* New York: W. W. Norton.

——— 1993. *Eight Little Piggies.* New York: W. W. Norton.

Grew, Nehemiah 1681. *Museum regalis societatis, or a description of the natural and artificial rarities belonging to the Royal Society, whereunto is subjoyned the comparative anatomy of stomachs and guts.* London: W. Rawlins.

Haldane, J. B. S. 1927. *Possible Worlds and Other Essays.* London: Chatto and Windus.

——— 1932. *The Causes of Evolution.* New York: Harper and Brothers.

Haywood, Alan 1985. *Creation and Evolution.* London: Triangle Books.

Hitler, Adolf 1939. *Mein Kampf,* complete and unabridged, fully annotated. Editorial sponsors: John Chamberlain, Sidney B. Fay, and others. New York: Reynal and Hitchcock.

Hodkinson, I. D., and D. Casson 1991. A lesser predilection for bugs: Hemiptera (Insecta) diversity in tropical rain forests. *Biological Journal of the Linnean Society* 43: 101–9.

Hummelinck, P. W. 1940. Mollusks of the genera *Cerion* and *Tudora. Studies on the fauna of Curaçao, Aruba, Bonaire, and the Venezuelan Islands,* no. 5.

Husson, A. M., and L. B. Holthuis 1969. On the type of *Antilope leucophaea* Pallas, 1776, preserved in the collection of the Rijksmuseum van Natuurlijke Historie Leiden. *Zoologische Mededelingen* 44: 147–57.

——— 1975. The earliest figures of the Blaauwbok, *Hippotragus Leucophaeus* (Pallas, 1766) and of the greater Kudu, *Tragelaphus Strepsiceros* (Pallas, 1766). *Zoologische Mededelingen* 49: 57–63.

Hutchinson, G. Evelyn 1959. Homage to Santa Rosalia, or why are there so many kinds of animals? *American Naturalist* 93: 145–59.

Huxley, Thomas Henry 1880. *The Crayfish: An Introduction to the Study of Zoology.* London: C. Kegan Paul.

——— 1893. *Evolution and Ethics.* London: Macmillan and Co.

Janzen, D. 1977. What are dandelions and aphids? *American Naturalist* 111: 586–89.

Jordan, David Starr 1922. *The Days of a Man, Being Memories of a Naturalist, Teacher, and Minor Prophet of Democracy.* Yonkers on Hudson, New York: World Books Co.

Keyes, Ralph 1992. *Nice Guys Finish Seventh: False Phrases, Spurious Sayings and Familiar Misquotations.* New York: Harper.

Kimbel, W. H., D. C. Johanson, and Y. Rak 1994. The first skull and other new discoveries of *Australopithecus afarensis* at Hadar, Ethiopia. *Nature* 368: 449–51.

Kuhn, Thomas S. 1962. *The Structure of Scientific Revolutions.* Chicago: University of Chicago Press.

Labandeira, C., and J. J. Sepkoski Jr. 1993. Insect diversity in the fossil record. *Science* 261: 310–15.

Laplace, Pierre Simon 1796 (An IV). *Exposition du système du monde.* Paris: L'imprimerie du Cercle–Social.

——— 1812. *Théorie analytique des probabilités.* Paris: Ve. Courcier.

Linnaeus, Carolus 1736. *Fundamenta botanica.* Amsterdam: Salomonem Schouten.

——— 1758. *Systema naturae per regna tria naturae, secundum classes, ordines, genera, species, cum characteribus, differentiis, synonymis, locis.* Stockholm: L.Salvii.

Lorenz, K. 1971 (originally published in 1950). Part and parcel in animal and human societies. In *Studies in Animal and Human Behavior,* vol. 2, 115–95. Cambridge, Mass.: Harvard University Press.

Lyell, Charles 1830–1833. *Principles of Geology.* 3 vols. London: John Murray.

MacDonald, J. D., R. B. Pike, and D. I. Williamson 1957. Larvae of the British species of *Diogenes, Pagurus, Anapagurus* and *Lithodes* (Crustacea, Decapoda). *Proceedings of the Zoological Society of London* 128: 209–57.

May, Robert M. 1989. An inordinate fondness for ants. *Nature* 341: 386–87.

——— 1989. High table tales. *Nature* 341: 695.

Mercati, Michael 1719. *Metallotheca vaticana.* Edited by J. M. Lancisi. Rome: Jo. Mariam Salvioni.

Merton, Robert K. 1965. *On the Shoulders of Giants: A Shandean Postcript.* New York: Free Press.

Meyers, Jeffry 1992. *Edgar Allan Poe: His Life and Legacy.* New York: Charles Scribner's Sons.

Mohr, Erna 1967. *Der Blaubock: Eine Documentation.* Hamburg: Paul Parey.

Mount, Jeffrey F., Stanley V. Margolis, William Showers, Peter Ward, and Eric Doehne 1986. Carbon and oxygen isotope stratigraphy of the Upper Maastrichtian, Zumaya, Spain: a record of oceanographic and biologic changes at the end of the Cretaceous Period. *Palaios* 1: 87–92.

Nettl, Paul 1956. *Beethoven Encyclopedia.* New York: Philosophical Library.

Niklas, K. J., R. M. Brown Jr., and R. Santos 1985. Ultrastructural states of preservation in Clarkia Angiosperm leaf tissues: implications on modes of fossilization in late Cenozoic history of the Pacific. In Charles J. Smiley (ed.), *Late Cenozoic History of the Pacific.* San Francisco: American Association for the Advancement of Science, 143–59.

Nixon, Richard 1988. *1999: Victory Without War.* New York: Simon and Schuster.

O'Riordan, C. E. 1983. *The Natural History Museum, Dublin.* Dublin: Stationery Office.

Osborn, Henry Fairfield 1924. Three new Theropoda, *Protoceratops* Zone, Central Mongolia. *Novitates American Museum of Natural History* 144: 1–12.

Pallas, Peter Simon 1766. *Miscellanea zoologica quibus novae imprimis atque obscurae animalium species describuntur et observationibus iconibusque illustrantur.* The Hague: P. van Cleef, 224.

Poe, Edgar Allan 1839. *The Conchologist's First Book: or, a System of Testaceous Malacology Arranged Expressly for the Use of Schools.* Philadelphia: Haswell, Barrington and Haswell.

Roberts, Mary 1834. *The Conchologist's Companion.* London: Whittaker.

——— 1846. *The Progress of Creation.* London: Smith, Elder and Co.

Rupke, Nicolaas A. 1983. *The Great Chain of History.* Oxford: University Press.

Russell, Bertrand 1967. *The Autobiography of Bertrand Russell.* Boston: Little, Brown.

Russell, J. B. 1991. *Inventing the Flat Earth.* New York: Praeger.

Schiebinger, Londa 1991. The private life of plants: sexual politics in Carl Linnaeus and Erasmus Darwin. In M. Benjamin (ed.), *Science and Sensibil-*

ity: Gender and Scientific Enquiry 1780–1945. New York: Oxford University Press, 121–43.

——— 1993. Why mammals are called mammals: gender politics in eighteenth century natural history. *The American History Review* 90: 382–411.

Schwartz, Hillel 1990. *Century's End*. New York: Doubleday.

Scilla, Augostino 1747. *De corporibus marinis lapidescentibus*. Rome: Antonii Rubeis.

Sheehan, P. M., D. E. Fastovsky, R. G. Hoffmann, C. D. Berghaus, and D. L. Gabriel 1991. Sudden extinction of the dinosaurs: Late Cretaceous, Upper Great Plains, U.S.A. *Science* 254: 835–39.

Shelley, Mary Wollstonecraft 1818. *Frankenstein, or, The Modern Prometheus*. London: Lackington, Hughes, Harding, Mavor and Jones.

Signor, P. W., and J. H. Lipps 1982. Sampling bias, gradual extinction patterns and catastrophes in the fossil record. *Geological Society of America Special Paper* (Regional Studies), No. 190: 291–96.

Sinclair, David 1977. *Edgar Allan Poe*. London: J. M. Dent and Sons.

Slater, A. E. 1951. Biological problems of space flight. *Journal of the British Interplanetary Society* 10: 154–58.

Smith, M. L., J. N. Bruhn, and J. P. Anderson 1992. The fungus *Armillaria bulbosa* is among the largest and oldest living organisms. *Nature* 356: 428–31.

Soltis, P. S., D. E. Soltis, and C. J. Smiley 1992. An *rbcl* sequence from a Miocene *Taxodium* (bald cypress). *Proceedings of the National Academy of Science USA* 89: 449–51.

Stoneking, Mark 1993. DNA and recent human evolution. *Evolutionary Anthropology* 2: 60–73.

Stork, Nigel E. 1988. Insect diversity: facts, fiction and speculation. *Biological Journal of the Linnean Society* 35: 321–37.

Swift, Jonathan 1704. A *Tale of a Tub. Written for the Universal Improvement of Mankind. To Which is Added, an Account of Battle Between the Ancient and Modern Books in St. James Library*. London: J. Nutt.

Symons, Julian 1978. *The Tell-Tale Heart: The Life and Works of Edgar Allan Poe*. New York: Harper & Row.

Tennyson, Alfred, Lord 1850. *In Memoriam*. Boston: Ticknor, Reed and Fields.

Thewissen, J. G. M., and S. A. Hasten 1993. Origin of underwater hearing in whales. *Nature* 361: 444–45.

——— and M. Aria 1994. Fossil evidence for the origin of aquatic locomotion in archaeocete whales. *Science* 263: 210–12.

Thompson, D. W. 1942. *On Growth and Form.* Cambridge: Cambridge University Press.

Van Valen, Leigh 1966. Deltatheridia, a new order of mammals. *Bulletin of the American Museum of Natural History* 132: 1–126.

Walossek, Dieter, and Klaus J. Müller 1994. Pentastomid parasites from the Lower Paleozoic of Sweden. *Transactions of the Royal Society of Edinburgh, Earth Science* 85: 1–37.

——— and R. M. Kristensen 1994. A more than half a billion years old stem-group tardigrade from Siberia. *Proceedings of the Sixth International Symposium on Tardigrada.* Cambridge, England.

Ward, Peter 1983. The extinction of the ammonites. *Scientific American* 249: 136–47.

——— 1992. *On Methuselah's Trail: Living Fossils and the Great Extinctions.* New York: W. H. Freeman.

Weissman, Paul 1994. Comet Shoemaker-Levy 9: the big fizzle is coming. *Nature* 370: 94–95.

Whewell, William 1832. Review of *Principles of Geology* by Charles Lyell. *Quarterly Review* 47: 103–32.

——— 1837. *History of the Inductive Sciences: From the Earliest to the Present Times.* London: J. W. Parker.

White, Andrew Dickson 1896. *A History of the Warfare of Science with Theology in Christendom.* New York: Appleton.

Whitney, William Dwight 1875. *Life and Growth of Language.* International Scientific Series. New York: D. Appleton and Co.

Williams, Raymond 1983. *The Year 2000.* New York: Pantheon Books.

Wilson, E. O. 1992. *The Diversity of Life.* Cambridge, Mass.: Harvard University Press.

Wolff, Torben 1961. Description of a remarkable deep sea hermit crab, with notes on the evolution of the Paguridea. *Galathea Report* 4: 11–32.

Zumbach, F. T. 1986. Edgar Allan Poe: Eine Biographie. Munich: Winkler Verlag.

Index